Aircraft instruments and integrated systems

E H J Pallett
IEng, AMRAeS

Foreword by
L F E Coombs
IEng, BSc, MPhil, AMRAeS, FRSA

PEARSON
Prentice
Hall

Harlow, England • London • New York • Boston • San Francisco • Toronto
Sydney • Tokyo • Singapore • Hong Kong • Seoul • Taipei • New Delhi
Cape Town• Madrid • Mexico City • Amsterdam • Munich • Paris • Milan

Pearson Education Limited
Edinburgh Gate
Harlow
Essex CM20 2JE
England

and Associated Companies throughout the world

Visit us on the World Wide Web at:
http://www.pearsoned.co.uk

First published 1992
ARP Impression 98

British Library Cataloguing in Publication Date
A catalogue record for this book is available from the British Library

ISBN 978-0-582-08627-2

Set in 4 in Compugraphic Times 11/13 pt

Printed and bound in Great Britain by Clays Ltd, Bungay, Suffolk.
EPC/15

Contents

Foreword

by L F E Coombs IEng, BSc, MPhil, AMRAeS, FRSA

The progress of all types of aviation has depended on providing the pilot with sufficient information to enable him or her to control the aircraft safely and to navigate it to its destination.

From 1903 onwards each advance in speed, range, altitude and versatility has had to be matched by instruments which enable the crew to maximize an aircraft's potential. In the beginning, i.e. the Wright 'Flyer' of 1903, the instrumentation was rudimentary, consisting of only an anemometer for airspeed, a stop watch and an engine revolution counter. Perhaps the piece of string attached to the canard structure in front of the pilot, to indicate aircraft attitude relative to the airflow, can also be classed as an instrument.

Limited instrumentation was a feature of the aircraft of the first decade of heavier-than-air powered flight. However, the demands of wartime flying accelerated the development of instruments, and by 1918 a typical cockpit would have an airspeed indicator, an altimeter, inclinometer, fuel pressure gauge, oil pressure indicator, rpm indicator, compass and a clock. Not until the end of the 1920s were instruments available by which a pilot could maintain attitude and heading when flying in cloud, or whenever the horizon was obscured. In the 1930s and '40s, considerable progress was made toward 'blind flying' instruments. In the 1950s came the 'director'-type attitude indicators and in the '60s more and more electromechanical instruments became available. By 1970 solid-state displays were edging their way on to the flight deck. In the past ten years the electronic tide has swept in to an extent that the modern flight deck is awash from wall to wall with solid-state displays such as the electronic instrument systems (EIS) and engine indication and crew alert systems (EICAS).

With a lifetime's interest in the man—machine interface of the cockpit, I have depended very much on the knowledge and advice of many in the aviation industry and associated publishing. Over the years a 'nose' is acquired which differentiates among the many authors and separates them into categories. I place E H J Pallett at the top of the list when it comes to the ability to grasp essentials from a mass of technological facts, to explain succinctly, and, above all, to write with that authority which can only be acquired from long practical experience.

Any author who writes on the subject of aircraft instrumentation, and who aims a book, as does Eddie Pallett, at licensed engineers and flight crew, has a difficult task, a task made more onerous by the wide range of instrument types in use. This extends from mechanical, through electromechanical, to electromechanical with some electronics, and finally to today's electronic solid-state displays. Both engineers and flight crews will come across all of these different technologies. Some aircraft types will include examples of each, while others will be 90 per cent all-electronic — the so-called 'glass cockpit'.

Another problem, well tackled in this book, arises from the fact that modern aircraft have systems rather than individual items of equipment, such as instrument display units; few instruments, therefore, can be considered in isolation. Many are just one part of a system: for example, the control and display unit (CDU) of a flight management system is the visual and tactile interface with the pilot. From this 'tip' depends the rest of the iceberg: the computers and data links, and the interface units with other systems.

We have come a long way from the time when the engineer had only to undo four bolts and two unions, and out came the airspeed indicator. The title *Aircraft instruments and integrated systems* is therefore most apposite.

Preface

The material for this book was initially intended to comprise a third edition of *Aircraft Instruments* which for the last twenty years has served as a reference to this area of the avionics field. Since the last revision however, many technological changes have taken place, and in deciding how information on such changes could be included in a new edition, it soon became evident that the existing 'framework' and title, would be totally inappropriate. Thus, after considerable restructuring the compilation of material proceeded on the basis that it would best serve as a complete replacement of *Aircraft Instruments*.

As far as the instrumentation requirements for aircraft are concerned the most significant of the technological changes has been that of processing data and presenting it in electronic display format. Instruments and systems utilizing such format are, by virtue of high levels of digital computer integration and signal distribution via data 'highway' busbar systems, able to project the same quantity of operational data which would otherwise have to be displayed by a very large number of conventional 'clockwork' type instruments. The scene was, therefore, set not only for making drastic reductions in the utilization of conventional instruments, but also for implementing full automation of the management of all aspects of in-flight operation of aircraft.

The development of electronic instrument systems ran parallel with the launching of the Boeing 757, 767 and Airbus A310 aircraft in 1978 as design projects, and these were to become the first of 'new technology' aircraft to enter commercial service in 1982−3. These aircraft and several of their descendant types, are now in service world-wide, together with many types of smaller aircraft, including helicopters, in which the foregoing technology has also satisfied an operational need.

Conventional instruments of course still fulfill an important role but the extent of their utilization now bears a more direct relationship to types of aircraft. For example, in those already referred to, a conventional airspeed indicator, altimeter and attitude indicator are provided to serve as 'standby' references. There are on the other hand many other types of aircraft in which conventional instruments still satisfy the full instrumentation requirements appropriate to their operational needs. The material for this book therefore, covers a representative selection from the wide range of instruments and systems that currently come within both areas of technology and the sequencing of the relevant chapters has been arranged in such a way as to reflect the transition from one area to the other.

Like its predecessor, the details emphasize fundamental principles and their applications to civil aircraft instruments and systems, and overall, are also intended to serve as a basic reference for those persons who, either independently or, by way of courses established by specialist training organizations, are preparing for Aircraft Maintenance Engineer Licence examinations. It is also hoped that the details will provide some support to the current technical knowledge requirements relevant to flight crew examinations. A large number of 'self-test' questions have been compiled and are set out in chapter sequence at the end of the book.

As with all books of this nature schematic diagrams and photographs are of great importance in supporting the written details and so it is hoped that the three hundred or so spread over the chapters which follow will achieve the desired objective. In reviewing some current aircraft installations together with the contents of *Aircraft Instruments*, and also of another of my books (*Microelectronics in Aircraft Systems*) I found that a number of diagrams and photographs were still appropriate and so it was expedient to make further use of these. The diagrams relating to 'new subject' material have, in many cases, been redrawn from my original 'roughs'. The remaining new diagrams and photographs (some of which are reproduced in colour) have been supplied to me from external sources, and in this connection I would particularly like to express my grateful thanks to Smith's Industries, Aer Lingus, and Boeing International Corporation for their assistance.

In conclusion, I wish to convey sincere thanks to Leslie Coombs not only for his help, past and present, in providing material on a subject of common interest, but in particular for having accepted my invitation to write the Foreword to this book. It necessitated his having to read through many pages of draft manuscript, but as this resulted in comments that required some changes of text, then I am sure that he too would agree that efforts were not wasted.

Copthorne   E.P.
W. Sussex

1 Instrument displays, panels and layouts

In flight, an aircraft and its operating crew form a 'man—machine' system loop which, depending on the size and type of aircraft, may be fairly simple or very complex. The function of the crew within the loop is that of controller, and the extent of the control function is governed by the simplicity or otherwise of the aircraft as an integrated whole. For example, in manually flying an aircraft, and manually initiating adjustments to essential systems, the controller's function is said to be a fully active one. If, on the other hand, the flight of an aircraft and system's adjustments are automatic in operation, then the controller's function becomes one of monitoring, with the possibility of reverting to the active function in the event of failure of systems.

Instruments, of course, play an extremely vital role in the control loop as they are the means of communicating data between systems and controller. Therefore, in order that a controller may obtain a maximum of control quality, and also to minimize the mental effort in interpreting data, it is necessary to pay the utmost regard to the content and format of the data displays.

The most common forms of data display are (a) *quantitative*, in which the variable quantity being measured is presented in terms of a numerical value and by the relative position between a pointer or index and a graduated scale, and (b) *qualitative*, in which the data is presented in symbolic or pictorial format.

Quantitative displays

There are three principal methods by which data may be displayed: (i) the *circular scale*, or more familiarly, the 'clock' type of scale, (ii) *straight scale*, and (iii) *digital*, or counter.

Circular scale

This may be considered as the classical method of displaying data in quantitative form and is illustrated in Fig. 1.1. The *scale base* refers to the graduated line, which may be actual or implied, running from end to end of the scale and from which the scale marks and line of travel of the pointer are defined.

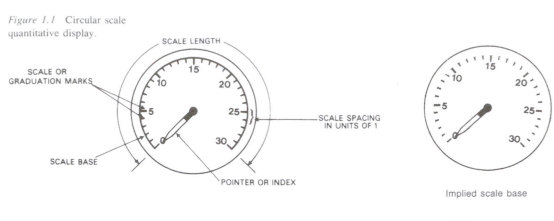

Figure 1.1 Circular scale quantitative display.

SCALE LENGTH

SCALE OR GRADUATION MARKS

SCALE SPACING IN UNITS OF 1

SCALE BASE

POINTER OR INDEX

Implied scale base

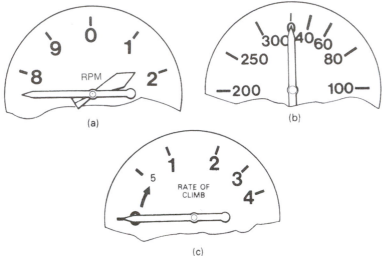

Figure 1.2 Linear and non-linear scales. (*a*) Linear; (*b*) square-law; (*c*) logarithmic.

(a)

(b)

(c)

Scale or *graduation marks* are those which constitute the scale of an instrument. For quantitative displays the number and size of marks are chosen in order to obtain quick and accurate interpretation of readings. In general, scales are divided so that the marks represent units of 1, 2 or 5, or decimal multiples thereof, and those marks which are to be numbered are longer than the remainder.

Spacing of marks is also governed by physical laws related to the quantity to be measured, but in general they result in spacing that is either *linear* or *non-linear*. Typical examples are illustrated in Fig. 1.2, from which it will also be noted that non-linear displays may be of the square-law or logarithmic-law type, the physical laws in this instance being related to airspeed and rate of altitude change respectively.

The sequence of numbering always increases in a clockwise direction, thus conforming to what is termed the 'visual expectation' of the observer. As in the case of marks, numbering is always in steps of 1, 2 or 5 or decimal multiples thereof. The numbers may be marked on the dial either inside or outside the scale base.

The distance between the centres of the marks indicating the minimum and maximum values of the chosen range of measurement, and measured along the scale base, is called the *scale length*. Governing factors in the choice of scale length for a particular range are the size of the instrument, the accuracy with which it needs to be read, and the conditions under which it is to be observed.

High-range long-scale displays

For the measurement of some quantities — for example, turbine engine speed, airspeed, and altitude — high measuring ranges are involved with the result that very long scales are required. This makes it difficult to display such quantities on single circular scales in standard-size cases, particularly in connection with the number and spacing of the marks. If a large number of marks are required their spacing might be too close to permit rapid reading, while, on the other hand, a reduction in the number of marks in order to 'open up' the spacing will also give rise to errors when interpreting values at points between scale marks.

Some of the displays developed as practical solutions are illustrated in Fig. 1.3. The display shown at (a) is perhaps the simplest way of accommodating a lengthy scale; by splitting it into two concentric scales, the inner one is made a continuation of the outer. A single pointer driven through two revolutions can be used to register against both scales, but as it can also lead to too frequent misreading, a

Figure 1.3 High-range long-scale displays. (*a*) Concentric scales; (*b*) fixed and rotating scales; (*c*) common scale and triple pointers.

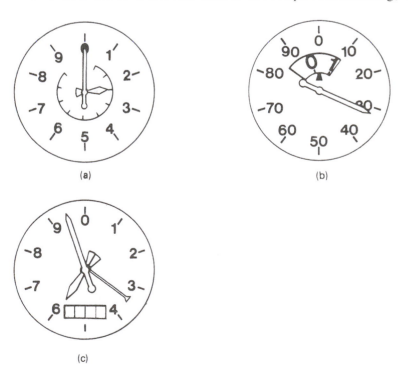

presentation by two concentrically-mounted pointers of different sizes is much better. A practical example of this is to be found in some types of engine speed indicator. In this instance, a large pointer rotates against an outer scale to indicate hundreds of rev/min, and at the same time it rotates a smaller pointer through appropriate ratio gearing, against an inner scale to indicate thousands of rev/min.

The method shown at (b) is employed in a certain type of pneumatic airspeed indicator; in its basic concept it is similar to the one just described. In this case, however, a single pointer rotates against a circular scale and drives a second scale plate instead of a pointer. This rotating plate, which records hundreds of knots as the pointer rotates through complete revolutions, is visible through an aperture in the main dial of the indicator.

Scale and operating ranges

Instrument scale lengths and ranges usually exceed that actually required for the operating range of the system with which an instrument is associated, thus leaving part of the scale unused. This may appear somewhat wasteful, but an example will show that it helps in improving the accuracy with which readings may be observed.

Let us consider a fluid system in which the operating pressure range is, say, $0-30$ lbf/in^2. It would be no problem to design a scale for the required pressure indicator which would be of a length equivalent to the system's total operating range, also divided into a convenient number of parts as shown in Fig. 1.4(a). However, under certain operating conditions of the system concerned, it may be essential to monitor pressures having such values as 17 or 29 lbf/in^2 and to do this accurately in the shortest possible time is not very easy, as a second look at the diagram will show.

If the scale is now redesigned so that its length and range exceed the system's operating range and also graduated in the manner noted

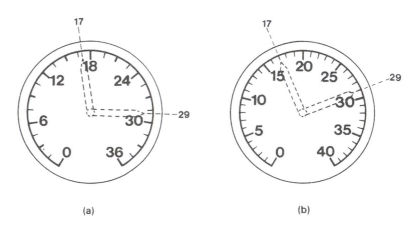

Figure 1.4 Reading accuracy.

(a)　　　　　　　　　　　　　(b)

earlier, then as shown at (b) the result makes it much easier to interpret and to monitor specific operating values.

Straight scale

In addition to the circular scale presentation, a quantitative display may also be of the straight scale (vertical or horizontal) type. For the same reason that the sequence of numbering is given in a clockwise direction on a circular scale, so on a straight scale the sequence is from bottom to top or from left to right.

Although such displays contribute to the saving of panel space and improved observational accuracy, their application to the more conventional types of mechanical and/or electro-mechanical instruments has been limited to those utilizing synchronous data-transmission principles. It is pertinent to note at this juncture that in respect of electronic CRT displays (see Chapter 11) there are no mechanical restraints, and so straight scales can, therefore, be more widely applied.

An example of a straight scale presentation of an indicator operating on the above-mentioned principles is illustrated in Fig. 1.5(a); it is used for indicating the position of an aircraft's landing flaps. The scales are graduated in degrees, and each pointer is operated by a synchro (see Chapter 5). The synchros are supplied with signals from transmitters actuated respectively by left and right outboard flap sections.

Another variation of this type of display is shown at (b) of Fig. 1.5. It is known as the *moving-tape* or *thermometer* display and was originally developed for the measurement of parameters essential to the operation of engines of large transport aircraft. Each display unit contains a servo-driven white tape in place of a pointer, which moves in a vertical plane and registers against a scale in a similar manner to the mercury column of a thermometer. As will be noted, there is one display unit for each parameter, the scales being common to each engine in the particular type of aircraft. When such displays are limited to only one or two parameters then, by scanning across the ends of the tapes, or columns, a much quicker and more accurate evaluation of changes in engine performance can be obtained as compared to 'clock' type displays. This fact, and the fact that panel space can be reduced, are clearly evident from the diagram.

Digital display

A digital, or counter, type of display is one that is generally to be found operating in conjunction with the circular type of display; two examples are shown in Fig. 1.6. In the application to an altimeter there are two counters: one presents a fixed pressure value which can

Figure 1.5 Straight scale displays. (*b*) gives a comparison between moving-tape and circular scale displays.

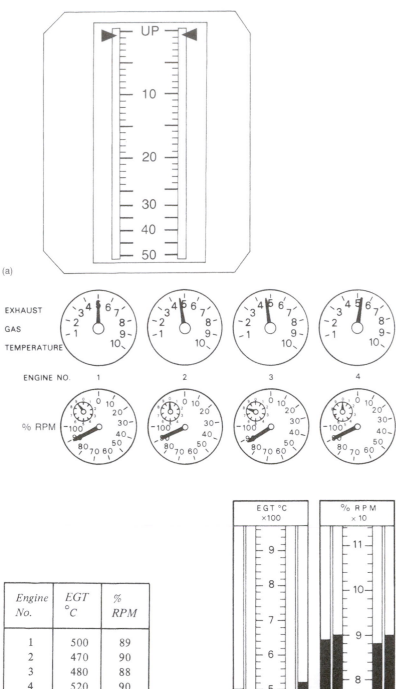

(a)

Engine No.	EGT °C	% RPM
1	500	89
2	470	90
3	480	88
4	520	90

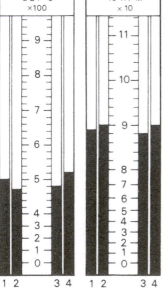

(b)

Figure 1.6 Application of digital counter displays.

DYNAMIC COUNTER DISPLAY

FEET

STATIC COUNTER DISPLAY

be mechanically set as and when required, and is known as a *static counter* display; while the other is geared to the altimeter mechanism and automatically presents changes in altitude, and is therefore known as a *dynamic counter* display. It is of interest to note that the presentation of altitude data by means of a scale and counter is yet another method of solving the long-scale problem already referred to on page 3. The counter of the turbine gas temperature (TGT) indicator is also a dynamic display since, in addition to the main pointer, it is driven by a servo transmission system (see also page 363).

Dual-indicator displays

These displays are designed principally as a means of conserving panel space, particularly where the measurement of the various quantities related to engines is concerned. They are normally of two basic forms: in one, two separate indicator mechanisms and scales are contained in one case, while in the other, which also has two mechanisms in one case, the pointers register against a common scale. Typical examples of display combinations are illustrated in Fig. 1.7.

Operational range markings

These markings take the form of coloured arcs, radial lines and sectors applied to the scales of instruments, their purpose being to highlight specific limits of operation of the systems with which the instruments are associated. The definitions of these marks are as follows:

RED *radial line*	Maximum and minimum limits
YELLOW *arc*	Take-off and precautionary ranges
GREEN *arc*	Normal operating range
RED *arc*	Range in which operation is prohibited

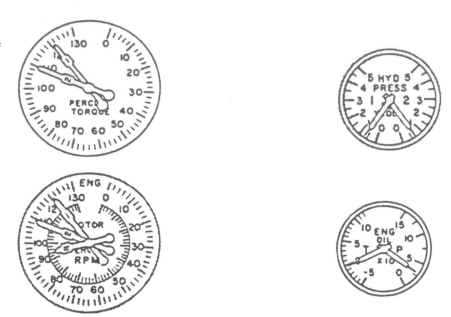

Figure 1.7 Dual-indicator displays. The display with three pointers has a helicopter application: it shows the speed of No. 1 and No. 2 engines, and of the main rotor.

In the example shown in Fig. 1.8(a), an additional WHITE arc is provided which serves to indicate the appropriate airspeed range over which an aircraft's landing flaps may be extended in the take-off, approach and landing configurations.

The application of sector-type markings is usually confined to those parts of an operating range in which it is sufficient to know that a certain condition has been reached rather than knowing actual quantitative values. For example, it may be necessary for an oxygen cylinder to be charged when the pressure has dropped to below, say, 500 lbf/in^2. The cylinder pressure gauge would therefore have a red sector on its dial embracing the marks from 0 to 500 as at (b) of Fig. 1.8. Thus, if the pointer should register within this sector, this alone is sufficient indication that recharging is necessary, and it is only of secondary importance to know what the actual pressure is.

Another method of indicating operating ranges is one that uses what are termed 'memory bugs'. These take the form of small pointers which, by means of an adjusting device, can be rotated around the dial plate of an instrument to pre-set them at appropriate operating values on the scale. An example of their application to a Mach/airspeed indicator (see page 47) is shown at (c) of Fig. 1.8.

Qualitative displays

These are of a special type in which the information is presented in a symbolic or pictorial form to show the condition of a system, whether the value of an output is increasing or decreasing, or to show the movement of flight control surfaces as in the example shown in Fig. 1.9.

Figure 1.8 Operational range markings.

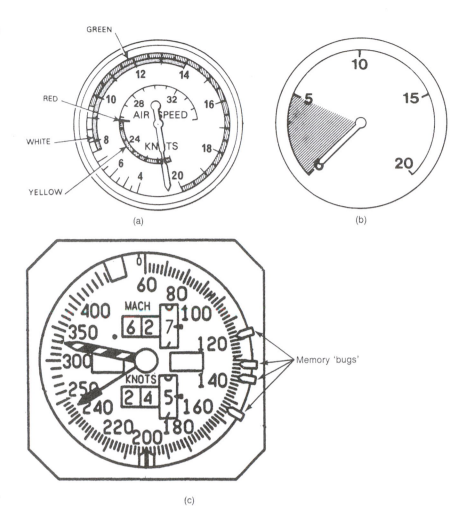

(a)

(b)

(c)

Figure 1.9 Qualitative display.

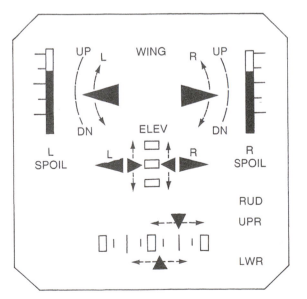

Director displays

These displays are associated principally with the monitoring of flight attitude and navigational data, and present it in a manner that indicates to the flight crew what control movements must be made, either to correct any departure from a desired flight path, or to cause an aircraft to perform a specific manoeuvre. It is thus apparent that in the development of such a display there must be a close relationship between the direction of control movements and the instrument pointer, or symbolic-type indicating element; in other words, movements should be in the 'natural' sense in order that the 'directives' or 'commands' of the display may be obeyed.

Displays of this nature are specifically applied to the two primary instruments which comprise conventional flight director systems and electronic flight instrument systems (see Chapters 9 and 12). One of the instruments (referred to as an Attitude Director Indicator) has its display origins in one of the oldest of flight attitude instruments, namely the gyro horizon (see Chapter 4), and so it serves as a basis for understanding the concept of director displays. As will be noted from Fig. 1.10, three elements make up the display of the instrument: a pointer registering against a bank-angle scale, an element symbolizing an aircraft, and an element symbolizing the natural horizon. Both the bank pointer and natural horizon element are stabilised by a gyroscope. As the instrument is designed for the display of attitude angles, and as also one of the symbolic elements can move with respect to the other, then it has two reference axes, that of the case which is fixed with respect to an aircraft, and that of the moving element.

Assuming that in level flight an aircraft's pitch attitude changes such as to bring the nose up, then the movement of the horizon element relative to the fixed aircraft symbol will be displayed as in diagram (a). This indicates that the pilot must 'get the nose down'. Similarly, if an aircraft's bank attitude should change whereby the left wing, say, goes down, then the display as at (b) would direct the pilot to 'bank the aircraft to the right'. In both cases the commands

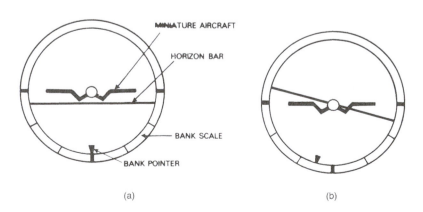

Figure 1.10 Director display (gyro horizon).

would be satisfied by the pilot moving the appropriate flight controls in the natural sense.

The display presentation of a typical Attitude Director Indicator is shown in Fig. 1.11(a), and as will be noted it is fundamentally similar to that of a gyro horizon. Details of its operation will be covered in a later chapter, but at this juncture it suffices to note that

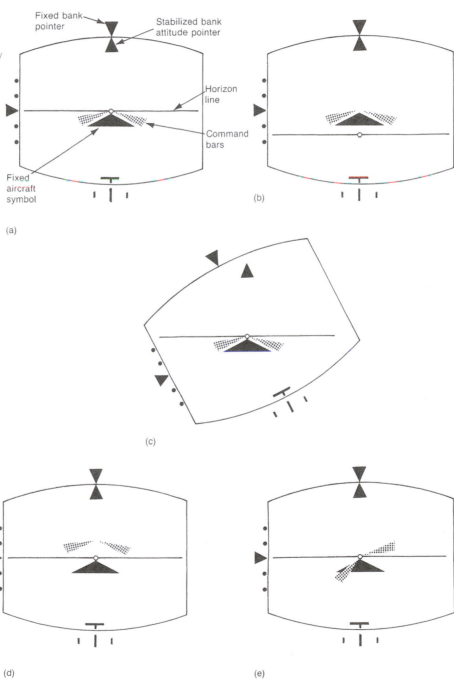

Figure 1.11 Attitude director display. (*a*) Aircraft straight and level; (*b*) aircraft nose up; (*c*) aircraft banked left; (*d*) 'fly up' command; (*e*) 'fly left' command.

the horizon symbolic element is driven by servomotors that receive appropriate attitude displacement signals from a remotely-located gyroscope unit. Thus, assuming as before a nose-up displacement of an aircraft, the signals transmitted by the gyroscope unit will cause the horizon symbolic element to be driven to a position below the fixed element symbolizing the aircraft, as shown at (b). The pilot is therefore directed to 'fly down' to the level flight situation as at (a).

If a change in the aircraft's attitude produces, say, a left bank, then in response to signals from the gyroscope unit the horizon symbolic element and bank pointer will be driven to the right as shown at (c). The pilot is therefore directed to 'fly right' to the level flight situation.

In addition to displaying the foregoing primary attitude changes, an indicator also includes what is termed a command bar display that enables a pilot to establish a desired change in aircraft attitude. If, for example, a climb attitude is to be maintained after take-off, then by setting a control knob the command bars are motor-driven to a 'fly up' position as shown at (d) of Fig. 1.11. During the climb the horizon symbolic element will be driven in the manner explained earlier, and the command bars will be recentred over the fixed element so that the display will be as shown in diagram (b).

Roll attitude, or turn commands, are established in a similar manner, the command bars in this case being rotated in the required direction; diagram (e) of Fig. 1.11 illustrates a 'fly left' command. As the aircraft's attitude changes the aircraft symbolic element moves with the aircraft, while the horizon symbolic element and bank pointer are driven in the opposite direction. When the command has been satisfied, the display will then be as shown in diagram (c).

The scales and pointers shown to the left and bottom of the indicator also form a director display that is utilized during the approach and landing sequence under the guidance of an Instrument Landing System. Details of the operation of this display and of the second indicator involved in a Flight Director System will be given in Chapter 9.

Electronic displays

With the introduction of digital signal-processing technology into the field colloquially known as 'avionics', and its application of micro-electronic circuit techniques, it became possible to make drastic changes to both quantitative and qualitative data display methods. In fact, the stage has already been reached whereby many of the conventional 'clock' type instruments which, for so long, have performed a primary role in data display, can be replaced entirely by a microprocessing method of 'painting' equivalent data displays on the screens of cathode ray tube (CRT) display units.

In addition to CRT displays (see Chapter 11), electronic display

Table 1.1 Applications of electronic displays

Display technology	Operating mode	Typical applications
Light-emitting diode	Active	Digital counter displays of engine performance
Liquid crystal	Passive	monitoring indicators; radio frequency selector indicators; distance measuring indicators; control display units of inertial navigation systems.
Electron CRT beam	Active	Weather radar indicators; display of navigational data; engine performance data; systems status; check lists.

techniques also include those of light-emitting diode and liquid crystal elements. Typical examples of their applications are given in Table 1.1. The operating mode of these displays may be either *active* or *passive*, the definitions of which are as follows:

Active: a display using phenomena potentially capable of producing light when the display elements are electrically activated.

Passive: a display which either transmits light from an auxiliary light source after modulation by the device, or which produces a pattern viewed by reflected ambient light.

Display configurations

Displays of the light-emitting diode and liquid crystal type are usually limited to applications in which a single register of alphanumeric values is required, and are based on what is termed a seven-segment matrix configuration or, in some cases, a dot matrix configuration.

Figure 1.12(a) illustrates the seven-segment configuration, the letters which conventionally designate each of the segments, and the patterns generated for displaying each of the decimal numbers 0−9. A segmented configuration may also be used for displaying alphabetic characters as well as numbers, but this requires that the number of segments be increased, typically from seven up to 13 and/or 16. Examples of these alphanumeric displays are illustrated at (b) of Fig. 1.12.

In a dot matrix display the patterns generated for each individual character are made up of a specific number of illuminated dots arranged in columns and rows. In the example shown at (c) of Fig. 1.12, the matrix is designated as a 4×7 configuration, i.e. it comprises four columns and seven rows.

Light-emitting diodes (LEDs)

An LED is a solid-state device comprising a forward-biased p-n junction transistor formed from a slice or chip of gallium arsenide

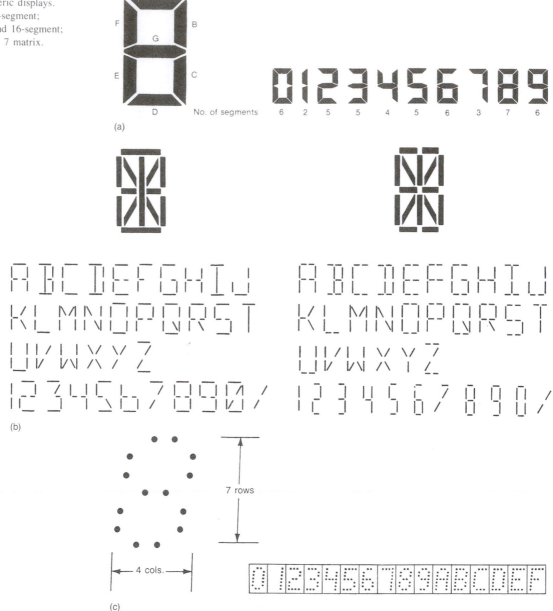

Figure 1.12 Electronic alphanumeric displays. (*a*) Seven-segment; (*b*) 13- and 16-segment; (*c*) a 4 × 7 matrix.

phosphide (GaAsP) moulded into a transparent covering as shown in Fig. 1.13. When current flows through the chip it emits light which is in direct proportion to the current flow. Light emission in different colours of the spectrum can, where required, be obtained by varying the proportions of the elements comprising the chip, and also by a technique of 'doping' with other elements, e.g. nitrogen.

In a typical seven-segment display format it is usual to employ one LED per segment and mount it within a reflective cavity with a

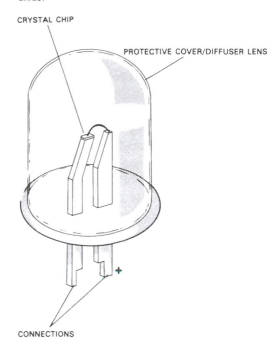

Figure 1.13 Light-emitting diode.

CRYSTAL CHIP

PROTECTIVE COVER/DIFFUSER LENS

CONNECTIONS

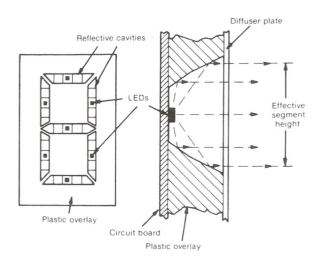

Reflective cavities

LEDs

Plastic overlay

Diffuser plate

Effective segment height

Circuit board

Plastic overlay

plastic overlay and a diffuser plate. The segments are formed as a sealed integrated circuit pack, the connecting pins of which are soldered to an associated printed circuit board. Depending on the application and the number of digits comprising the appropriate quantitative display, independent digit packs may be used, or combined in a multiple digit display unit.

LEDs can also be used in a dot-matrix configuration, and an example of this as applied to a type of engine speed indicator is shown in Fig. 1.14. Each dot making up the decimal numbers is an individual LED and they are arranged in a 9 × 5 matrix. The counter is of unique design in that its signal drive circuit causes an apparent 'rolling' of the digits which simulates the action of a mechanical drum-type counter as it responds to changes in engine speed.

Liquid crystal display (LCD)

The basic structure of a seven-segment LCD is shown in Fig. 1.15. It consists of two glass plates coated on their inner surfaces with a thin film of transparent conducting material (referred to as polarizing film) such as indium oxide. The material on the front plate is etched to form the seven segments, each of which forms an electrode. A mirror image is also etched into the oxide coating of the back glass plate, but this is not segmented since it constitutes a common return for all segments. The space between the plates is filled with a liquid crystal

15

Figure 1.14 Engine speed indicator with a dot matrix LED. (Courtesy of Smith's Industries Ltd.)

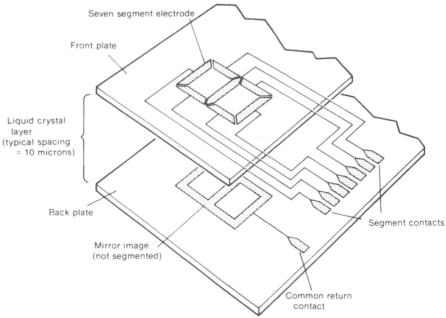

Figure 1.15 Structure of an LCD.

compound, and the complete assembly is hermetically sealed with a special thermoplastic material to prevent contamination.

When a low-voltage, low-current signal is applied to the segments, the polarization of the compound is changed together with a change in its optical appearance from transparent to reflective. The

Figure 1.16 Application of an LCD.

magnitude of the optical change is basically a measure of the light reflected from, or transmitted through, the segment area to the light reflected from the background area. Thus, unlike an LED, it does not emit light, but merely acts on light passing through it. Depending on polarizing film orientation, and also on whether the display is reflective or transmissive, the segments may appear dark on a light background (as in the case of digital watches and pocket calculators) or light on a dark background. An example of LCD application is shown in Fig. 1.16.

Head-up displays

A head-up display (HUD) is one in which vital in-flight data are presented at the same level as a pilot's line of sight when he is viewing external references ahead of the aircraft, i.e. when he is maintaining a 'head-up' position. This display technique is one that has been in use for many years in military aviation, and in particular it has been essential for those aircraft designed for carrying out very high-speed low-level sorties over all kinds of terrain.

As far as civil aviation is concerned, HUD systems have been designed specifically for use in public transport category aircraft during the approach and landing phase of flight, but thus far it has been a matter of choice on the operators' part whether or not to install systems in their aircraft. This has resulted principally from the differing views held by operators, pilot representative groups, and aviation authorities on the benefits to be gained, notably in respect of a system's contribution to the landing of an aircraft, either automatically or manually, in low-visibility conditions.

The principle adopted in a HUD system is to display the required data on the face of a CRT and to project them through a collimating lens as a symbolic image on to a transparent reflector plate, such that the image is superimposed on a pilot's normal view, through the windscreen, of the terrain ahead. The display is a combined alphanumeric and symbolic one, and since it is focused at infinity it permits simultaneous scanning of the 'outside world' and the display without refocusing the eyes. The components of a typical system are shown in Fig. 1.17.

Figure 1.17 Head-up display.

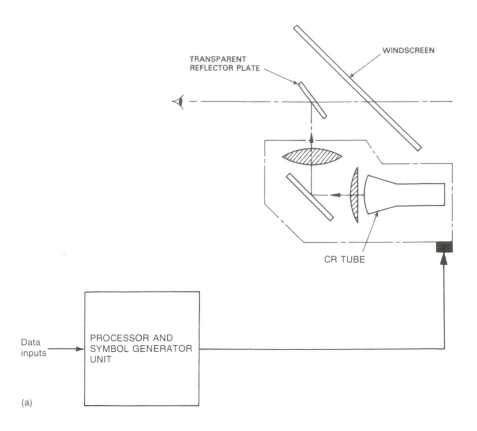

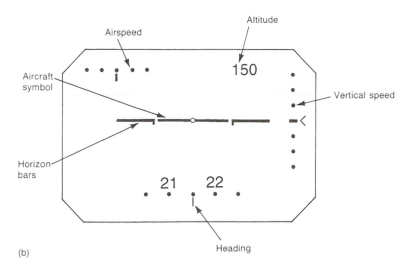

The amount of data required for display is governed by the requirements of the various flight phases and operational role of an aircraft, i.e. military or civil, but the parameters shown in diagram (b) are common to all. The format and disposition of the displays corresponding to required parameters can vary between systems; for example, a heading display may be in the form of a rotating arc at the upper part of the reflector plate, and altitude may be indicated by the registering of moving dots with a fixed index at one side of the plate instead of the changing digital counter readout located as shown in the diagram. Additional data such as decision height, radio altitude and runway outlines may also be displayed.

Panels and layouts

All instruments essential to the operation of an aircraft are accommodated on panels, the number and disposition of which vary in accordance with the number of instruments required for the appropriate type of aircraft and its cockpit or flight deck layout. A main instrument panel positioned in front of pilots is, of course, a feature common to all types of aircraft since instruments displaying primary data must be within the pilot's normal line of vision. The panel may be mounted in the vertical position or, as is now more common practice, sloped forward at about 15° from the vertical to minimize parallax errors. Typical positions of other panels are: overhead, at the side, and on a control pedestal located centrally between pilots. Figure 1.18 illustrates the foregoing arrangement appropriate to the Boeing 737-300 series aircraft. Where a flight engineer is required as a member of an operating crew, then panels would also be located at the station specifically provided on the flight deck.

Instrument grouping

Flight instruments

Basically there are six flight instruments whose indications are so co-ordinated as to create a 'picture' of an aircraft's flight condition and required control movements; they are the airspeed indicator, altimeter, gyro horizon, direction indicator, vertical speed indicator and turn-and-bank indicator. It is, therefore, most important for these instruments to be properly grouped to maintain co-ordination and to assist a pilot in observing them with the minimum of effort.

The first real attempt at establishing a standard method of grouping was the 'blind flying panel' or 'basic six' layout shown in Fig. 1.19(a). The gyro horizon occupies the top centre position, and since it provides positive and direct indications of attitude, and attitude changes in the pitching and rolling planes, it is utilized as the master instrument. As control of airspeed and altitude are directly related to attitude, the airspeed indicator, altimeter and vertical speed indicator flank the gyro horizon and support the interpretation of pitch attitude. Changes in direction are initiated by banking an aircraft, and the degree of heading change is obtained from the direction indicator; this instrument therefore supports the interpretation of roll attitude

Figure 1.18 Flight deck
layout: Boeing 737—300 series
aircraft.

Figure 1.19 Flight instrument
grouping. (*a*) Basic six;
(*b*) basic 'T' (with flight
director system indicators).

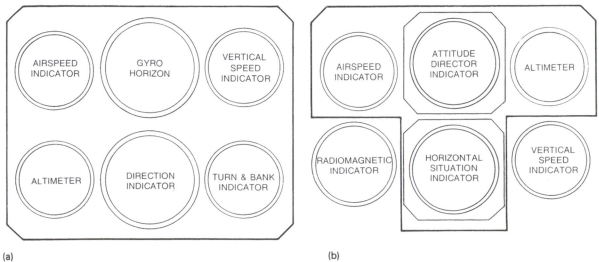

(a)

(b)

and is positioned directly below the gyro horizon. The turn-and-bank indicator serves as a secondary reference instrument for heading changes, so it too supports the interpretation of roll attitude.

With the development and introduction of new types of aircraft, and of more comprehensive display presentations afforded by the indicators of flight director systems, a review of the functions of certain of the instruments and their relative positions within the group resulted in the adoption of the 'basic T' arrangement as the current standard. As will be noted from diagram (b) of Fig. 1.19, there are now four 'key' indicators: airspeed, pitch and roll attitude, an altitude indicator forming the horizontal bar of the 'T', and a horizontal situation (direction) indicator forming the vertical bar. As far as the positions flanking the latter indicator are concerned, they are taken up by other less specifically essential flight instruments which, in the example shown, are the vertical speed indicator and a radiomagnetic indicator (RMI). In some cases a turn-and-bank indicator, or an indicator known as a turn co-ordinator, may take the place shown occupied by the RMI. In many instances involving the use of flight director system indicators and/or electronic flight instrument system display units, a turn-and-bank indicator is no longer used.

In the case of electronic flight instrument systems, the two CRT display units (EADI and EHSI) are also used in conjunction with four conventional-type indicators to form the basic 'T', as shown in Fig. 1.20(a). In displays of more recent origin, and now in use in such aircraft as the Boeing 747−400 (see also Fig. 12.11), the CRT screens are much larger in size, thus making it possible for the EADI to display airspeed, altitude and vertical speed data instead of conventional indicators. The presentation, which also corresponds to the basic 'T' arrangement is illustrated at (b) of Fig. 1.20.

Power plant instruments

The specific grouping of instruments required for the monitoring of power plant operation is governed primarily by the type of power plant, the size of aircraft, and therefore the space available for location of instruments. In a single-engined aircraft this does not present too much of a problem since the small number of instruments required may flank the flight instruments, thus keeping them within a small 'scanning range'.

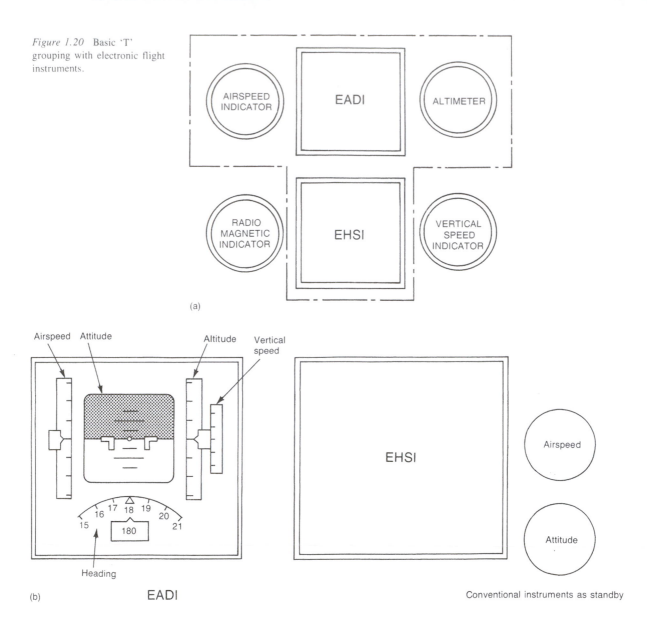

Figure 1.20 Basic 'T' grouping with electronic flight instruments.

(a)

(b)   EADI   Conventional instruments as standby

The problem is more acute in multi-engined aircraft, the number of instruments for all essential parameters doubling up with each engine. For twin-engined aircraft, and for certain medium-size four-engined aircraft, the practice is to group the instruments at the centre of the main instrument panel and between the two groups of flight instruments.

In those aircraft having a flight engineer's station, the instruments are grouped on the control panels at this station. Those instruments measuring parameters required to be known by the pilot during take-off, cruising and landing, e.g. rev/min and turbine gas temperature, are duplicated at the centre of the main instrument panel.

The positions of the instruments within a group are arranged so that those relating to each power plant correspond to the power plant positions as seen in plan view. It will be apparent from the layout of Fig. 1.21 that by scanning a row of instruments a pilot or engineer can easily compare the readings of a given parameter, and by scanning a column of instruments can assess the overall performance

Figure 1.21 Power plant instrument grouping.

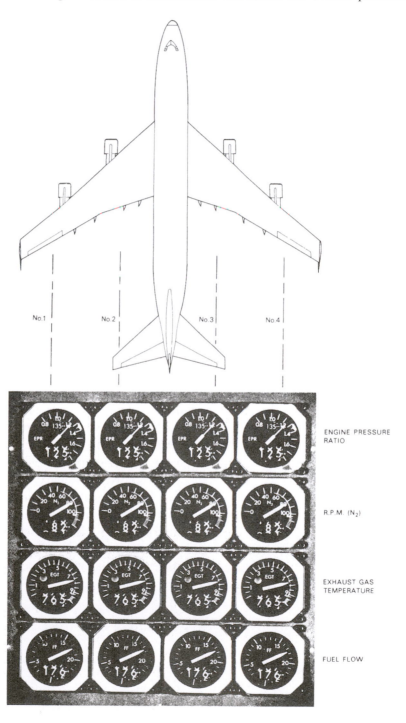

Figure 1.22 Power plant
instrument grouping (solid-state
displays): Primary parameters

pattern of a particular power plant. Another advantage of this
grouping method is that all the instruments for one power plant are
more easily associated with the controls for that power plant.

Figure 1.22 illustrates the grouping arrangement currently adopted
in Boeing 737–400 series aircraft for the display of the primary
parameters associated with its power plants.

The numeric values corresponding to each parameter are indicated
by LEDs arranged in a dot matrix 'rolling digit' configuration, and
located at the centre of permanently defined scale bases, graduations
and coloured range markings. In addition to the counter displays,
LEDs are also located around the periphery of each scale base, and
in their active state they simulate the rotation of conventional
indicator pointers.

2 Air data instruments

An air data (or manometric) system of an aircraft is one in which the total pressure created by the forward motion of an aircraft, and the static pressure of the atmosphere surrounding it, are sensed and measured in terms of speed, altitude and rate of altitude change (vertical speed). The measurement and indication of these three parameters may be done by connecting the appropriate sensors, either directly to mechanical-type instruments, or to a remotely-located air data computer which then transmits the data in electrical signal format to electro-mechanical or servo-type instruments.

Since the primary source of air for these measurements is the earth's atmosphere itself, then it is necessary to have some understanding of its characteristics before going into the operating principles of the measuring instruments and systems involved.

The earth's atmosphere

The earth's atmosphere is the surrounding envelope of air, which is a mixture of a number of gases, the chief of which are nitrogen and oxygen. By convention, this gaseous envelope is divided into several concentric layers extending from the earth's surface, each with its own distinctive features; these are shown in Fig. 2.1.

The lowest layer, and the one in which conventional types of aircraft are flown, is termed the *troposphere*, and extends to a boundary height termed the *tropopause*.

Above the tropopause, the next layer, termed the *stratosphere*, also extends to a boundary height called the *stratopause*.

At greater heights the remaining atmosphere is divided into further layers which are termed the *chemosphere, ozonosphere, ionosphere* and *exosphere*.

Throughout all these layers the atmosphere undergoes a gradual transition from its characteristics at sea-level to those at the fringes of the exosphere where it merges with the completely airless outer space.

Atmospheric pressure

The atmosphere is held in contact with the earth's surface by the force of gravity, which produces a pressure within the atmosphere. Gravitational effects decrease with increasing distances from the

Figure 2.1 Earth's
atmosphere.

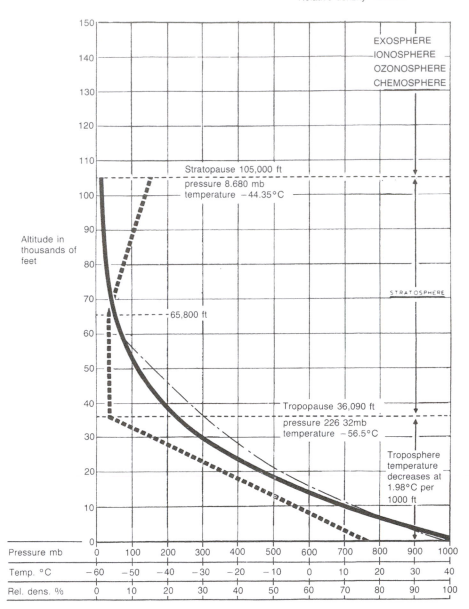

Pressure ▬▬▬
Temperature ▪▪▪▪
Relative density – – –

EXOSPHERE
IONOSPHERE
OZONOSPHERE
CHEMOSPHERE

Stratopause 105,000 ft
pressure 8.680 mb
temperature −44.35°C

Altitude in
thousands of
feet

STRATOSPHERE

65,800 ft

Tropopause 36,090 ft
pressure 226 32mb
temperature −56.5°C

Troposphere
temperature
decreases at
1.98°C per
1000 ft

Pressure mb	0	100	200	300	400	500	600	700	800	900	1000
Temp. °C	−60	−50	−40	−30	−20	−10	0	10	20	30	40
Rel. dens. %	0	10	20	30	40	50	60	70	80	90	100

earth's centre, and this being so, atmospheric pressure decreases
steadily with increases of height above the earth's surface.

The units in which atmospheric pressure is expressed are: *pounds
per square inch* (psi), *inches of mercury* (in Hg) and *millibars* (mb).
Conversion factors for these units are given in Appendix 1.

The steady fall in atmospheric pressure has a dominating effect on the density of the air, which changes in direct proportion to changes of pressure.

Atmospheric temperature

Another important factor affecting the atmosphere is its temperature. The air in contact with the earth is heated by conduction and radiation, and as a result its density decreases as the air starts rising. In doing so, the pressure drop allows the air to expand, and this in turn causes a fall in temperature from a known sea-level value. It falls steadily with increasing height up to the tropopause, and the rate at which it falls is termed the *lapse rate* (from the Latin *lapsus*, meaning slip). In the stratosphere the temperature at first remains constant at some reduced value, then increases again to a maximum.

Standard atmosphere

In order to obtain indications of airspeed, altitude and vertical speed, it is of course necessary to know the relationship between the pressure, temperature and density variables, and altitude. If such indications are to be presented with absolute accuracy, direct measurements of the three variables would have to be taken at all altitudes and fed into the appropriate instruments as correction factors. Such measurements, while not impossible, would, however, demand some rather complicated sensor mechanisms. It has therefore always been the practice to base all measurements and calculations on what is termed a *standard atmosphere*, or one in which the values of pressure, temperature and density at different altitudes are *assumed* to be constant. These assumptions have in turn been based on established meteorological and physical observations, theories and measurements, and so the standard atmosphere is accepted internationally. As far as airspeed indicators, altimeters and vertical speed indicators are concerned, the inclusion of the assumed values of the relevant variables in the calibration laws permits the use of sensing elements that respond solely to pressure changes.

The assumptions are: (i) the atmospheric pressure at mean sea-level is equal to 14.7 psi, 1013.25 mb, or 29.921 in Hg; (ii) the temperature at mean sea-level is 15°C (59°F); (iii) the air temperature decreases by 1.98°C (3.556°F) for every 1000 ft increase in altitude (the lapse rate already referred to) from 15°C at mean sea-level to −56.5°C (−69.7°F) at 36 090 ft. Above this altitude the temperature is assumed to remain constant at −56.5°C.

It is from the above mean sea-level values that all other corresponding values have been calculated and presented in what is

27

termed the *International Standard Atmosphere* (ISA). Altitudes and values are given in Table 1.

Basic air data system

In its basic form the system consists of a pitot-static probe, the three primary air data instruments (airspeed indicator, altimeter and vertical speed indicator) and pipelines and drains interconnected as shown diagrammatically in Fig. 2.2. Sensing of the total, or pitot, pressure (p_t) and of the static pressure (p_s) is effected by the probe, which is

Figure 2.2 Basic air data system.

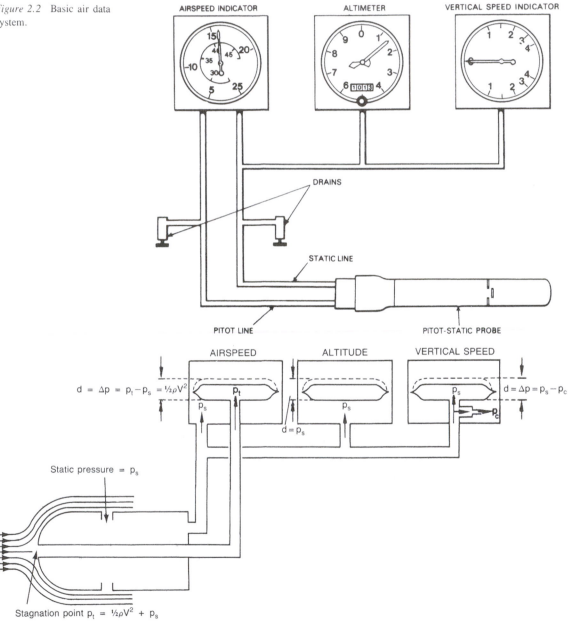

suitably located in the airstream and transmits these pressures to the sensing elements within the indicators.

The pressure transmission produces small displacements of the sensing elements in such a manner that displacements corresponding to (a) *airspeed* are proportional to the difference between p_t and p_s, (b) *altitude* are directly proportional to p_s, and (c) *vertical speed* are proportional to the difference between p_s and a 'case' pressure p_c produced by a calibrated metering unit. The displacements are, in turn, transmitted to an indicating element via an appropriate magnifying system.

The complexity of an air data system depends primarily upon the type and size of aircraft, the number of locations at which primary air data are to be displayed, the types of instrument installed, and the number of other systems requiring air data inputs. The point about complexity may be particularly noted from Figs 2.3 and 2.4, which show in schematic form the systems used in two types of public transport aircraft currently in service.

Probes

Probes may be either of the combined pitot-static tube type, or of the single pitot tube type, the latter being used in air data systems that utilize remotely-located static vents or ports (see also page 35).

Figure 2.3 Typical pitot probe and static vent system (1).

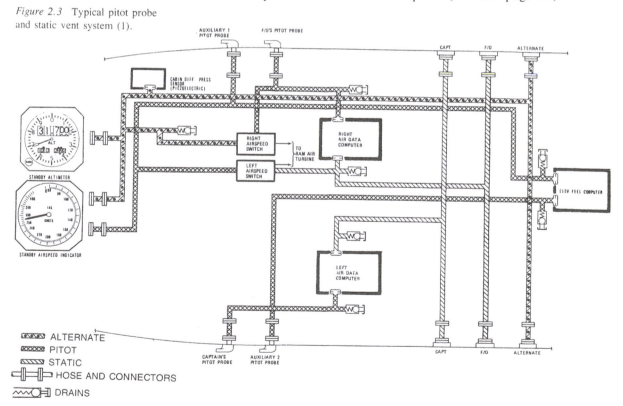

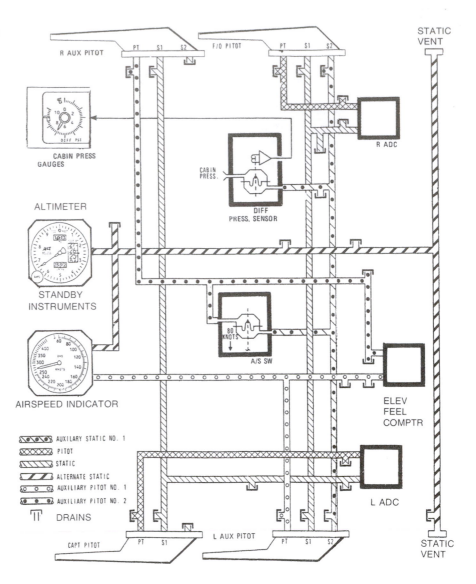

Figure 2.4 Typical pitot probe and static vent system (2)

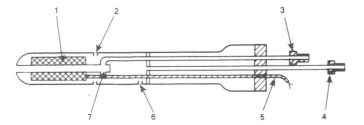

Figure 2.5 Basic form of pitot-static probe. 1 Heating element, 2 static slots, 3 pitot tube connection, 4 static tube connection, 5 heater element cable, 6 external drain hole, 7 pitot tube drain hole.

A probe of the combined tube type is shown in basic form in Fig. 2.5. The tubes are mounted concentrically, the pitot tube being inside the static tube which also forms the casing of the probe. Static pressure is admitted through small ports around the casing. The pressures are transmitted from their respective tubes by means of

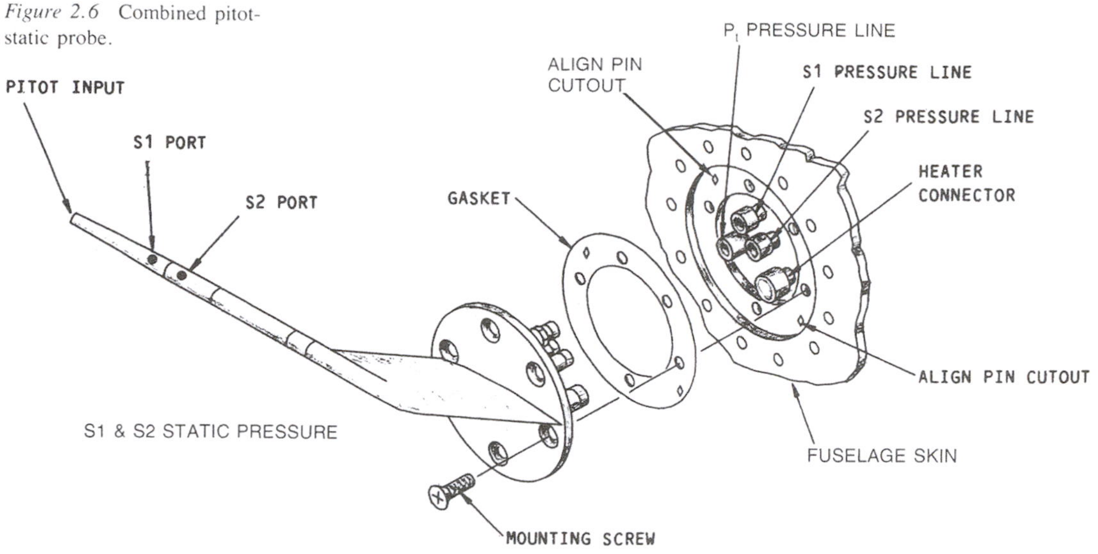

Figure 2.6 Combined pitot-static probe.

PITOT INPUT

S1 PORT

S2 PORT

GASKET

S1 & S2 STATIC PRESSURE

MOUNTING SCREW

ALIGN PIN CUTOUT

P_t PRESSURE LINE

S1 PRESSURE LINE

S2 PRESSURE LINE

HEATER CONNECTOR

ALIGN PIN CUTOUT

FUSELAGE SKIN

metal pipes which may extend to the rear of the probe, or at right angles to it, depending on whether the probe is to be mounted at the leading edge of an aircraft's wing, under a wing, or at the side of a fuselage. Locations of probes will be covered in more detail under the heading of 'Position error'.

A chamber is normally formed between the static holes and the pipe connection to smooth out any turbulent air flowing into the holes which might occur when the probe is yawed, before transmitting it to the instruments.

Protection against icing is provided by a heating element fitted around the pitot tube, or, as in some designs, around the inner circumference of the probe casing, and in such a position that the maximum heating effect is obtained at points where ice build-up is most likely to occur. The temperature/characteristics of some elements are such that the current consumption is automatically regulated according to the temperature conditions to which the probe is exposed.

Figure 2.6 illustrates a type of combined tube probe; it is supported on a faired casing which is secured to the side of an aircraft's fuselage by means of a mounting flange. The assembly incorporates two sets of static holes (S1 and S2) connected to individual pipes terminating at the mounting flange; the use of both sets is shown in more detail in Fig. 2.4. Pitot pressure is transmitted via an appropriate connecting union and pipe terminating at the mounting flange.

An example of a 'pitot tube only' type of probe is shown in Fig. 2.7, and from this it will be noted that its mounting arrangements are similar to those adopted for the combined tube type. A typical application of the probe is shown in more detail in Fig. 2.3.

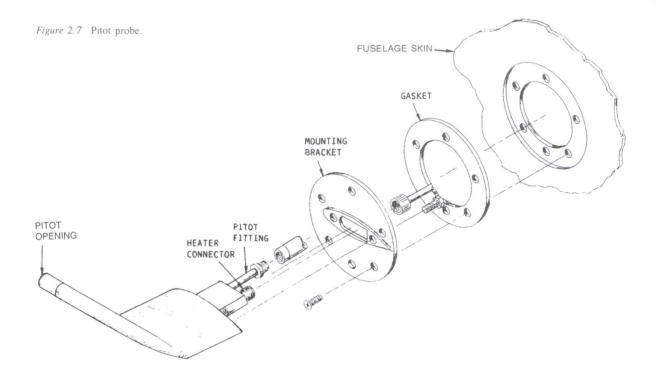

Figure 2.7 Pitot probe.

FUSELAGE SKIN

GASKET

MOUNTING
BRACKET

PITOT
OPENING

PITOT
FITTING

HEATER
CONNECTOR

Heating circuit arrangements

The heating elements of some probes require a 28 V dc supply for operation, while others are designed to operate from a 115 V ac supply, their application to any one type of aircraft being governed principally by the primary power supply system adopted.

In any heating circuit it is of course necessary to have a control switch, and it is also usual to provide some form of indication of whether or not the circuit is functioning correctly. Two typical dc-powered circuit arrangements are shown in Fig. 2.8.

In the arrangement shown at (*a*) the control switch, when in the 'ON' position, allows current to flow to the heater via the coil of a relay which will be energized when there is continuity between the switch and the grounded side of the heater. If a failure of the heater, or a break in another section of its circuit occurs, the relay will de-energize and its contacts will then complete the circuit from the second pole of the switch to illuminate the red light which gives warning of the failed circuit condition. The broken lines show an alternative arrangement of the light circuit whereby illumination of an amber light indicates that the heater circuit is in operation.

In the arrangement shown at (b) an ammeter is connected in series with the heater element so that not only is circuit continuity indicated, but also the amount of current being consumed by the element.

An example of an ac-powered heater circuit is shown in Fig. 2.9;

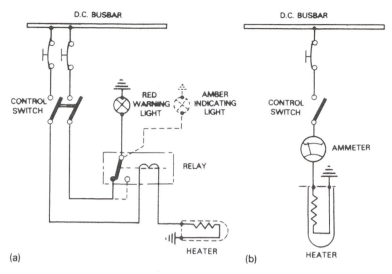

Figure 2.8 Typical probe heating circuit arrangement (dc). (*a*) Light and relay; (*b*) ammeter.

Figure 2.9 Ac-powered heating circuit.

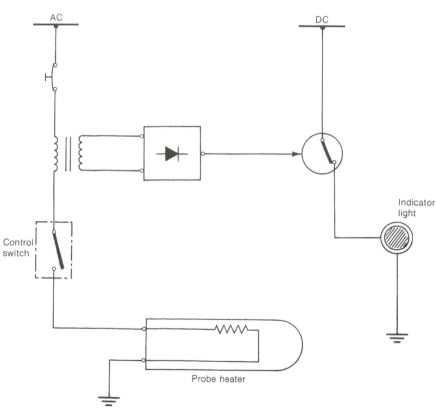

dc power is also used for indicator light circuit operation. With both power supplies available, and the system control switch in the 'ON' position, single-phase ac is supplied to the heater element via the primary winding of a transformer. The dc power to the amber indicator lights passes to ground via a normally closed solid-state switch, so they will initially remain illuminated.

33

When current is supplied to the heater element, a current is also induced in the secondary winding of the transformer and is supplied to a bridge rectifier. The rectified output is, in turn, supplied to the solid-state switch. When the heating current has reached a sufficient level, the increased rectified output causes the solid-state switch to interrupt the indicator light circuit; a 'light out' thus indicates that 'probe heat is on'. Functioning of the indicator lights can be checked by a press-to-test switch within the body of the light unit.

Position error

The accurate measurement of airspeed and altitude by means of a combined tube type of probe presents two main difficulties: one, to design a probe which will not cause any disturbance to the airflow over it, and the other, to find a suitable location on an aircraft where the airflow over it will not be affected by attitude changes of the aircraft. The effects of such disturbances are greatest on the static pressure sensing section of an air data system, giving rise to what is termed a *position* or *pressure error* (PE). This error may be more precisely defined as 'the amount by which the local static pressure at a given point in the flow field differs from the free-stream static pressure'. As a result of PE, an airspeed indicator and an altimeter can develop errors in their indications. The indications of a vertical speed indicator remain unaffected by PE.

As far as airflow over a probe is concerned, we may consider it, and the aircraft to which it is fitted, as being alike because some of the factors determining airflow are: shape, size, speed and angle of attack. The shape and size of a probe are dictated by the speed at which it is moved through the air; a large-diameter casing, for example, can present too great a frontal area which at very high speeds can initiate the build-up of a shock wave which will break down the airflow over the probe. This shock wave can have an appreciable effect on the static pressure, extending as it does for a distance equal to a given number of diameters from the nose of a probe. One way of overcoming this is to decrease the casing diameter and to increase the distance of the static holes from the nose of the probe. Furthermore, a number of holes may be provided along the length of the casing of a probe spaced in such a way that some will always be in the region of undisturbed airflow.

A long, small-diameter probe is an ideal one from an aerodynamic point of view, but it can present certain practical difficulties; for example, its 'stiffness' may not be sufficient to prevent vibration at high speed, and it could also be difficult to accommodate the heating elements necessary for anti-icing. Thus, in establishing the ultimate relative dimensions of a probe, a certain amount of compromise must be accepted.

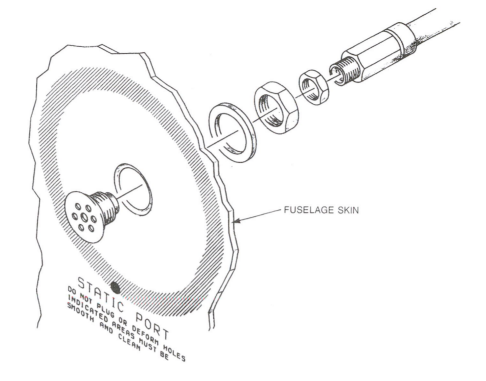

Figure 2.10 Static vent or port.

FUSELAGE SKIN

STATIC PORT
DO NOT PLUG OR DEFORM HOLES
INDICATED AREAS MUST BE
SMOOTH AND CLEAN

When a probe is at some angle of attack to the airflow, it causes air to flow into the static holes which creates a pressure above that of the prevailing static pressure, and a corresponding error in static pressure measurement. The pressures developed at varying angles of attack depend on such factors as the axial location of the static holes along the probe casing, their positions around the circumference, and on their size.

Static vents

From the foregoing, it would appear that, if all these problems are created by pressure effects at the static holes of a probe, they might as well be separated from it and positioned elsewhere on an aircraft. This is, in fact, a solution put into practice on many types of aircraft by using a single tube type of probe (see Fig. 2.7) in conjunction with a static vent located in the side of the fuselage. A typical static vent is shown in Fig. 2.10.

Location of probes and static vents

The choice of probe locations is largely dependent on the type of aircraft, speed range and aerodynamic characteristics, and as a result there is no common standard for all aircraft. Typical locations are: ahead of a wing tip, under a wing, ahead of a vertical stabilizer tip,

35

at the side of a fuselage nose section, and ahead of a fuselage nose section.

Independent static vents, when fitted, are always located one on each side of a fuselage and interconnected so as to balance out dynamic pressure effects resulting from any yawing or sideslip motion of an aircraft.

The actual PE due to a chosen location is determined for the appropriate aircraft type during the initial flight-handling trials of a prototype, and is finally presented in tabular or graphical format, thus enabling corrections to be directly applied to the readings of the relevant air data instruments, and as appropriate to various operating configurations. In most cases, corrections are performed automatically and in a variety of ways. One method is to employ aerodynamically-compensated probes, i.e. probes which are so contoured as to create a local pressure field which is equal and opposite to that of the aircraft, so that the resultant PE is close to zero. Other methods more commonly adopted utilize correction devices either within separate transducers, or within central air data computers (see page 165).

Alternate pressure sources

If failure of the primary pitot and static pressure sources should occur, e.g. complete icing-up of a probe due to a heating element failure, then of course errors will be introduced in instruments' indications and in other systems dependent on such pressures. As a safeguard against failure, therefore, a standby system may be installed in aircraft employing combined tube type probes whereby static atmospheric pressure and/or pitot pressure from alternate sources can be selected and connected into the primary system.

The required pressure is selected by means of selector valves connected between the appropriate pressure sources and the air data instruments, and located within easy reach of the flight crew. Figure 2.11 illustrates diagrammatically the method adopted in a system utilizing an alternate static pressure source only. The valves are shown in the normal operating position, i.e. the probes supply pitot and static pressures to the instruments on their respective sides of the aircraft.

In the event of failure of static pressure from one or other probe, the instruments can be connected to the alternate source by manually changing over the position of the relevant selector valve.

The layout shown in Fig. 2.12 is one in which an alternate source of both pitot and static pressures can be selected. Furthermore, it is an example of a system which utilizes the static holes of a combined tube type of probe as the alternate static pressure source. The valves are shown in their normal position, i.e. the probes supply pitot pressure to the instruments on their respective sides of an aircraft,

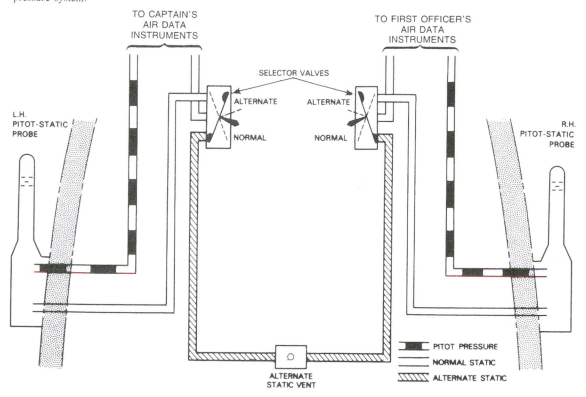

Figure 2.11 Alternate static pressure system.

TO CAPTAIN'S
AIR DATA
INSTRUMENTS

TO FIRST OFFICER'S
AIR DATA
INSTRUMENTS

SELECTOR VALVES

ALTERNATE   ALTERNATE

L.H.
PITOT-STATIC
PROBE

NORMAL   NORMAL

R.H.
PITOT-STATIC
PROBE

ALTERNATE
STATIC VENT

▬▬ PITOT PRESSURE

── NORMAL STATIC

▨▨ ALTERNATE STATIC

and the static pressure is supplied from static vents. In the event of failure of pitot pressure from one or other probe, the position of the relevant selector valve must be manually changed over to connect the air data instruments to the opposite probe. The alternate static source is selected by means of a valve similar to that employed in the pitot pressure system, and, as will be seen from Fig. 2.12, it is a straightforward change-over function.

The probes employed in the system just described are of the type illustrated in Fig. 2.6, reference to which shows that two sets of static holes (front and rear) are connected to separate pipes at the mounting base. In addition to being connected to their respective selector valves, the probes are also coupled to each other by a cross-connection of the static holes and pipes; thus, the front set of holes are connected to the rear set on opposite probes. This balances out any pressure differences which might be caused by the location of the static holes along the fore-and-aft axis of the probes.

Pipelines and drains

Pitot and static pressures are transmitted through seamless and corrosion-resistant metal (light alloy and/or tungum) pipelines, and,

Figure 2.12 Alternate pitot pressure and static pressure system.

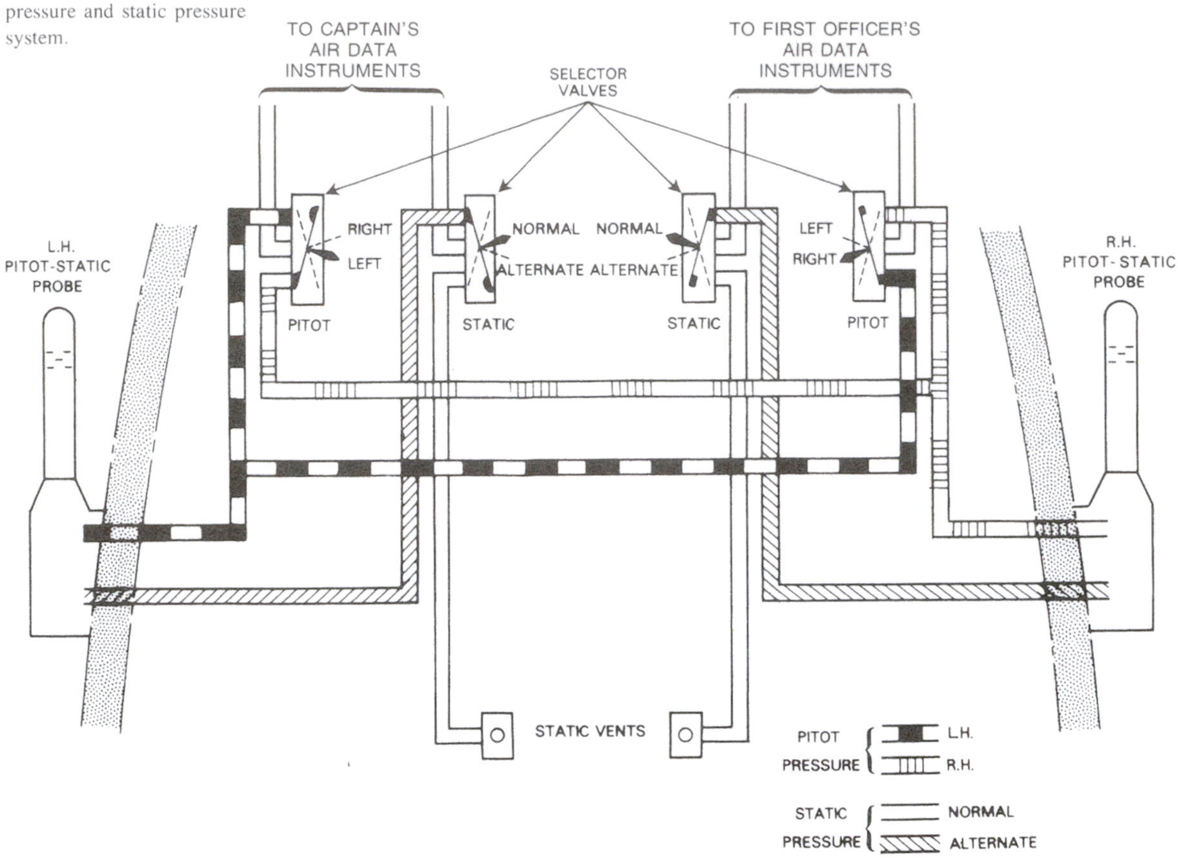

where connections to components mounted on anti-vibration mountings are required, flexible pipelines are used. The diameter of pipelines is related to the distance from the pressure sources to the instruments in order to eliminate pressure drop and time-lag factors.

In order for an air data system to operate effectively under all flight conditions, provision must also be made for the elimination of water that may enter the system as a result of condensation, rain, snow, etc., thus reducing the probability of 'slugs' of water blocking the lines. Such provision takes the form of drain holes in probes, drain traps and valves in the system's pipelines. Drain holes provided in probes are of such a diameter that they do not introduce errors in instrument indications.

The method of draining the pipelines varies between aircraft types, and some examples are shown in Fig. 2.13. Drain traps are designed to have a capacity sufficient to allow for the accumulation of the maximum amount of water that could enter a system between servicing periods. Valves are of the self-closing type so that they cannot be left in the open position after drainage of accumulated water.

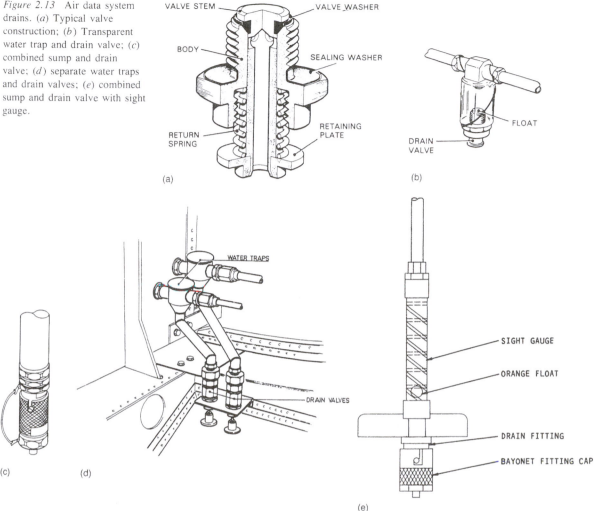

Figure 2.13 Air data system drains. (*a*) Typical valve construction; (*b*) Transparent water trap and drain valve; (*c*) combined sump and drain valve; (*d*) separate water traps and drain valves; (*e*) combined sump and drain valve with sight gauge.

Air data instruments

The three primary air data instruments may be either of the pure 'pneumatic' type, or the servo-operated type. Pneumatic-type instruments are those which are connected to probes and/or static vents, and therefore respond to the pressures transmitted directly to them. They are commonly used in the more basic air data systems installed in many types of small aircraft, while in the more complex systems adopted in large public transport aircraft, they are used only in a standby role.

Servo-operated instruments are, on the other hand, of the indirect type in that they respond to electrical signals generated by pressure transducers within *central air data computers* (CADCs) to which probes and static vents are connected. The fundamental principles of these instruments will be described in a later chapter.

Figure 2.14 Pitot pressure.

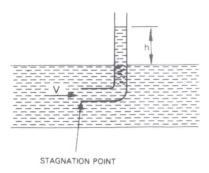

STAGNATION POINT

Airspeed indicators (pneumatic)
These indicators measure speed in terms of the difference between
the pitot and static pressures detected by either a combined pitot-static
probe, or a pitot probe and static vent, as appropriate.

Pitot pressure This may be defined as the additional pressure
produced on a surface when a flowing fluid is brought to rest, or
stagnation, at the surface.

Let us consider a pitot probe placed in a fluid with its open end
facing upstream as shown in Fig. 2.14. When the fluid flows at a
certain velocity V over the probe it will be brought to rest at the
nose; this point is known as the *stagnation point*. If the fluid is an
ideal one, i.e. is not viscous, then the total energy is equal to the
sum of the potential energy, the kinetic energy and pressure energy,
and remains constant. In connection with this probe, however, the
potential energy is neglected, thus leaving the sum of the remaining
two terms as the constant.

In coming to rest at the stagnation point, kinetic energy of the fluid
is converted into pressure energy. This means that work must be
done by the mass of fluid and this raises an equal volume of fluid
above the level of the fluid stream. The work done in raising the
fluid is equal to the product of its mass, the height through which it
is raised, and acceleration due to gravity. It is also equal to the
product of the ratio of the mass (m) to density (ρ) and pressure (p);
thus,

$$\text{Work done} = \frac{m}{\rho} p$$

The kinetic energy of a mass m before being brought to rest is
equal to $\frac{1}{2} mV^2$, where V is the speed, and since this is converted into
pressure energy,

$$\frac{m}{\rho} p = \tfrac{1}{2} mV^2.$$

Therefore

$$p = \tfrac{1}{2} \rho V^2.$$

The quantity $\tfrac{1}{2} \rho V^2$ is additional to the static pressure in the region of the fluid flow, and is usually referred to as the *dynamic pressure*, denoted by the letter Q.

The factor $\tfrac{1}{2}$ assumes that the fluid is an ideal one and so does not take into account the fact that the shape of a body subject to fluid flow may not bring the fluid to rest at the stagnation point. This coefficient is, however, determined by experiment and for pitot pressure probes it has been found that its value corresponds almost exactly to the theoretical one.

The $\tfrac{1}{2} \rho V^2$ law, as it is usually called in connection with airspeed measurement, does not allow for the effects of compressibility of air as speed increases. In order therefore to minimize 'compressibility errors' in indication, the calibration law is modified as follows:

$$p = \tfrac{1}{2} \rho V^2 \left(1 + \tfrac{1}{4} \frac{V^2}{a_0{}^2} \right)$$

where p = pressure difference (mmH$_2$O)
ρ = density of air at sea-level
V = speed of aircraft (mph or knots)
a_0 = speed of sound at sea-level (mph)

Airspeed terminology
Indicated airspeed (IAS)   The readings of an airspeed indicator corrected only for instrument error, i.e. the difference between the true value and the indicated value. Errors and appropriate corrections to be applied are determined by comparison against calibration equipment having high standards of accuracy.

Computed airspeed   Basically, this is IAS with corrections for position error (PE) applied (see page 34). The term 'computed' applies specifically to air data computer systems in which PE corrections are automatically applied to an airspeed sensing module via an electrical correction network.

Calibrated airspeed (CAS)   This is also associated with air data computer systems and is the computed airspeed compensated for the non-linear, or square-law, response of the airspeed sensing module.

Equivalent airspeed (EAS)   This is the airspeed calculated from the measured pressure difference p when using the constant sea-level value of density ρ. In air data computer systems, CAS is automatically compensated for compressibility of air at a pitot probe to obtain EAS at varying speeds and altitudes.

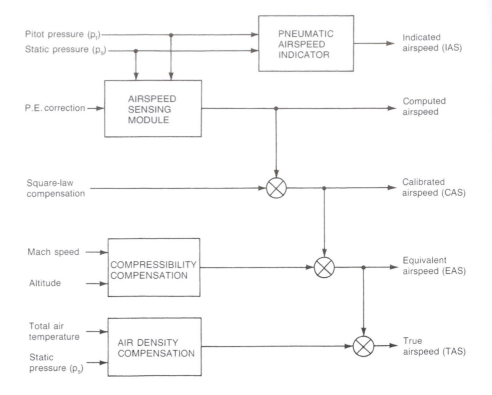

Figure 2.15 Airspeed terminology.

True airspeed (TAS)   This is EAS compensated for changes in air temperature and density at various flight levels. This is also done automatically in air data computer systems.

The foregoing airspeeds are summarized pictorially in Fig. 2.15.

Limiting speeds

V_{mo} Maximum operating speed in knots.
M_{mo} Maximum operating speed in terms of Mach number.

Typical indicator   The mechanism of a typical pneumatic-type airspeed indicator is illustrated in Fig. 2.16. The pressure-sensing element is a metal capsule, the interior of which is connected to the pitot pressure connector via a short length of capillary tube which damps out pressure surges. Static pressure is exerted on the exterior of the capsule and is fed into the instrument case via the second connector. Except for this connector the case is sealed.

Displacements of the capsule in accordance with what is called the 'square-law' are transmitted via a magnifying lever system, gearing, and a square-law compensating device to the pointer, which moves over a linear scale calibrated in knots. Temperature compensation is achieved by a bimetallic strip arranged to vary the magnification of

Figure 2.16 Typical pneumatic airspeed indicator.

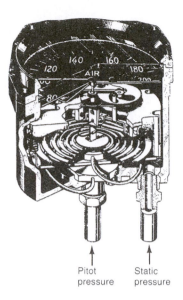

Figure 2.17 Square-law characteristics. (*a*) Effect of linear deflection/pressure response; (*b*) effect of direct magnification.

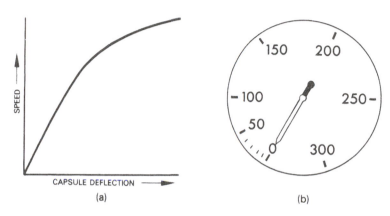

the lever system in opposition to the effects of temperature on system and capsule sensitivity.

Square-law compensation Since airspeed indicators measure a differential pressure which varies with the square of the airspeed, it follows that, if the deflections of the capsules responded linearly to the pressure, the response characteristic in relation to speed would be similar to that shown in Fig. 2.17(*a*). If also the capsule were coupled to the pointer mechanism so that its deflections were directly magnified, the instrument scale would be of the type indicated at (*b*).

The non-linearity of such a scale makes it difficult to read accurately, particularly at the low end of the speed range; furthermore, the scale length for a wide speed range would be too great to accommodate conveniently in the standard dial sizes.

Therefore, to obtain the desired linearity a method of controlling either the capsule characteristic, or the dimensioning of the coupling

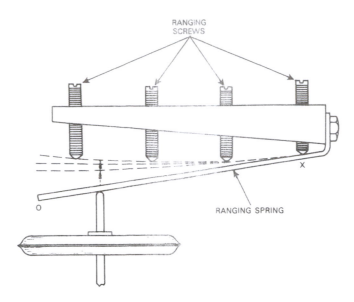

Figure 2.18 Tuning spring compensator. *OX* = effective spring length diminishing as spring makes contact with screws.

RANGING SCREWS

RANGING SPRING

O

X

element conveying capsule deflections to the pointer, is necessary. Of the two methods the latter is the more practical because means of adjustment can be incorporated to overcome the effects of capsule 'drift' plus other mechanical irregularities as determined during calibration.

The principle of a commonly used version of the foregoing method is one in which the length of a lever is altered as progressive deflections of the capsule take place, causing the mechanism, and pointer movement, to be increased for small deflections and decreased for large deflections. In other words, it is a principle of variable magnification.

Another type of square-law compensating device is shown in Fig. 2.18. It consists of a special ranging or 'tuning' spring which bears against the capsule and applies a controlled retarding force to capsule expansion. The retarding force is governed by sets of ranging screws which are pre-adjusted to contact the spring at appropriate points as it is lifted by the expanding capsule. As speed and differential pressure increase, the spring rate increases and its effective length is shortened; thus linearity is obtained directly at the capsule and eliminates the need for a variable magnifying lever system. In some types of servo-operated indicators, a specially profiled cam provides square-law compensation (see also page 171).

Machmeters and Mach/airspeed indicators
In order for aircraft to operate at speeds approaching and exceeding that of sound, their aerodynamic profiles and structural design must be such that they minimize, or ideally overcome, the limiting effects that high-velocity airflow and its associated forces could otherwise have on in-flight behaviour of aircraft. Since the speed of sound

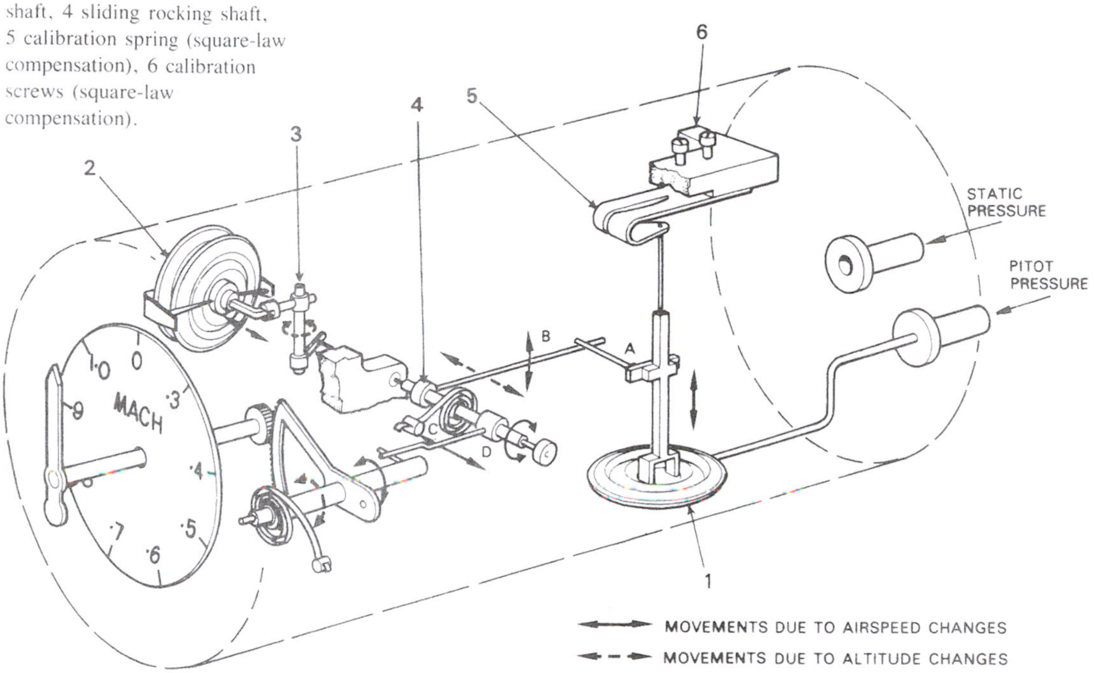

Figure 2.19 Machmeter. 1 Airspeed capsule, 2 altitude capsule, 3 altitude rocking shaft, 4 sliding rocking shaft, 5 calibration spring (square-law compensation), 6 calibration screws (square-law compensation).

depends on atmospheric pressure and density, it will vary with altitude, and this suggests that for an aircraft to operate within speed limits commensurate with structural safety, a different speed would have to be maintained for each altitude. This obviously is not a practical solution, and so it is therefore necessary to have a means whereby the ratio of an aircraft's speed, V, and the speed of sound, a, can be computed from pressure measurement and indicated in a conventional manner. This ratio, V/a, is termed the *Mach number (M)*, and the instrument which measures it is termed a *Machmeter*.

A Machmeter is a compound air data instrument which, as may be seen from Fig. 2.19, accepts two variables and uses them to compute the required ratio. The first variable is *airspeed* and therefore a mechanism based on that of a conventional airspeed indicator is adopted to measure this in terms of the pressure difference $p_t - p_s$, where p_t is the total or pitot pressure, and p_s is the static pressure. The second variable is *altitude*, and this is also measured in the conventional manner, i.e. by means of an aneroid capsule sensitive to p_s. Deflections of the capsules of both mechanisms are transmitted to the indicator pointer by rocking shafts and levers, the dividing function of the altitude unit being accomplished by an intermediate sliding rocking shaft.

Let us assume that the aircraft is flying under standard sea-level conditions at a speed V of 500 mph. The speed of sound at sea-level is approximately 760 mph, therefore the Mach number is 500/760 = 0.65. Now, the speed measured by the airspeed mechanism is, as we have already seen, equal to the pressure difference $p_t - p_s$, and so the sliding rocking shaft and levers A, B, C and D will be set to angular positions determined by this difference. The speed of sound cannot be measured by the instrument, but since it is governed by static pressure conditions, the altimeter mechanism can do the next best thing and that is to measure p_s and feed this into the indicating system, thereby setting a datum position for the point of contact between the levers C and D. Thus a Machmeter indicates the Mach number V/a in terms of the pressure ratio $(p_t - p_s)/p_s$, and for the speed and altitude conditions assumed the pointer will indicate 0.65.

What happens at altitudes above sea-level? As already pointed out, the speed of sound decreases with altitude, and if an aircraft is flown at the same speed at all altitudes, it gets closer to and can exceed the speed of sound. For example, the speed of sound at 10 000 ft decreases to approximately 650 mph, and if an aircraft is flown at 500 mph at this altitude, the Mach number will be 500/650 = 0.75, a 10 per cent increase over its sea-level value. It is for this reason that critical Mach numbers (M_{crit}) are established for the various types of high-speed aircraft, and being constant with respect to altitude it is convenient to express any speed limitations in terms of such numbers.

We may now consider how the altitude mechanism of the Machmeter functions in order to achieve this, by taking the case of an aircraft having an M_{crit} of, say, 0.65. At sea-level and as based on our earlier assumption, the measured airspeed would be 500 mph to maintain $M_{crit} = 0.65$. Now, if the aircraft is to climb to and level off at a flight altitude of 10 000 ft, during the climb the decrease of p_s causes a change in the pressure ratio. It affects the pressure difference $p_t - p_s$ in the same manner as a conventional indicator is affected, i.e. the measured airspeed is decreased. The airspeed mechanism therefore tends to make the pointer indicate a lower Mach number. However, the altitude mechanism simultaneously responds to the decrease in p_s, its capsule expanding and causing the sliding rocking shaft to carry lever C towards the pivot point of lever D.

The magnification ratio between the two levers is therefore altered as the altitude mechanism divides $p_t - p_s$ by p_s, lever D being forced down so as to make the pointer maintain a constant Mach number of 0.65.

The critical Mach number for a particular type of aircraft is indicated by a pre-adjusted lubber mark located over the dial of the Machmeter.

Mach/airspeed indicator This indicator is one which combines the functions of both a conventional airspeed indicator and a Machmeter,

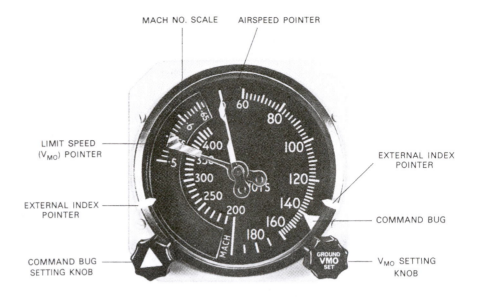

Figure 2.20 Mach/airspeed indicator.

and presents the requisite information in the manner shown in Fig. 2.20. The mechanism consists of two measuring elements which drive their own indicating elements, i.e. a pointer and a fixed scale to indicate airspeed, and a rotating dial and scale calibrated to indicate Mach number. A second pointer, known as the *velocity maximum operating* (V_{mo}) pointer, is also provided for the purpose of indicating the maximum safe speed of an aircraft over its operating altitude range; in other words, it is an indicator of critical Mach number. The pointer is striped red and white and can be pre-adjusted to the desired limiting speed value, by pulling out and rotating the setting knob in the bottom right-hand corner of the indicator bezel. The adjustment is made on the ground against charted information appropriate to the operational requirements of the particular type of aircraft. The purpose of the setting knob in the bottom left-hand corner of the bezel is to enable the pilot to position a command 'bug' with respect to the airspeed scale, thereby setting an airspeed value which may be used as a datum for an autothrottle control system, or as a fast/slow speed indicator. Two external index pointers around the bezel may be manually set to any desired reference speed, e.g. the take-off speeds V_1 and V_R.

In operation, the airspeed measuring and indicating elements respond to the difference between pitot and static pressures in the conventional manner, and changes in static pressure with changes in altitude cause the Mach number scale to rotate (anti-clockwise with increasing altitude) relative to the V_{mo} pointer. When the limiting speed is reached, and the corresponding Mach number graduation coincides with the V_{mo} pointer setting, mechanical contact is made between the scale and pointer actuating assemblies so that continued rotation of the scale will also cause the pointer to rotate in unison.

The pointer rotates against the tension of a hairspring which returns the pointer to its originally selected position when the Mach speed decreases to below the limiting speed. It will be noted from Fig. 2.20 that at the high end of the speed range, the airspeed pointer can also register against the Mach scale, thereby giving a readout of speed in equivalent units. The necessary computation is effected by calibrating the scales to logarithmic functions of pitot and static pressures.

In addition to their basic indicating function, Mach/airspeed indicators can also be designed to acutate switch units coupled to visual or audio devices which give warning when such speeds as Mach limiting, or landing gear extension are reached. In aircraft having an autothrottle system, certain types of Mach/airspeed indicator are designed to provide a speed error output which is proportional to the difference between the reading indicated by the airspeed pointer and the setting of the command 'bug'. This is accomplished by means of a CT/CX synchro (see page 140) combination which senses the positions of the airspeed pointer and the command bug, and produces an output error signal which, after amplification, is then supplied to the autothrottle system.

Indicated/computed airspeed indicator
An example of this type of indicator is shown in Fig. 2.21. It is very similar in construction and presentation to the Mach/airspeed indicator in that it employs pitot and static pressure-sensing elements which position the appropriate pointers. It has, however, the additional feature of indicating the airspeed computed by a central air data computer (see Chapter 7). The indicating element for this purpose is a servomotor-driven digital counter, the motor being supplied with signals from a synchronous transmission system. In the

Figure 2.21 Indicated/computed airspeed indicator.

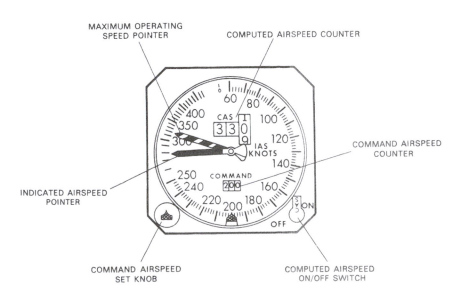

event of failure of such signals a yellow warning flag obscures the counter drums. A check on the operation of the failure monitoring and flag circuits can be made by moving the calibrated airspeed (CAS) switch from its normally 'ON' position to 'OFF'.

As in the case of certain types of Mach/airspeed indicators, provision is made for setting in a command airspeed signal and for transmitting it to an autothrottle system which will adjust the engine power to attain a commanded speed. In the example illustrated, the command set knob mechanically adjusts a synchrotel (see page 149) which also senses indicated airspeed. Thus, the synchrotel establishes the airspeed error signal output required by the autothrottle computer. A readout of the command speed is given on a digital counter which is also mechanically set by the command speed knob.

Altimeters (pneumatic)

These instruments operate on the aneroid barometer principle; in other words they respond to changes in atmospheric pressure, and in accordance with appropriate calibration laws they indicate these changes in terms of equivalent altitude values.

The dial presentation and mechanical features of a typical pneumatic type of altimeter are shown in Fig. 2.22. The pressure-sensing element consists of twin capsules, which transmit their

Figure 2.22 Pneumatic-type altimeter.

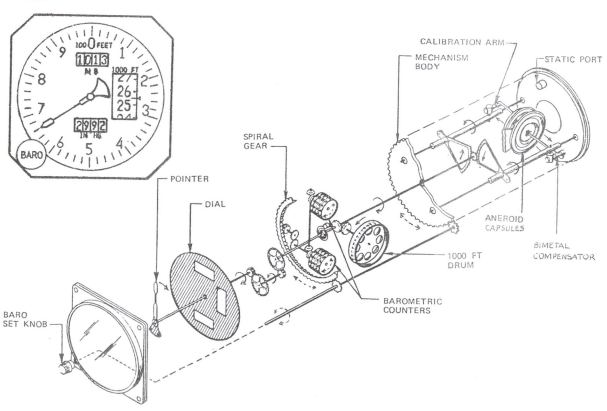

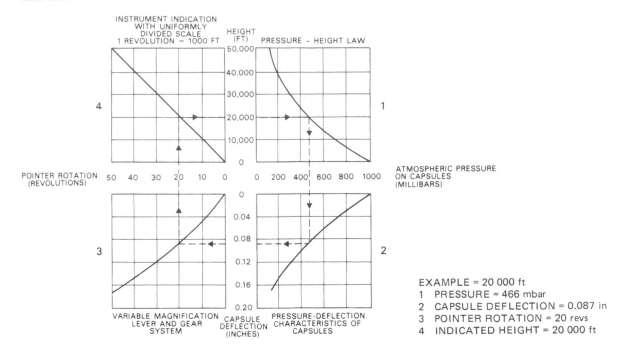

Figure 2.23 Conversion of pressure/height relationship to a linear scale.

INSTRUMENT INDICATION WITH UNIFORMLY DIVIDED SCALE
1 REVOLUTION = 1000 FT

HEIGHT (FT)

PRESSURE – HEIGHT LAW

4

1

POINTER ROTATION (REVOLUTIONS)

ATMOSPHERIC PRESSURE ON CAPSULES (MILLIBARS)

3

2

VARIABLE MAGNIFICATION LEVER AND GEAR SYSTEM

CAPSULE DEFLECTION (INCHES)

PRESSURE-DEFLECTION CHARACTERISTICS OF CAPSULES

EXAMPLE = 20 000 ft
1   PRESSURE = 466 mbar
2   CAPSULE DEFLECTION = 0.087 in
3   POINTER ROTATION = 20 revs
4   INDICATED HEIGHT = 20 000 ft

deflections in response to pressure changes, to a single pointer and altitude drum via sector gears and pinions. The direction of the solid arrows shown in the diagram corresponds to the movements obtained under increasing altitude conditions. The complete mechanism is contained within a casing which, with the exception of the static pressure connection, forms a sealed unit.

In order to derive a linear altitude scale from the non-linear pressure/altitude relationship (see Fig. 2.1 again) it is necessary to incorporate some form of conversion within the altimeter mechanism. This conversion is represented by the graphical example shown in Fig. 2.23. Typically, linearity is obtained by a suitable choice of material for the capsules and their corresponding deflections (curve 2) and also choice of deflection characteristics of the variable magnification and lever system for transmitting the relevant deflections to the pointer (curve 3). The resultant of both curves produces the linear scale as at curve 4. To cater for variations between deflection characteristics of individual capsules, and so allow for calibration, adjustments are always provided whereby the lever and gear system magnification may be matched to suit the capsule characteristics.

The pressure-sensing element of the altimeter is compensated for changes in temperature of the air supplied to it by a bi-metal compensator device connected in the magnification lever system. The temperature coefficient of the instrument is chiefly due to the change of elasticity of the capsule material with change of temperature; this, in turn, varies the degree of deflection of the capsules in relation to the pressure acting external to them. For example, if at sea-level the temperature should decrease, the elasticity of the capsules would increase; in other words, and from the definition of elasticity, the capsules have a greater tendency 'to return to their original size' and so would expand and cause the altimeter to over-read. At higher altitudes the same effects on elasticity will take place, but since the pressures acting on the capsules will also have decreased, then, by comparison, the expansion of the capsules becomes progressively greater. The bi-metal compensator is simultaneously affected by the decrease in ambient temperature, but by virtue of its characteristics it exerts forces through the lever system to oppose the error-producing deflection of the capsules.

Barometric pressure setting As pointed out earlier in this chapter, the basis for the calibration of air data instruments is the ISA and its assumed values. As far as altimeters are concerned, they will, under ISA conditions, indicate what is termed *pressure altitude*. In practice, however, atmospheric pressure and temperature are continually changing, and so under these 'non-standard' conditions altimeters would be in error and would then display what is termed *indicated altitude*.

We may consider these errors by taking the case of a simple altimeter situated at various levels. In standard conditions, and at a sea-level airfield, an altimeter would respond to a pressure of 1013.25 mb (29.92 in Hg) and indicate the pressure altitude of zero feet. Similarly, at an airfield level of 1000 ft, it would respond to a standard pressure of 977.4 mb (28.86 in Hg) and indicate a pressure altitude of 1000 ft. Assuming that at the sea-level airfield the pressure falls to 1012.2 mb (29.89 in Hg), the altimeter will indicate that the airfield is approximately 30 ft above sea-level; in other words, it will be in error by +30 ft. Again, if the pressure increases to 1014.2 mb (29.95 in Hg), the altimeter in responding to the pressure change will indicate that the airfield is approximately 30 ft below sea-level, an error of −30 ft.

In a similar manner, errors would be introduced in the readings of such an altimeter in flight and whenever the atmospheric pressure at any particular altitude departed from the assumed standard value. For example, when an aeroplane flying at 5000 ft enters a region in which the pressure has fallen from the standard value of 842.98 mb to, say, 837 mb, the altimeter will indicate an altitude of approximately 5190 ft.

Since the ISA also assumes certain temperature values at all altitudes, then consequently non-standard values can also cause errors

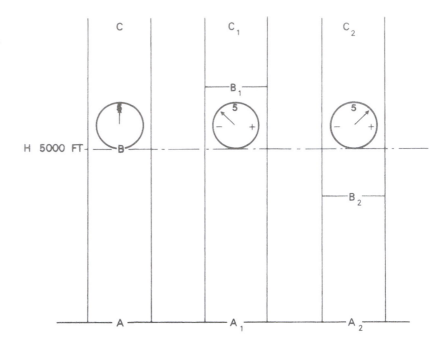

Figure 2.24 Effect of atmospheric temperature on an altimeter.

in altimeter readings. Variations in temperature cause differences of air density and therefore differences in weight and pressure of the air. This may be seen from the three columns shown in Fig. 2.24. At point A the altimeter measures the pressure of the column AC. At point H, which is, say, at an altitude of 5000 ft above A, the pressure on the altimeter is less by that of the part AB below it. If the temperature of the air in part AB increases, the column of air will expand to A_1B_1, and so the pressure on the altimeter at H will now be less by that of A_1H. The pressure of A_1B_1 is, however, still the same as that of AB, and so the pressure of A_1H must be less than that of AH. Thus the altimeter, in rising from A_1 to 5000 ft, will register a smaller reduction of pressure than when it rose from point A to 5000 ft. In other words, it will read less than 5000 ft. Similarly, when the temperature of the air between points A and H decreases, the part AB of the column reduces to A_2B_2 and the change of pressure on the altimeter in rising from A_2 to 5000 ft will be not only the pressure of A_2B_2 (which equals AB) but also the pressure of B_2H. The altimeter will thus read a greater pressure drop and will indicate an altitude greater than 5000 ft. The relationship between the various altitudes associated with flight operations is presented graphically in Fig. 2.25.

It will be apparent from the foregoing that, although the simple form of altimeter performs its basic function of measuring changes in atmospheric pressure accurately enough, the corresponding altitude indications are of little value unless they are corrected to standard pressure data. In order, therefore, to compensate for altitude errors

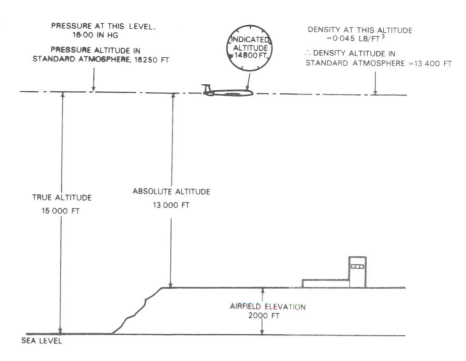

Figure 2.25 Relation between various altitudes.

PRESSURE AT THIS LEVEL,
16·00 IN HG

PRESSURE ALTITUDE IN
STANDARD ATMOSPHERE, 16250 FT

INDICATED
ALTITUDE
14800 FT.

DENSITY AT THIS ALTITUDE
=0·045 LB/FT³

∴ DENSITY ALTITUDE IN
STANDARD ATMOSPHERE =13 400 FT

TRUE ALTITUDE
15 000 FT

ABSOLUTE ALTITUDE
13 000 FT

AIRFIELD ELEVATION
2000 FT

SEA LEVEL

due to atmospheric pressure changes, altimeters are provided with a manually-operated setting device which allows prevailing ground pressure values to be preset.

In the altimeter shown in Fig. 2.22, the adjustment device consists of two drum counters (one calibrated in in. Hg and the other in mb) interconnected through gearing to a setting knob. When the knob is rotated then, as shown by the dotted arrows, both counters can be set to indicate the prevailing barometric pressure, i.e. the static pressure, in the equivalent units of measurement. Likewise it will be noted that the setting knob is also geared to the sensing element mechanism body, so that this mechanism can also be rotated. The deflected position of the capsules appropriate to whatever pressure is acting on them at the time will not be disturbed by rotation of the mechanism. However, in order to maintain the correct pressure/altitude relationship, rotation of the setting knob will cause the altitude pointer and drum to rotate and so indicate the altitude corresponding to the pressures set on the counters. The underlying principle of this may be understood by considering the setting device to be a millibar scale having a simple geared connection to the altitude pointer as shown in Fig. 2.26.

At (a) the altimeter is assumed to be subjected to standard conditions; thus the millibar scale, in this case, when set to 1013.25 mb, positions the pointer at the 0 ft graduation. If the setting is then changed to, say, 1003 mb as at (b), the scale will be rotated clockwise, causing the altitude pointer to rotate anti-clockwise and to indicate −270 ft. If now the altimeter is raised through 270 ft as at

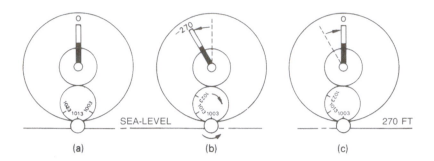

Figure 2.26 Principle of barometric pressure setting.

(a) (b) (c)

SEA-LEVEL   270 FT

(c) a pressure decrease of 10 mb will be sensed by the capsule and its corresponding deflection will cause the altitude pointer to return to the zero graduation. Thus, whatever pressure is set on the millibar scale, the altimeter will indicate zero when subjected to that pressure. Similarly, any setting of the altitude pointer automatically adjusts the millibar scale reading to indicate the pressure at which the altitude indicated will be zero.

'Q' code for altimeter setting The setting of altimeters to the barometric pressures prevailing at various flight levels and airports is part of flight operating techniques, and is essential for maintaining adequate separation between aircraft, and also terrain clearance during take-off and landing. In order to make the settings flight crew are dependent on observed meteorological data which are requested and transmitted from air traffic control. The requests and transmissions are adopted universally and form part of the ICAO 'Q' code of communication.

There are two code letter groups commonly used in connection with altimeter setting procedures, and they are defined as follows:

QFE Setting the barometric pressure prevailing at an airport to make the altimeter read zero on landing at, and taking off from, that airport. The zero reading is regardless of the airport's elevation above sea-level.

QNH Setting the barometric pressure to make the altimeter read airport elevation above sea-level on landing and take-off. The pressure set is a value reduced to mean sea-level in accordance with ISA. When used for landing and take-off, the setting is generally known as '*airport QNH*'. Any value is only valid in the immediate vicinity of the airport concerned.

Since an altimeter with a QNH setting reads altitude above sea-level, the setting is also useful in determining terrain clearance when an aircraft is en route. For this purpose, the UK and surrounding seas are divided into fourteen *Altimeter Setting Regions*, each transmitting an hourly '*regional QNH*' forecast.

There is also a third setting and this is referred to as the *Standard Altimeter Setting* (SAS), in which the barometric pressure counters

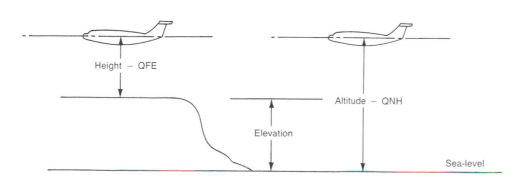

Figure 2.27 Altitude, elevation and height.

Transition altitude

Flight levels — SAS

Height — QFE

Altitude — QNH

Elevation

Sea-level

are set to the ISA values of 1013.2 mb or 29.92 in Hg. It is used for flights above a prescribed *transition altitude* and has the advantage that with all aircraft using the same airspace and flying on the same altimeter setting, the requisite limits of separation between aircraft can more readily be maintained. The transition altitude within UK airspace is usually 3000 ft to 6000 ft, and from these data altitudes are quoted as *flight levels*: e.g. 4000 ft is FL 40 and 15 000 ft is FL 150.

The following definitions, together with Fig. 2.27, show how the terms 'altitude', 'elevation' and 'height' are used in relation to altimeter setting procedures.

Altitude is the vertical distance of a level, point or object considered as a point above mean sea-level. Thus, an altimeter indicates an altitude when a QNH is set.

Elevation is the vertical distance of a fixed point above or below mean sea-level. For altimeter settings the QFE datum used is the airport elevation which is the highest usable point on the landing area. Where a runway is below the airport elevation, the QFE datum used is the elevation of the touchdown point, referred to as touchdown elevation.

Height is the vertical distance of a level, point or object considered as a point measured from a specified datum. Thus, an altimeter indicates a height above airport elevation (the specified datum) when a QFE is set.

Vertical speed indicators
These indicators (also known as rate-of-climb indicators) are the third of the primary group of air data instruments, and are very sensitive

Figure 2.28 Vertical speed indicator presentation.

differential pressure gauges, designed to indicate the rate of altitude change from variations in static pressure alone.

Since the rate at which the static pressure changes is involved in determining vertical speed, a time factor has to be introduced as a pressure function. This is accomplished by incorporating a special air metering unit in the sensing system, its purpose being to create a lag in static pressure across the system and so establish the required pressure differences.

A pneumatic type of indicator consists basically of three principal components: a capsule-type sensing element, an indicating element and a metering unit, all of which are housed in a sealed case connected to the static pressure source. The dial presentation is such that zero is at the 9 o'clock position; thus the pointer is horizontal in the straight and level flight attitude, and can move from this position to indicate climb and descent in the correct sense. Certain types of indicator employ a linear scale, but in the majority of applications, indicators having a scale calibrated to indicate the logarithm of the rate of pressure change are preferred. The reason for this is that a logarithmic scale is more open near the zero graduation, and so provides for better readability and for more accurate observation of variations from level flight conditions. A typical example of this presentation is shown in Fig. 2.28.

An indicator mechanism is shown in schematic form in Fig. 2.29, from which it will be noted that the metering unit forms part of the static pressure connection and is connected to the interior of the capsule by a length of capillary tube. This tube serves the same purpose as the one employed in a pneumatic type of airspeed indicator, i.e. it prevents pressure surges reaching the capsule. It is, however, of greater length due to the fact that the capsule is much more flexible and sensitive to pressure changes. The other end of the metering unit is open to the interior of the case to apply static pressure to the exterior of the capsule. Let us now see how the instrument operates under the three flight conditions shown in the diagram.

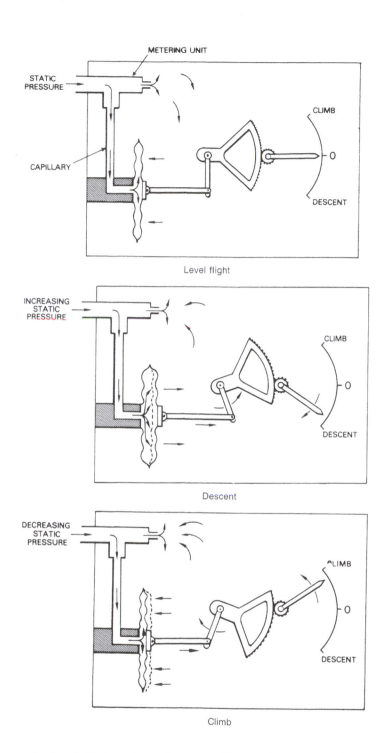

Figure 2.29 Principle of vertical speed indicator.

Level flight

Descent

Climb

In level flight, air at the prevailing static pressure is admitted to the interior of the capsule, and also to the instrument case via the metering unit. Thus, there is no difference of pressure across the capsule and the pointer indicates zero.

At the instant of commencing a descent, there will still be no

57

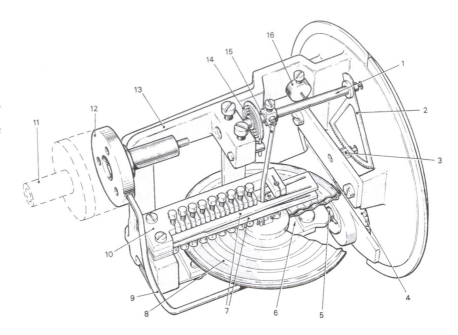

Figure 2.30 Construction of a typical vertical speed indicator. 1 Rocking shaft assembly, 2 sector, 3 hand-staff pinion, 4 gearwheel, 5 eccentric shaft assembly, 6 capsule plate assembly, 7 calibration springs, 8 capsule, 9 capillary tube, 10 calibration bracket, 11 static connection, 12 metering unit, 13 mechanism body, 14 hairspring, 15 link, 16 balance weight.

pressure difference, but as the aircraft is now descending into conditions of higher static pressure then such pressure will be directly sensed inside the capsule. The pressure inside the case, however, will not build up at the same rate as the capsule pressure, because in having to pass through the metering unit the airflow into the case is restricted. Thus, a differential pressure is created across the capsule, causing it to distend and so make the pointer indicate a descent.

During a climb, an aircraft will of course pass through conditions of decreasing static pressure, but as the metering unit will then restrict the airflow out of the case, a differential pressure is created, as a result of case pressure now being greater than that inside the capsule, causing it to collapse and so make the pointer indicate a climb.

Apart from the changes of static pressure with changes of altitude; air temperature, density and viscosity changes are other very important variables which must be taken into account, particularly as the instrument depends on rates of airflow. In addition, the volumetric capacities of cases and capsules must also be considered in order to obtain the constant differential pressures necessary for the indication of specific rates of climb and descent. Metering units are designed to compensate for the effects of the variables over the ranges normally encountered, and so from the theoretical point of view a vertical speed indicator is a somewhat sophisticated instrument.

The construction of a typical indicator is shown in Fig. 2.30. It consists of a cast aluminium-alloy body which forms the support for all the principal components with the exception of the metering unit, which is secured to the rear of the instrument case. Displacements of the capsule in response to differential pressure changes are transmitted to the pointer via a balanced link and rocking-shaft assembly, and a quadrant and pinion. The flange of the metering unit connects with the static pressure connection of the indicator case, and with the capsule via the capillary tube.

In order to achieve the correct relationship between the capsule's pressure/deflection characteristics and the pointer position at all points of the scale, forces are exerted on the capsule by two pre-adjusted calibration springs. The upper spring and its adjusting screws control the rate of descent calibration, while the lower spring and screws control that of rate of climb.

An adjustment device is provided at the front of the indicator for setting the pointer to zero, and when operated it moves the capsule assembly up or down to position the pointer via the magnifying lever and gearing system. The range of adjustment around zero depends on the scale range of any one type of indicator, but ±200 and ±400 ft/min are typical values.

Instantaneous vertical speed indicators (IVSI) These indicators consist of the same basic elements as conventional VSIs, but in addition they employ an accelerometer unit which is designed to create a more rapid differential pressure effect, specifically at the initiation of a climb or descent. The basic principle is illustrated in Fig. 2.31.

The accelerometer comprises a small cylinder, or dashpot, containing a piston held in balance by a spring and by its own mass. The cylinder is connected in a capillary tube leading to the capsule, and is thus open directly to the static pressure source. When a change in vertical speed is initiated, the piston is immediately displayed under the influence of a vertical acceleration force, and this creates an immediate pressure change inside the capsule. For example, at initiation of a descent, the piston moves up and thereby decreases the volume of chamber 'A' to produce an immediate increase of pressure inside the capsule. The capsule displacement in turn produces instantaneous deflection of the indicator pointer over the descent portion of the scale. At initiation of an ascent, the converse of the foregoing responses would apply. The accelerometer response decays in each case after a few seconds, but by this time the change in actual static pressure becomes effective, so that a pressure differential is produced by the metering unit in the conventional manner. The purpose of the restrictor in the bypass line is to prevent any loss of pressure change effects created by displacements at the acceleration pump.

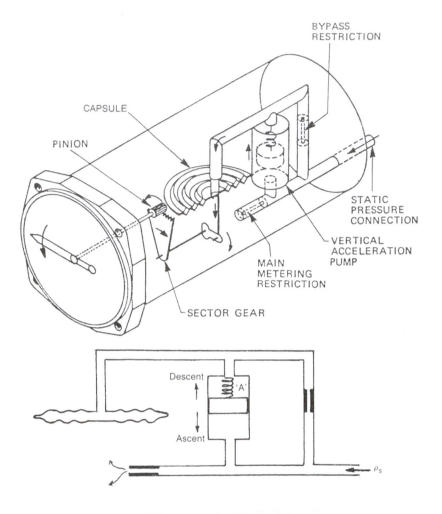

Figure 2.31 Instantaneous vertical speed indicator.

BYPASS RESTRICTION

CAPSULE

PINION

STATIC PRESSURE CONNECTION

VERTICAL ACCELERATION PUMP

MAIN METERING RESTRICTION

SECTOR GEAR

Descent

'A'

Ascent

ρ_s

Air temperature sensing

Air temperature is another of the basic parameters used to establish data vital to the performance monitoring of aircraft and engines, e.g. true airspeed measurement, temperature control, thrust settings, fuel/air ratio settings, etc. of turbine engines, and it is therefore necessary to provide a means of in-flight measurement.

The temperature which would overall be the most ideal is that of air under pure static conditions at the flight levels compatible with the operating range of any particular type of aircraft concerned. The measurement of *static air temperature* (SAT) by direct means is, however, not possible for all types of aircraft for the reason that measurements can be affected by the adiabatic compression of air resulting from increases in airspeed. At speeds below 0.2 Mach, the air temperature is very close to static conditions, but at higher speeds, and as a result of changes in boundary layer behaviour and the effects of friction, the temperature is raised to a value appreciably higher than SAT; this increase is referred to as *ram rise*.

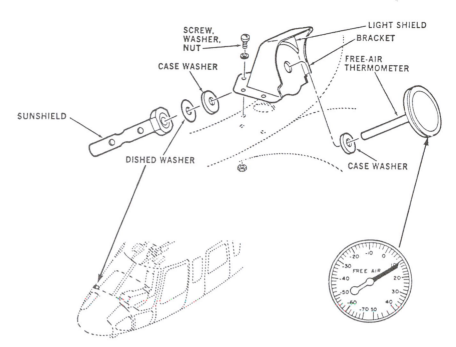

Figure 2.32 Direct-reading air temperature indicator.

For use in aircraft capable of high Mach speeds, and for efficient control and management of the overall performance of their engines, it is customary to sense and measure the maximum temperature rise possible. This parameter is referred to as *total air temperature* (TAT) and is derived when the air is brought to rest (or nearly so) without further addition or removal of heat. If the corresponding SAT value is to be determined and indicated, it is necessary to calculate the value of ram rise and then subtract it from that of TAT. Details of the method by which this is normally accomplished will be given in Chapter 7.

Various types of sensor may be adopted for the sensing of air temperature. The simplest type, and one which is used in some types of small low-speed aircraft for the indication of SAT, is a direct-reading indicator which operates on the principle of expansion and contraction of a bi-metallic element when subjected to temperature changes. The element is arranged in the form of a helix anchored at one end of a metal sheath or probe; the opposite, or free end of the helix, is attached to the spindle of a pointer. As the helix expands or contracts, it winds or unwinds causing the pointer to rotate over the scale of a dial fixed to the probe. The thermometer is secured through a fixing hole in the side window of a cockpit, or in the wrap-around portion of a windscreen, so that the probe protrudes into the airstream. An example of this thermometer and its installation in one type of helicopter is shown in Fig. 2.32.

The measurement of TAT requires a more sophisticated measuring technique, and because the proportion of ram rise due to adiabatic

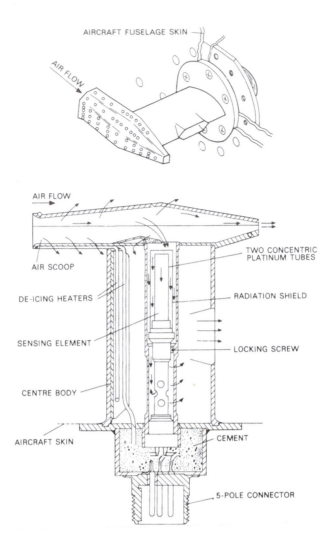

Figure 2.33 Total air temperature probe (1).

AIRCRAFT FUSELAGE SKIN

AIR FLOW

AIR FLOW

TWO CONCENTRIC PLATINUM TUBES

AIR SCOOP

DE-ICING HEATERS

RADIATION SHIELD

SENSING ELEMENT

LOCKING SCREW

CENTRE BODY

AIRCRAFT SKIN

CEMENT

5-POLE CONNECTOR

compression is dependent on the ability of a sensor to sense and recover this rise, then a TAT sensor must itself be of a more sophisticated design. In this context, the sensitivity of a sensor is normally expressed as a percentage termed the *recovery factor*. Thus, a sensor having a factor of 0.80 would measure SAT plus 80 per cent of the ram rise.

TAT sensors are of the probe type, and one example is shown in Fig. 2.33. The probe is in the form of a small strut and air intake made of nickel-plated beryllium copper which provides good thermal conductivity and strength. It is secured at a pre-determined location in the front fuselage section of an aircraft (typically at the side, or upper surface of the nose) and outside of any boundary layer which may exist. In flight, the air flows through the probe in the manner indicated; separation of any water particles from the air is effected by the airflow being caused to turn through a right angle before passing round the sensing element. The bleed holes in the intake casing

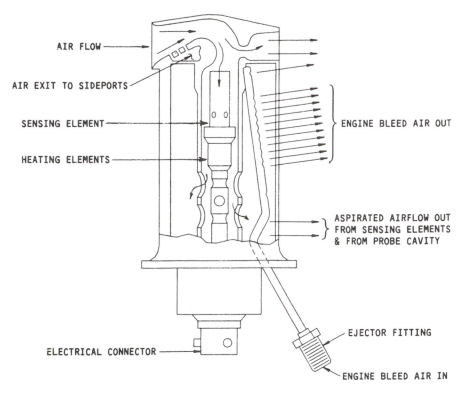

Figure 2.34 Total air temperature probe (2).

AIR FLOW

AIR EXIT TO SIDEPORTS

SENSING ELEMENT

HEATING ELEMENTS

ENGINE BLEED AIR OUT

ASPIRATED AIRFLOW OUT FROM SENSING ELEMENTS & FROM PROBE CAVITY

EJECTOR FITTING

ENGINE BLEED AIR IN

ELECTRICAL CONNECTOR

permit boundary layer air to be drawn off under the influence of the higher pressure that is created within the intake and casing of the probe.

A pure platinum wire resistance-type sensing element is used and is hermetically sealed within two concentric platinum tubes. The element is wound on the inner tube, and since they are both of the same metal, a close match of thermal expansion and minimizing of thermal strain is ensured. The probe has an almost negligible time-lag, and a high recovery factor of approximately 1.00. An axial wire heating element, supplied with 115 V ac at 400 Hz, is mounted integral with the probe to prevent the formation of ice, and is of the self-compensating type in that as the temperature rises so does the element resistance rise, thereby reducing the heater current. The heater dissipates a nominal 260 W under in-flight icing conditions, and can have an effect on indicated air temperature readings. The errors involved, however, are small, some typical values obtained experimentally being 0.9°C at 0.1 Mach decreasing to 0.15°C at Mach 1.0.

A second type of TAT probe is shown in Fig. 2.34. The principal differences between it and the one just described relate to the air intake configuration and the manner in which airflow is directed through it and the probe casing. The purpose of the engine bleed air injector fitting and tube is to create a negative differential pressure within the casing so that outside air is drawn through it at such a rate that the heating elements have a negligible effect on the temperature/resistance characteristics of the sensing element.

In some cases, an auxiliary sensing element is provided in a probe. The purpose of this element is to transmit a signal to other systems requiring air temperature information. An example of this would be the airspeed measuring circuit of an ADC for the computing of true airspeed (see Chapter 7).

Air temperature indicators   As in the case of other instruments, TAT indicators can, as a result of the instrumentation arrangements adopted for each particular type of aircraft, vary in the manner in which they display the relevant data. Some of the variations are illustrated in Fig. 2.35.

The circuit of a probe and a basic conventional pointer and scale type of indicator is shown in Fig. 2.36. The system is supplied with 115 V ac which is then stepped down and rectified by a power supply module within the indicator. The probe element forms one part of a resistance bridge circuit, and as the element's resistance changes with temperature, the bridge is unbalanced, causing current to flow through the moving coil of the indicator.

Figure 2.37 illustrates the circuit arrangement of a servo-operated indicator employing a mechanical drum-type digital counter display. The generation of the appropriate temperature signals is also accomplished by means of a dc bridge circuit, but in this case unbalanced conditions are monitored by a solid-state chopper circuit which produces an error signal to drive an ac servomotor via an operational amplifier. The motor then drives the counter drums, and at the same time positions the wiper contact of a potentiometer to start rebalancing of the circuit, until at some constant temperature condition the circuit is 'nulled'.

In order to indicate whether temperatures are either positive or negative, the rebalancing/feedback system also activates a 'sign changer' and an indicator drum, and a switch which reverses the polarity of the bridge circuit when the temperature indications pass through zero.

Detection of failure of the 26 V ac power to the indicator, and sensing of an excessive null voltage in the rebalancing/feedback system, is provided by a failure monitor circuit module. This controls an 'OFF' flag which under normal conditions is held out of view by an energized solenoid.

The internal arrangement of an LCD (see page 15) type of indicator is schematically shown in Fig. 2.38. The temperature data signals are transmitted from a digital type of ADC (see Chapter 7) via a data bus and receiver to a microcomputer. The power supply to the computer is connected via supply, low voltage and failure monitor modules. In addition to TAT, the indicator can also display SAT and

Figure 2.35 TAT displays. (*a*) Conventional pointer and scale; (*b*) servo/digital counter; (*c*) LCD; (*d*) electronic (CRT).

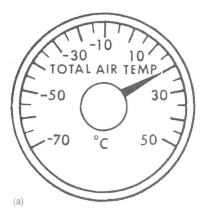

(a)

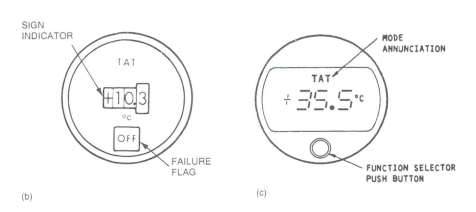

SIGN INDICATOR

TAT

°C

FAILURE FLAG

(b)

MODE ANNUNCIATION

TAT

FUNCTION SELECTOR PUSH BUTTON

(c)

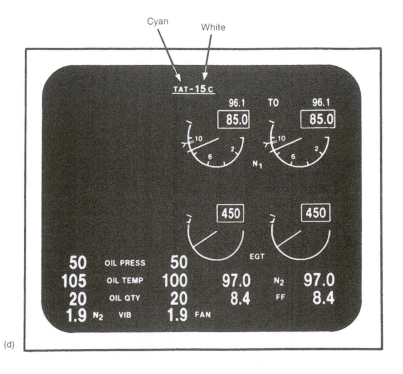

Cyan   White

(d)

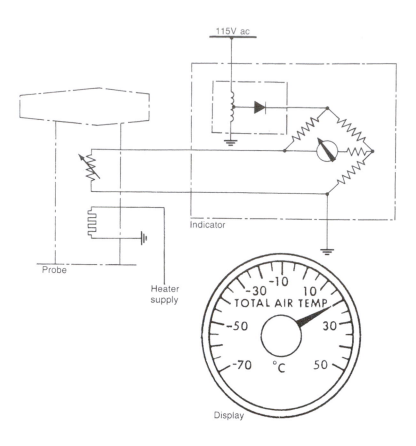

Figure 2.36 TAT indicator system.

115V ac

Indicator

Probe

Heater supply

TOTAL AIR TEMP
-30 -10 10
-50 30
-70 °C 50

Display

Figure 2.37 Servo-operated TAT indicator.

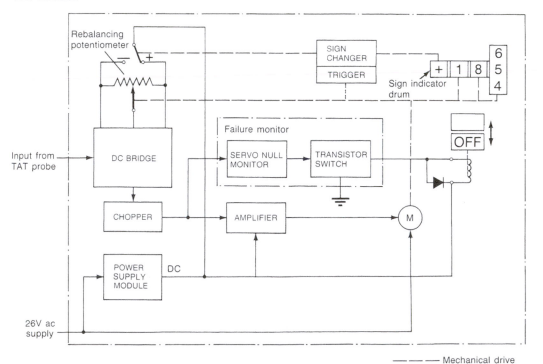

Rebalancing potentiometer

SIGN CHANGER

TRIGGER

Sign indicator drum

+ 1 8 6 5 4

Failure monitor

Input from TAT probe

DC BRIDGE

SERVO NULL MONITOR

TRANSISTOR SWITCH

OFF

CHOPPER

AMPLIFIER

M

POWER SUPPLY MODULE

DC

26V ac supply

— — — Mechanical drive

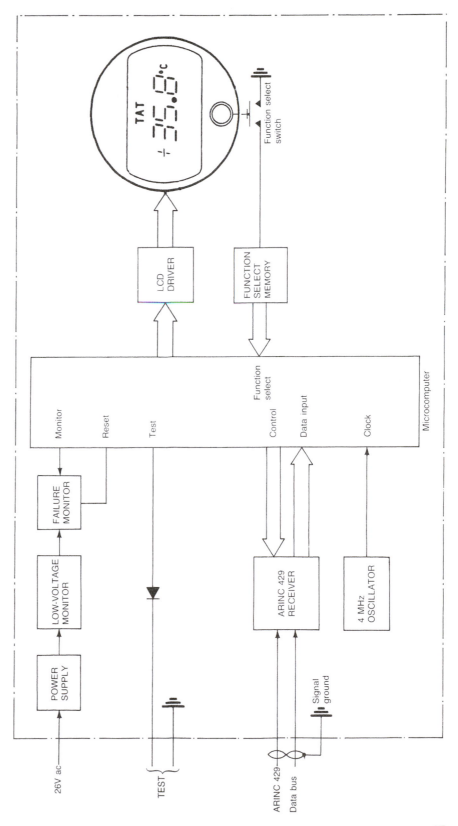

Figure 2.38 LCD-type air temperature indicator.

67

Figure 2.39 Function mode selection sequence.

TAT

$\pm\ 35.8\ \text{°C}$

Function selector push-button

$\pm\ 30.0\ \text{°C}$

SAT

$450\ \text{KT}$

TAS

TAS, each of which can be selected in sequence by a push-button function select switch. When power is first applied, the indicator displays TAT, as in Fig. 2.39; to select SAT the switch is pushed in, and then pushed in again to display TAS. Pushing the switch in for a third time returns the display to TAT. A test input facility is provided, and when activated it causes the display to alternate between all seven segments (of each of the three digits), 'ON' for two seconds, and blank for one second.

As noted earlier, indications of SAT can be derived by subtracting the ram rise from the measured values of TAT. Since this is normally done by also supplying TAT signals to the speed-measuring module of an ADC, the operating principles of SAT indication will be covered in Chapter 7.

Details of the coloured display shown at (d) of Fig. 2.35 will be given in Chapter 16.

Air data alerting and warning systems

In connection with the in-flight operation of aircraft, it is necessary to impose limitations in respect of certain operating parameters compatible with the airworthiness standards to which each type of aircraft is certificated. It is also necessary for systems to be provided which will, both visually and aurally, alert and warn a flight crew whenever the imposed operational limitations are being exceeded.

The number of parameters to be monitored in this way varies in relation to the type of aircraft and the number of systems required for its operation overall. As far as air data measuring systems are concerned, the principal parameters are airspeed and altitude, so let us now consider the operating principles of associated alerting and warning systems typical of those used in some of the larger types of public transport aircraft.

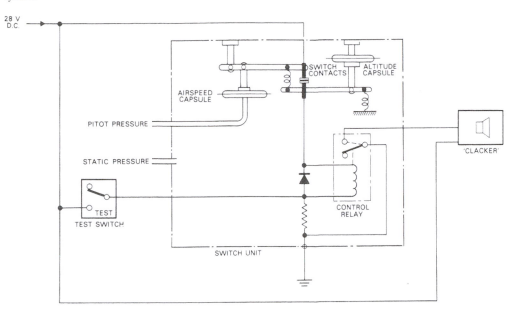

Figure 2.40 Mach warning system.

Mach warning system

This system provides an aural warning when an aircraft's speed reaches the maximum operating value in terms of Mach number, i.e. M_{mo} (a typical value is $0.84M$). The warning is in addition to any limiting speed reference pointers or 'bugs' that are provided in Mach/airspeed indicators (see Fig. 2.20 again).

The system consists of a switch unit which, as can be seen from Fig. 2.40, comprises airspeed and altitude sensing units connected to an aircraft's pitot probe and static vent system in a manner similar to that of a Machmeter. It will also be noted that in lieu of a pointer actuating system, the sensing units actuate the contacts of a switch which is connected to a 28 V dc power source.

At speeds below the limiting value, the switch contacts remain closed and the dc passing through them energizes a control relay. The contacts of this relay interrupt the ground connection to an aural warning device generally referred to as a 'clacker' because of the sound it emits when in operation. When the limiting Mach speed at any given altitude is reached, the airspeed sensing unit causes the switch contacts to open, thereby de-energizing the control relay so that its contacts now complete a connection from the 'clacker' to ground. Since the 'clacker' is directly supplied with dc, then it will be activated to provide the appropriate warning, which is emitted at a specific frequency (typically 7 Hz).

A toggle switch that is spring-loaded to 'OFF' is provided for the

69

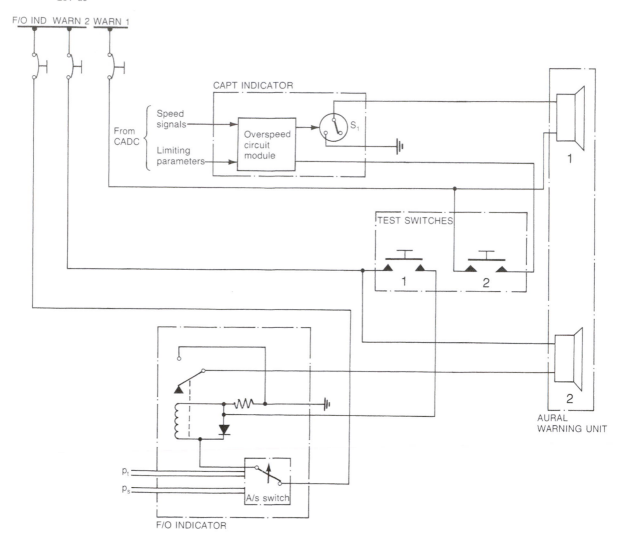

Figure 2.41 Combined indicator and switch unit system.

purpose of functional checking the system. When placed in the 'TEST' position, it allows dc to flow to the ground side of the switch unit control relay, thereby providing a bias sufficient to de-energize the relay and so cause the 'clacker' to be activated.

In some aircraft systems, Mach/airspeed indicators with 'built-in' warning switch units may be used and so arranged that they operate two independent 'clackers'. In the example shown in Fig. 2.41, the indicator in the captain's group of flight instruments is servo-operated by signals from an ADC. The other indicator, which is in the first officer's group of flight instruments, is also of the servo-operated type, but contains a switch unit that is connected directly to the pitot probe and static vent system. The 'clacker' units associated with the indicators are respectively designated as 'aural warning 1' and 'aural warning 2'.

The captain's indicator contains an overspeed circuit module that is supplied by the ADC with prevailing speed data and also the limiting V_{mo} and M_{mo} values appropriate to the type of aircraft. The circuit module is, in turn, connected to a solid-state switch (S_1) that is powered 'open' at speeds below V_{mo} and M_{mo}. If, however, these speeds are reached, then S_1 is relaxed to provide a ground connection for the dc supply to 'aural warning 1' clacker unit which thus gives the necessary warning.

The contacts of the switch unit in the first officer's indicator are connected to a relay, and since these contacts remain closed at speeds below maximum values, the relay is de-energized. When the maximum speed is reached, the relay coil circuit is interrupted and its contacts then change over to provide a ground connection for the dc supply which activates 'aural warning 2' clacker unit.

Test switches are provided for checking the operation of each clacker by simulation of overspeed conditions. When switch 1 is operated dc is applied to the overspeed circuit module in the captain's indicator, and causes the switch S_1 to relax. The operation of switch 2 applies dc to the relay coil such that it is shorted out against the standing supply from the closed airspeed switch; the relay is therefore de-energized to provide a ground connection for 'aural warning 2' clacker unit.

The indicators themselves provide visual indications of overspeed and these are discernible when the airspeed pointers become positioned coincident with pre-set maximum limit pointers (see Figs 2.20 and 2.21).

Altitude alerting system

This system is designed to alert a flight crew, by aural and visual means, of an aircraft's approach to, or deviation from, a pre-selected altitude. The components of a typical system are shown in Fig. 2.42(a).

An aircraft's pressure altitude is provided as a signal input to the alert controller unit from an altimeter via a coarse/fine synchro system. The selected altitude is set by means of a knob on the controller, and is indicated by a digital counter which is geared to the rotors of control and resolver synchros, so that they produce a corresponding signal. The signal is compared with the pressure altitude signal, and the resulting difference is supplied to level detection circuits within the controller. At predetermined values of rotor voltages of both synchros, two signals are produced and

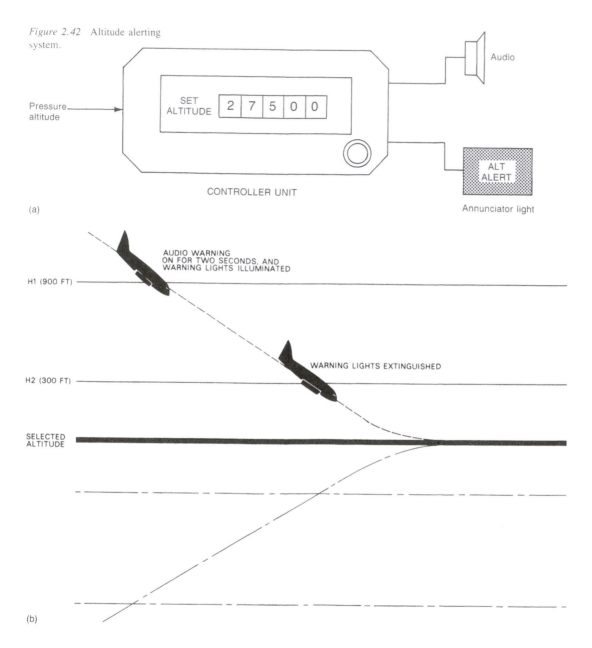

Figure 2.42 Altitude alerting system.

SET ALTITUDE | 2 | 7 | 5 | 0 | 0

Pressure altitude

Audio

ALT ALERT

CONTROLLER UNIT

(a)

Annunciator light

AUDIO WARNING ON FOR TWO SECONDS, AND WARNING LIGHTS ILLUMINATED

H1 (900 FT)

WARNING LIGHTS EXTINGUISHED

H2 (300 FT)

SELECTED ALTITUDE

(b)

supplied as inputs to a logic circuit and timing network which controls the aural and visual alerting devices.

The sequence of alerting is shown at (b) of Fig. 2.42. As an aircraft descends or climbs to the preselected altitude the difference signal is reduced, and the logic circuit so processes the input signals that, at a pre-set outer limit H_1 (typically 900 ft) above or below preselected altitude, one signal activates the aural alerting device which remains on for two seconds; the annunciator light is also illuminated. The light remains on until at a further pre-set inner limit H_2 (typically 300 ft) above or below preselected altitude, the second signal causes the annunciator light to be extinguished. As an aircraft approaches the preselected altitude, the synchro system approaches the 'null' position, and no further alerting takes place. If an aircraft should subsequently depart from the preselected altitude, the controller logic circuit changes the alerting sequence such that the indications correspond to those given during the approach through outer limit H_1, i.e. aural alert on for two seconds, and annunciator light illuminated.

Angle of attack sensing

The angle of attack (AoA), or *alpha* (α) *angle*, is the angle between the chord line of the wing of an aircraft and the direction of the relative airflow, and is a major factor in determining the magnitude of lift generated by a wing. Lift increases as α increases up to some critical value at which it begins to decrease due to separation of the slow-moving air (the boundary layer) from the upper surface of the wing, which, in turn, results in separation and turbulence of the main airflow. The wing, therefore, assumes a stalled condition, and since it occurs at a particular angle rather than a particular speed, the critical AoA is also referred to as the *stalling angle*. The angle relates to the design of aerofoil section adopted for the wings of any one particular type of aircraft, and so, of course, its value varies accordingly; typically it is between 12° and 18°.

The manner in which an aircraft responds as it approaches and reaches a stalled condition depends on many other factors, such as wing configuration, i.e. high, low, swept-back, and also on whether the horizontal stabilizer is in the 'T'-tail configuration. Other factors relate to the prevailing speed of an aircraft, which largely depends on engine power settings, flap angles, bank angles and rates of change of pitch. The appropriate responses are pre-determined for each type of aircraft in order to derive specifically relevant procedures for recovering from what is, after all, an undesirable situation.

An aircraft will, in its own characteristic manner, provide warning of a stalled condition, e.g. by buffeting, gentle or severe pitch-down attitude change, and/or 'wing drop', and although recoverable, in a situation such as an approach when an aircraft is running out of airspace beneath itself, these inherent warnings could come too late! It is, therefore, necessary to provide a means whereby α can be sensed directly, and at some value just below that at which a stalled condition can occur it can provide an early warning of its onset.

Stall warning systems

The simplest form of system, and one which is adopted in several types of small aircraft, consists of a hinged-vane-type sensor mounted

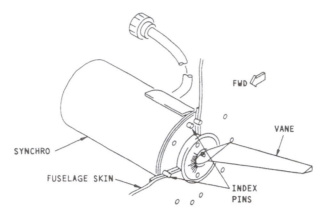

Figure 2.43 Alpha sensor.

FWD

VANE

SYNCHRO

FUSELAGE SKIN

INDEX
PINS

in the leading edge of a wing so that the vane protrudes into the airstream. In normal level flight conditions, the airstream maintains the vane in a parallel position. If the aircraft's attitude changes such that α increases, then, by definition, the airflow will meet the leading edge at an increasing angle, and so cause the vane to be deflected. When α reaches that at which the warning unit has been preset, the vane activates a switch to complete a circuit to an aural warning unit in the cockpit.

In larger types of aircraft, stall warning systems are designed to perform a more active function, in that they are either of the 'stick-shaker' or 'stick push or nudger' type; for some aircraft configurations they are used in combination.

Figure 2.43 illustrates the type of sensor normally used for these systems. It consists of a precision counter-balanced aerodynamic vane which positions the rotor of a synchro. The vane is protected against ice formation by an internal heater element. The complete unit is accurately aligned by means of index pins at the side of the front fuselage section of an aircraft. Since the pitch attitude of an aircraft is also changed by the extension of its flaps, the sensor synchro is also interconnected with a synchro within the transmitter of the flap position indicating system, in order to modify the α signal output as a function of flap position.

Stick-shaking is accomplished by a motor which is secured to a control column and drives a weighted ring that is deliberately unbalanced to set up vibrations of the column, to simulate the natural buffeting associated with a stalled condition.

Sensor signals, and signals for the testing of a system, are processed through a circuit module unit located on a flight deck panel. Control switches for normal operation and for testing are also provided in this unit. Sensing relays and shock strut microswitches on the nose landing gear are included in the circuit of a system to permit operational change-over from ground to air.

The circuit of a typical system is shown in basic form in Fig. 2.44. When the aircraft is on the ground and electrical power is on, the

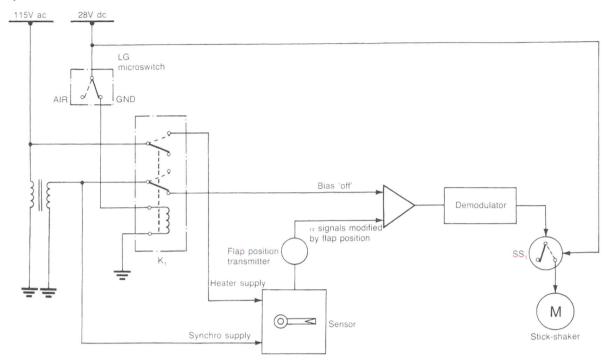

Figure 2.44 Stick-shaker system.

contacts of the landing gear microswitches complete a dc circuit to a sensing relay K_1 which, on being energized, supplies an ac voltage (in this case 11.8 V) to the circuit module amplifier. The output is then supplied to a demodulator whose circuit is designed to 'bias off' the ac voltage from the contacts of K_1, so that the solid-state switch SS_1 remains open to isolate the stick-shaker motor from its dc supply. The vane heater element circuit is also isolated from its ac supply by the opening of the second set of contacts of K_1. The sensor synchro is supplied directly from the ac power source.

During take-off, and when the nose gear 'lifts off', the microswitches operate to de-energize relay K_1, and with the system control switch at 'NORMAL', the system is fully activated. The only signal now supplied to the amplifier and demodulator is the modified α signal.

In normal flight, the signal produced and supplied as input to the amplifier is less than a nominal value of 20 mV, and in phase with the ac voltage supplied as a reference to the demodulator. If the aircraft's attitude should approach that of a stalled condition, the α signal will exceed 20 mV and become out-of-phase. The demodulator then produces a resultant voltage which triggers the switch SS_1 to connect a 28 V dc supply direct to the stick-shaker motor, which then starts vibrating the control column.

A confidence check on system operation may be carried out by placing the circuit module control switch in the 'TEST' position. This energizes a relay which switches the sensor signal to the motor of an indicator, the dial of which will be rotated by the motor if there is circuit continuity. Since the switch isolates the sensor circuit from the amplifier, the reference voltage to the demodulator triggers the switch SS_1 to operate the stick-shaker motor. The control switch also has a 'HEATER OFF' position which isolates the vane heater circuit from its power supply, thus enabling the vane to be manipulated manually without inflicting burns.

In most cases, two systems are installed in an aircraft, so that a sensor is located on each side of the front fuselage section, and a stick-shaker motor on each pilot's control column.

In certain types of aircraft the sensor signals are transmitted to an air data computer, which then supplies an output, corresponding to actual α angle, to a comparator circuit within an electronic module of the stall warning system. The comparator is also supplied with signals from a central processor unit (also within the module) which processes a programme to determine maximum α angles based on the relationship between flap position and three positions of the leading edge slats. The positions are: retracted, partially extended and fully extended, and so signals corresponding to three different computed angles are processed for comparison with an actual α angle signal. If the latter is higher than a computed maximum, the circuit to the stick-shaker motor is completed.

Stick-pushers

In some types of aircraft, particularly those with rear-mounted engines and a 'T'-tail configuration, it is possible for what is termed a 'deep' or 'super' stall situation to develop. When such aircraft first get into a stalled condition then, as in all cases, the air flowing from the wings is of a turbulent nature, and if the α angle is such that the engines are subjected to this airflow, loss of power will occur as a result of surging and possible 'flame-out'. If, then, the stall develops still further, the horizontal stabilizer will also be subjected to the turbulent airflow with a resultant loss of pitch control. The aircraft then sinks rapidly in the deep stalled attitude, from which recovery is difficult, if not impossible. This was a lesson that was learned, with tragic results, during the flight testing of two of the earliest types of commercial aircraft configured as mentioned, namely, the BAC 1−11 and HS 'Trident'.

In order to prevent the development of a deep stall situation, warning systems are installed which, in addition to stick-shaking, utilize the α sensor signals to cause a forward push on the control columns and downward deflection of the elevators. The manner in which this is accomplished varies; in some aircraft, the signals are transmitted to a linear actuator which is mechanically connected to the feel and centring unit of the elevator control system. In aircraft having computerized flight control systems, α sensor signals are transmitted to the elevator control channel of the flight control computer. Whenever stick-push is activated, the elevator control channels of automatic flight control systems are automatically disengaged via an interlock system.

Indicators

There is no standard requirement for angle of attack indicators to be installed in aircraft, with the result that the adoption of any one available type is left as an option on the part of an aircraft manufacturer and/or operator. When selected for installation, however, they must not be used as the only means of providing stall warning, but as a supplement to an appropriate type of stick-shake and stick-push system.

Indicators are connected to the alpha sensors of a stall warning system, and display the relevant data in a variety of ways, depending on their design. In some cases a conventional pointer and scale type of display is used, while in aircraft having electronic flight instrument display systems, the data can be programmed into computers such that it is displayed against a vertical scale, usually located adjacent to that indicating vertical speed, on the attitude director indicator. Another type of indicator currently in use has a pointer which is referenced against horizontal yellow, green and red bands; a dividing line between the yellow and green bands signifies the angle at which the stick-shaker operates.

3 Direct-reading compasses

Compasses of this type were the first of the many airborne flight and navigational aids ever to be introduced in aircraft, their primary function being to show the direction in which an aircraft is heading with respect to the earth's magnetic meridian.

As far as present-day aircraft are concerned, the use of direct-reading compasses as a *primary* directional reference source is confined to small types of aircraft whose design and operating requirements are at a fairly basic level. In the more sophisticated types of aircraft, directional references are derived from flight instrument systems and navigational aids based on advanced technology, and although airworthiness requirements still necessitate the installation of direct-reading compasses, they are relegated to a *secondary* role.

The operating principle of a direct-reading compass is based on established fundamentals of magnetism, and on the reaction between the magnetic field of a suitably suspended magnetic element, and that of terrestrial magnetism.

Terrestrial magnetism   The surface of the earth is surrounded by a weak magnetic field which culminates in two internal *magnetic poles* situated near the North and South *true* or *geographic poles*. That this is so is obvious from the fact that a magnet freely suspended at various parts of the earth's surface will be found to settle in a definite direction, which varies with locality. A plane passing through the magnet and the centre of the earth would trace out on the earth's surface an imaginary line called the *magnetic meridian* as shown in Fig. 3.1.

It would thus appear that the earth's magnetic field is similar to that which would be expected at the surface if a short but strongly magnetized bar magnet were located at the centre. This partly explains the fact that the magnetic poles are relatively large areas, due to the spreading out of the lines of force, and it also gives a reason for the direction of the field being horizontal in the vicinity of the equator. The origin of the earth's field is still not precisely known, but, for purposes of explanation, the supposition of a bar magnet at its centre is useful in visualizing the general form of the magnetic field as it is known to be.

The field differs from that of an ordinary magnet in several

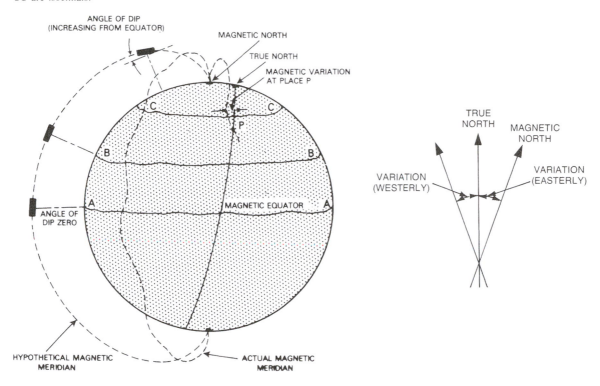

Figure 3.1 Terrestrial magnetism. Lines AA, BB and CC are isoclinals.

respects. Its points of maximum intensity, or strength, are not at the magnetic poles (theoretically they should be) but occur at four other positions, two near each pole, known as *magnetic foci*. Moreover, the poles themselves are continually changing their positions, and at any point on the earth's surface the field is not symmetrical and is subject to changes both periodic and irregular.

Magnetic variation

As meridians and parallels are constructed with reference to the true or geographic North and South poles, so can magnetic meridians be constructed with reference to the magnetic poles. If a map were prepared to show both true and magnetic meridians, it would be observed that these intersect each other at angles varying from 0° to 180° at different parts of the earth, diverging from each other sometimes in one direction and sometimes in the other. The horizontal angle contained between the true and the magnetic meridian at any place is known as the *magnetic variation* or *declination*.

When the direction of the magnetic meridian inclines to the left of the true meridian, the variation is said to be westerly, while an inclination to the right produces easterly variation. It varies in amount from 0° along those lines where the magnetic and true meridians run together, to 180° in places between the true and magnetic poles. At some places on the earth, where the ferrous nature of the rock disturbs the main magnetic field, local attraction exists and abnormal variation occurs which may cause large changes in its value over very short distances. While the variation differs all over the world, it does not maintain a constant value in any one place, and the following changes, themselves not constant, may be experienced: (i) *Secular change*, which takes place over long periods due to the changing positions of the magnetic poles relative to the true poles; (ii) *Annual change*, which is a small seasonal fluctuation superimposed on the secular change; (iii) *Diurnal* or *daily change*.

Information regarding variation and its changes are given on special charts. Lines are drawn on the charts, and those which join places having equal variation are called *isogonal lines*, while those drawn through places where the variation is zero are called *agonic lines*.

Magnetic dip

As stated earlier, a freely suspended magnet will settle in a definite direction at any point on the earth's surface and will lie parallel to the magnetic meridian at that point. It will not, however, lie parallel to the earth's surface at all points for the reason that the lines of force themselves are not horizontal, as may be seen from Fig. 3.2. These lines emerge vertically from the North magnetic pole, bend over and descend vertically into the South magnetic pole; it is only at what is known as the *magnetic equator* that they pass horizontally along the earth's surface. If, therefore, a suspended magnet is carried along a meridian from north to south, it will be on end, red end down, at the start, horizontal near the equator, and finish up again on end but with the blue end down.

The angle the lines of force make with the earth's surface at any

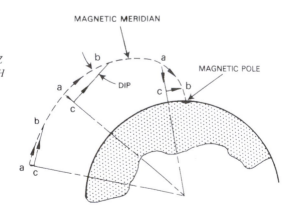

Figure 3.2 Relationship between magnetic components and dip.

$a - c$ = Vertical component Z
$c - b$ = Horizontal component H
$a - b$ = total force T
Given angle of dip θ and H,

$\dfrac{Z}{H} = \tan \theta$ and $Z = H \tan \theta$

$\dfrac{H}{T} = \cos \theta$ and $T = H \cos \theta$

$T^2 = H^2 + Z^2$

MAGNETIC MERIDIAN

MAGNETIC POLE

DIP

given place is called the *angle of dip* or *magnetic inclination*, and varies from 0° at the magnetic equator to 90° at the magnetic poles. The angle of dip at all places undergoes changes similar to those described for variation and is also shown on charts of the world. Places on these charts having the same dip angle are joined by lines known as *isoclinals*, while those at which the angle is zero are joined by a line known as the *aclinic line* or magnetic equator, of which mention has already been made.

Earth's total force

When a magnet freely suspended in the earth's field comes to rest, it does so under the influence of the total force of the field. This total force is resolved into its horizontal and vertical components, termed H and Z respectively. The relationship between these components and dip is shown in Fig. 3.2.

As in the case of variation and dip, charts of the world are published showing the values of the components for all places on the earth's surface, together with the mean annual change. Lines of equal H and Z forces are referred to as *isodynamic lines*.

The earth's magnetic force may be stated either as a relative value or an absolute value. If stated as a relative value, and in the case of compasses this is the case, it is given relative to the H force at Greenwich.

Compass construction

Direct-reading compasses have the following common principal features: a magnet system housed in a bowl; liquid damping; liquid expansion compensation; and deviation compensation. The majority of compasses currently in use are of the card type, and the construction of two examples is illustrated in Fig. 3.3.

The magnet system of the example shown at (a) comprises an annular cobalt-steel magnet to which is attached a light-alloy card, graduated in increments of 10°, and referenced against a lubber line fixed to the interior of the bowl. The system is pendulously suspended by an iridium-tipped pivot resting in a sapphire cup supported in a holder or stem. The pivot point is above the centre of gravity of the magnet system which is balanced in such a way as to minimize the effects of angle of dip over as wide a range of latitudes North and South as possible.

The bowl is of plastic (Diakon) and so moulded that it has a magnifying effect on the card and its graduations. It is filled with a silicone fluid to make the compass *aperiodic*, i.e. to ensure that after the magnet system has been deflected, it returns to equilibrium directly without oscillating or overshooting. The fluid also provides

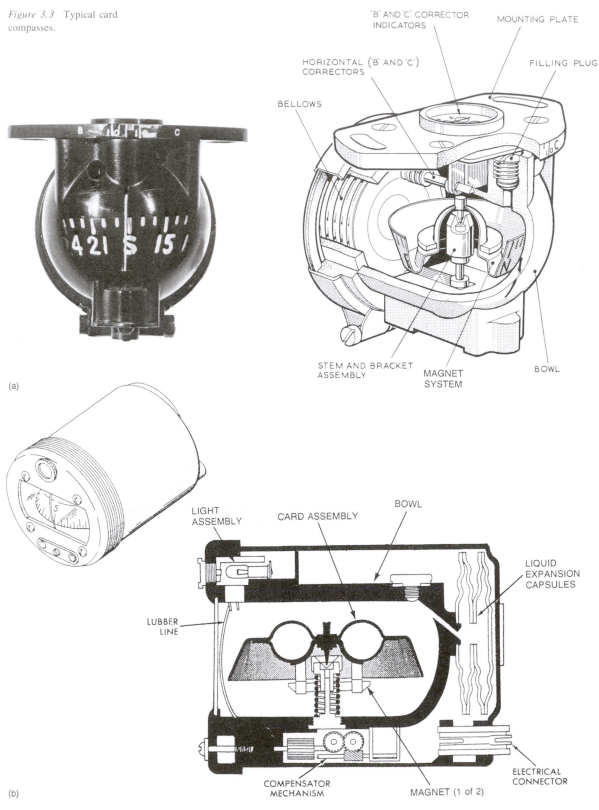

Figure 3.3 Typical card compasses.

'B' AND 'C' CORRECTOR INDICATORS

MOUNTING PLATE

HORIZONTAL ('B' AND 'C') CORRECTORS

FILLING PLUG

BELLOWS

STEM AND BRACKET ASSEMBLY

MAGNET SYSTEM

BOWL

(a)

LIGHT ASSEMBLY

CARD ASSEMBLY

BOWL

LIQUID EXPANSION CAPSULES

LUBBER LINE

COMPENSATOR MECHANISM

MAGNET (1 of 2)

ELECTRICAL CONNECTOR

(b)

82

the system with a certain buoyancy, thereby reducing the weight on the pivot and so diminishing the effects of friction and wear. Changes in volume of the fluid due to temperature changes, and their resulting effects on damping efficiency, are compensated by a bellows type of expansion device secured to the rear of the bowl.

Compensation of the effects of deviation due to longitudinal and lateral components of aircraft magnetism (see page 87) is provided by permanent magnet coefficient 'B' and 'C' corrector assemblies secured to the compass mounting plate.

The compass shown at (b) of Fig. 3.3 is designed for direct mounting on a panel. Its magnet system is similar to the one described earlier except that needle-type magnets are used. The bowl is in the form of a brass case which is sealed by a front bezel plate. Changes in liquid volume are compensated by a capsule type of expansion device. A permanent-magnet deviation compensator is located at the underside of the bowl, the coefficient 'B' and 'C' spindles being accessible from the front of the compass. A small lamp is provided for illuminating the card of the magnet system.

Compass location

The location of a compass in any one type of aircraft is of importance, and is pre-determined during the design stage by taking into account the effects which mechanical and electrical equipment in cockpit or flight deck areas may have on indications. In this connection it is usual to apply the *compass safe distance* rule which, precisely defined, is 'the minimum distance at which equipment may be safely positioned from a compass without specified design values of maximum deviation being exceeded under all operating conditions'. The distance is measured from the centre of a compass magnet system to the nearest point on the surface of equipment. Values are quoted by manufacturers as part of the operating data appropriate to their equipment.

Errors in indication

The pendulous suspension of a magnet system, although satisfactory from the point of view of counteracting dip, is unfortunately a potential source of errors under in-flight operating conditions in which certain force components are imposed on the system. There are two main errors that result from such components, namely *acceleration error* and *turning error*.

Acceleration error

This may be broadly defined as the error, caused by the effect of the earth's field component Z, in the directional properties of the magnet

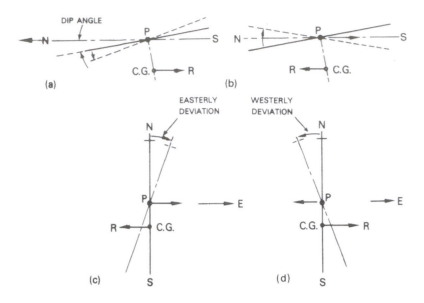

Figure 3.4 Acceleration errors. (a) Acceleration on northerly heading in northern hemisphere; (b) deceleration on northerly heading in northern hemisphere; (c) acceleration on easterly heading in northern hemisphere; (d) deceleration on easterly heading in northern hemisphere.

system when its centre of gravity is displaced from its normal position, such errors being governed by the heading on which acceleration or deceleration takes place.

When accelerating or decelerating on any fixed heading, a force is applied to the magnet system at the point of suspension P, this being its only connection. The reaction to this force will be equal and opposite and must act through the centre of gravity, which is below and offset from P due to the slight dip of the magnet system. The two forces constitute a couple which, dependent on the heading being flown, causes the magnet system merely to change its dip offset angle, or to rotate in azimuth.

Consider now an acceleration on a northerly heading in the northern hemisphere. The forces brought into play will be as shown in Fig. 3.4(a). Since both the point P and centre of gravity are in the plane of the magnetic meridian, the reaction R causes the 'N' end of the magnet system to go down, thus increasing the dip offset angle without any azimuth rotation. Conversely, when decelerating, the reaction R tilts the 'S' end of the magnet system as shown at (b).

In either the northern or southern hemispheres, acceleration or deceleration on headings other than the N–S meridian will produce azimuth rotation of the magnet system and consequent errors.

When an acceleration occurs on an easterly heading in the northern hemisphere, as at (c) of Fig. 3.4, a force will again act through point P, and the reaction R through the centre of gravity. In this case, however, they are acting away from each other and the couple produced tends to rotate the magnet system in a clockwise direction, thus indicating an apparent turn to the north, or what is termed easterly deviation. The reverse effects occur during a deceleration, producing an apparent turn to the south or westerly deviation.

Hence, in the northern hemisphere, acceleration causes easterly deviation on easterly headings, and westerly deviation on westerly headings, whilst deceleration has the reverse effect. In the southern hemisphere the results will be reversed in each case.

As northerly or southerly headings are approached, the magnitude of the apparent deviation decreases, the acceleration error varying as the sine of the compass heading.

One further point may be mentioned in connection with these errors, and that is the effect of aircraft attitude changes. If an aircraft flying level is put into a climb at the same speed, the effect on its compass magnet system will be the same as if the aircraft had decelerated. If the change in attitude is also accompanied by a change in speed, the apparent deviation may be quite considerable.

Turning errors

During a turn, the point P of a compass magnet system is carried with the aircraft along the curved path of the turn. The system's centre of gravity, being offset, is subjected to the centrifugal acceleration force produced by the turn, causing the system to swing outwards and to rotate so that apparent deviations, or turning errors, will be observed. In addition, the magnet system tends to maintain a position parallel to the transverse plane of the aircraft, thus giving it a lateral tilt the angle of which is governed by the aircraft's bank angle. For a correctly banked turn, the tilt angle would be maintained equal to the bank angle, because the resultant of centrifugal force and gravity lies normal to the aircraft's transverse plane, and also to the plane through the point P and centre of gravity of the magnet system. In this case, centrifugal force itself would have no effect other than to exert a pull on the centre of gravity and so decrease the offset dip angle of the magnet system.

As soon as the system is tilted, however, and regardless of whether or not the aircraft is correctly banked, the system is free to move under the influence of the earth's component Z which will then have a component in the lateral plane of the system, causing it to rotate and further increase the turning error.

The extent and direction of the error is dependent upon the aircraft's heading, the magnet system tilt angle, and the dip. In order to form a clearer understanding of its effects, we may consider a few examples of heading changes from the magnetic meridian, and in both the northern and southern hemispheres.

Turning from a northerly heading towards east or west
Figure 3.5(a) represents the magnet system of a compass in an aircraft flying on a northerly heading in the northern hemisphere. Let us assume that a change in heading to the eastward is required. As

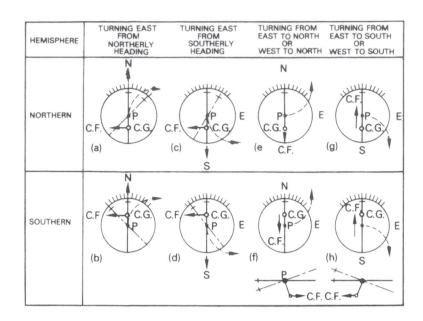

Figure 3.5 Turning errors.

HEMISPHERE	TURNING EAST FROM NORTHERLY HEADING	TURNING EAST FROM SOUTHERLY HEADING	TURNING FROM EAST TO NORTH OR WEST TO NORTH	TURNING FROM EAST TO SOUTH OR WEST TO SOUTH
NORTHERN	(a)	(c)	(e)	(g)
SOUTHERN	(b)	(d)	(f)	(h)

soon as the turn commences, the centrifugal acceleration acts on the centre of gravity causing the system to rotate in the same direction as the turn, and since the system is tilted, the earth's component Z exerts a pull on the N end causing further rotation of the system. Now, the magnitude of system rotation is dependent on the rate at which turning and banking of the aircraft is carried out, and resulting from this three possible indications may be registered: (i) a turn of the correct sense, but smaller than that actually carried out when the magnet system turns at a slower rate than the aircraft; (ii) no turn at all when the system and aircraft are turning at the same rate; (iii) a turn in the opposite sense when the system turns at a rate faster than the aircraft. The same effects will occur if the heading changes from N to W whilst flying in the northern hemisphere.

In the southern hemisphere (diagram (b)) the effects are somewhat different. The south magnetic pole is now the dominant pole and so the offset dip angle of the magnet system changes to displace the centre of gravity to the north of point P. We may again consider the case of an aircraft turning eastward from a northerly heading. Since the centre of gravity is now north of point P, the centrifugal acceleration acting on it causes the magnet system to rotate more rapidly in the opposite direction to the turn, i.e. indicating a turn in the correct sense but of greater magnitude than is actually carried out.

Turning from a southerly heading towards east or west
If the turns are executed in the northern hemisphere (Fig. 3.5(c)) then because the magnet system's centre of gravity is still south of point

P, the rotation of the system and the indications registered will be the same as turning from a northerly heading in the northern hemisphere.

In turning from a southerly heading in the southern hemisphere (Fig. 3.5(d)) the magnet system's centre of gravity is north of the point P and produces the same effects as turning from a northerly heading in the southern hemisphere.

In all the above cases, the greatest effect on compass indications will be found when turns commence near to northerly or southerly headings, being most pronounced when turning through north. For this reason the term *northerly turning error* is often used when describing the effects of centrifugal acceleration on compass magnet systems.

Turning through east or west

When turning from an easterly or westerly heading in either the northern or southern hemispheres (diagrams (e)−(h)) no errors will result because the centrifugal acceleration acts in a vertical plane through the magnet system's centre of gravity and point P. The centre of gravity is merely deflected to the north or south of point P, thus increasing or decreasing the magnet system's pendulous resistance to dip.

A point which may be noted in connection with turns from E or W is that when the N or S end of the magnet system is tilted up, the line of the system is nearer to the direction where the directive force is zero, i.e. at right angles to the line of dip. Thus, if a compass has not been accurately adjusted during a 'swing', any uncorrected deviating force will become dominant and so cause indications of apparent turns.

Aircraft magnetism and its effects on compasses

Magnetism is unavoidably present in aircraft in varying amounts, and can therefore also produce errors in the indications of compasses. However, by analysis it is divided into two main types and also resolved into components acting in definite directions, so that steps can be taken to minimize the errors, or *deviations* as they are called, resulting from such components.

The two types of magnetism can be further divided in the same way that magnetic materials are classified according to their ability to be magnetized, namely *hard-iron* and *soft-iron*.

Hard-iron magnetism is of a peramanent nature and is caused, for example, by the presence of magnetically 'hard' materials in an aircraft's structure, in power plants and other equipment, the earth's field building itself into such materials during the many varied manufacturing and assembly processes involved in the overall construction of an aircraft.

Soft-iron magnetism is of a temporary nature and is caused by the metallic materials of an aircraft which are magnetically 'soft' becoming magnetized due to induction by the earth's field. The effect of this type of magnetism is dependent on an aircraft's heading, attitude and its geographical position.

There is also a third type of magnetism, due to the sub-permanent magnetism of what is called 'intermediate' iron, which can be retained for varying periods. Such magnetism depends, not only on heading, attitude and geographical position of an aircraft, but also on the nature of its previous motion, vibrations, lightning strikes and other external effects.

The various magnetic components which cause deviations are designated by letters, those for permanent hard-iron magnetism being capitals, and those for soft-iron magnetism being small letters. The resulting deviations are termed easterly when positive, and westerly when negative.

Components of hard-iron magnetism

The total effect of this type of magnetism at a compass position may be considered as having originated from equivalent bar magnets lying longitudinally, laterally and vertically, as shown in Fig. 3.6. The components are respectively denoted as *P, Q* and *R*, and are either positive or negative depending on the locations of the blue poles of the equivalent magnets. The strength of these components does not vary with heading or change of latitude, but may do so with time due to a weakening of aircraft magnetism. The deviations caused by each of the components are set out in Table 3.1.

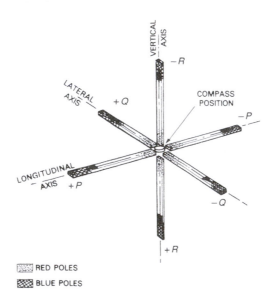

Figure 3.6 Components of hard-iron magnetism.

Table 3.1

Component	Axis	Component polarity	Aircraft heading			
			North	East	South	West
			Deviations			
P	Horizontal	+ve	0	max. +ve	0	max. −ve
		−ve	0	max. −ve	0	max. +ve
Q	Lateral	+ve	max. +ve	0	max. −ve	0
		−ve	max. −ve	0	max. +ve	0
			Aircraft nose up			
		+ve	0	max. +ve	0	max. −ve
		−ve	0	max. −ve	0	max. +ve
			Aircraft nose down			
		+ve	0	max. −ve	0	max. +ve
R	Vertical	−ve	0	max. +ve	0	max. −ve
			Aircraft banked to port			
		+ve	max. +ve	0	max. −ve	0
		−ve	max. −ve	0	max. +ve	0
			Aircraft banked to stb'd			
		+ve	max. −ve	0	max. +ve	0
		−ve	max. +ve	0	max. −ve	0

Notes: 1. +ve and −ve deviations are termed easterly and westerly respectively.
2. Component R effective only in the aircraft attitudes indicated.

Components of soft-iron magnetism

The effect of this type of magnetism may be considered as originating from a piece of soft-iron in which magnetism has been induced by the earth's field. This field, as we already know, has two components designated H and Z, but in the analysis of soft-iron magnetism H is resolved into two additional components X and Y. These, together with component Z, are also related to the three principal axes of an aircraft, namely X — longitudinal, Y — lateral and Z — vertical.

The polarities and strengths of components X and Y vary with changes in aircraft heading relative to the fixed direction of the earth's component H. Components X, Y and Z also change with geographical location because this results in changes in the earth's field strength and direction. A change in the polarity of component Z will only occur with a change in magnetic hemisphere.

Table 3.2

Component	Axis	Component polarity	Aircraft heading							
			North	NE	East	SE	South	SW	West	NW
						Deviations				
aX		+ve	0	max. +ve	0	max. −ve	0	max. +ve	0	max. −ve
		−ve	0	max. −ve	0	max. +ve	0	max. −ve	0	max. +ve
bY	Longitudinal	+ve	0	−ve	max. −ve	−ve	0	−ve	max. −ve	−ve
		−ve	0	+ve	max. +ve	+ve	0	+ve	max. +ve	+ve
cZ		+ve −ve	* Same as corresponding polarities of component P.							
dX		+ve	max. +ve	+ve	0	+ve	max. +ve	+ve	0	+ve
		−ve	max. −ve	−ve	0	−ve	max. −ve	−ve	0	−ve
eY	Lateral	+ve	Same as −ve component aX.							
		−ve	Same as +ve component aX.							
fZ		+ve −ve	* Same as corresponding polarities of component Q.							
gX		+ve −ve								
hY	Vertical	+ve −ve	* Same as corresponding polarities of component R.							
kZ		+ve −ve								

* See Table 3.1.

Note: 1. +ve and −ve deviations are termed easterly and westerly respectively.

2. The polarities and direction of components cZ and fZ depend on whether an aircraft is in the northern or southern hemisphere.

Each of the three components produce three soft-iron components that are designated aX, bY, cZ; dX, eY, fZ; and gX, hY, kZ. The deviations caused by such components are set out in Table 3.2.

Total magnetic effect

The total effect of the magnetic fields that produce deviating forces relative to each of the three axes of an aircraft is determined by algebraically summing the quantities appropriate to each of the related components. Thus:

$$\text{TOTAL } X^1 \text{ (longitudinal)} = X+P+aX+bY+cZ$$
$$\text{TOTAL } Y^1 \text{ (lateral)} = Y+Q+dX+eY+fZ$$
$$\text{TOTAL } Z^1 \text{ (vertical)} = Z+R+gX+hY+kZ$$

Figure 3.7 Relationship between aircraft magnetism and deviation coefficients.

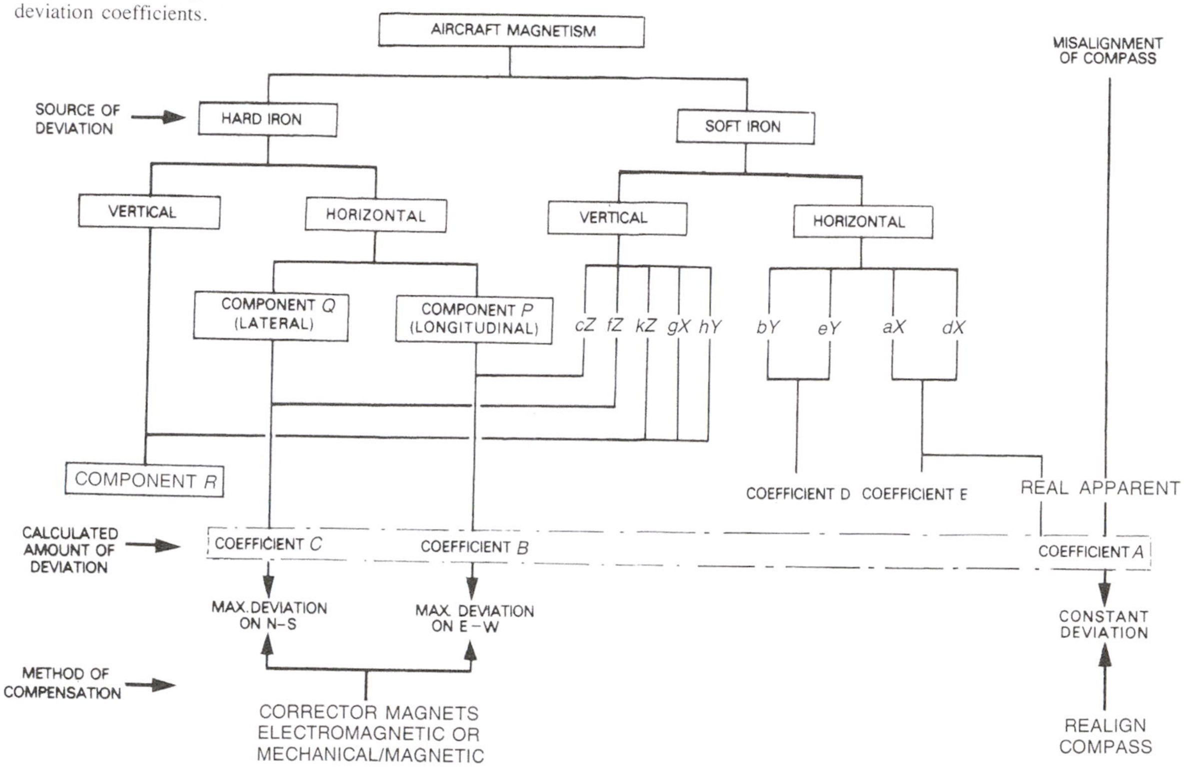

Deviation coefficients

Before steps can be taken to minimize the deviations caused by hard-iron and soft-iron components of aircraft magnetism, their values on each heading must be obtained and quantitatively analysed into *coefficients of deviation*. There are five coefficients designated *A, B, C, D* and *E*, termed positive or negative as the case may be, and expressed in degrees. The relationship between them and the components of aircraft magnetism is shown in Fig. 3.7.

Coefficient A

This represents a constant deviation and may be termed as either *real A*, which is caused by the soft-iron components *bY* and *dX*, or *apparent A*, which is a deviation produced by non-magnetic causes such as misalignment of a direct-reading compass or of a flux detector unit where appropriate (see page 182) with respect to an aircraft's longitudinal axis. In practice it is not necessary to distinguish between them, since they are both understood to be included in the term coefficient *A*.

The coefficient is calculated by taking the average of the algebraic differences between deviations measured on a number of equidistant compass headings; this also applies to the other four coefficients. In the case of A, the average may be determined from deviations on either the four cardinal headings or, for greater accuracy, on these headings plus the four quadrantal headings. Thus:

$$A = \frac{\text{Deviation on N} + \text{E} + \text{S} + \text{W}}{4}$$

or

$$A = \frac{\text{Deviation on N} + \text{NE} + \text{E} + \text{SE} + \text{S} + \text{SW} + \text{W} + \text{NW}}{8}$$

The coefficient is positive or negative, depending on whether the constant deviation which it represents is easterly or westerly.

Coefficient B

This represents the resultant deviation due to the presence, either together or separately, of hard-iron component P and soft-iron component cZ. When these components are of like signs, they cause deviation in the same direction, but when of unlike signs they tend to counteract each other. The coefficient is calculated from the formula:

$$B = \frac{\text{Deviation on E} - \text{Deviation on W}}{2}$$

Since components P and cZ cause deviation which varies as the sine of an aircraft's heading θ, then deviation due to coefficient B may also be expressed as $B \times \sin \theta$.

Coefficient C

This represents the resultant deviation due to the presence, either together or separately, of hard-iron component Q and soft-iron component fZ. When of like and unlike signs these components cause deviations whose directions are the same as those caused by components P and cZ. The coefficient is calculated from the formula:

$$C = \frac{\text{Deviation on N} - \text{Deviation on S}}{2}$$

Since components Q and fZ cause deviation which varies as the cosine of an aircraft's heading, then deviation due to coefficient C may also be expressed as $C \times \cos \theta$.

Coefficient D

This represents the deviation due to the presence, either together or separately, of components aX and eY which cause deviations of the same direction when they are of unlike signs and counteract each other when of like signs. When a $+aX$ or a $-eY$ predominates, or when they are present together, the coefficient is said to be positive, whilst a $-aX$ or a $+eY$ predominating or together cause a negative coefficient D. It is calculated from the formula:

$$D = \frac{(\text{Dev. on NE} + \text{Dev. on SW}) - (\text{Dev. on SE} + \text{Dev. on NW})}{4}$$

The deviations caused by components aX and eY vary as the sine of twice an aircraft's heading; therefore deviations may also be expressed as $D \times \sin 2\theta$.

Coefficient E

This coefficient represents the deviation due to the presence of components bY and dX of like signs. When a $+bY$ and a $+dX$ are combined, coefficient E is said to be positive, whilst a combination of a $-by$ and a $-dX$ gives a negative coefficient; the two components must in each case be equal in magnitude. The coefficient is calculated from the formula:

$$E = \frac{(\text{Dev. on N} + \text{Dev. on S}) - (\text{Dev. on E} + \text{Dev. on W})}{4}$$

The deviations caused by the components bY and dX vary as the cosine of twice an aircraft's heading; therefore deviations may also be expressed as $E \times \cos 2\theta$.

The total deviation on an uncorrected compass for any given direction of an aircraft's heading by compass may be expressed by the equation:

$$\text{Total deviation} = A + B\sin\theta + C\cos\theta + D\sin2\theta + E\cos2\theta$$

Deviation compensation

In order to determine by what amount compass readings are affected by hard- and soft-iron magnetism, a special calibration procedure known as 'swinging' is carried out so that adjustments can be made to compensate for the deviations.

These adjustments are effected by compensator or corrector magnet devices which, in the case of direct-reading compasses, always relate only to deviation coefficients B and C. Adjustment for coefficient A is effected by repositioning the compass in its mounting by the requisite number of degrees.

A compensator forms an integral part of a compass (see Fig. 3.3) and in common with the majority of those in current use it contains two pairs of permanent magnets which can be rotated through gearing as shown in Fig. 3.8.

One pair of magnets is positioned laterally to provide a variable longitudinal field, thereby permitting adjustment for coefficient B,

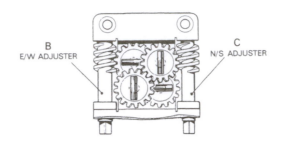

Figure 3.8 Deviation compensator device.

B
E/W ADJUSTER

C
N/S ADJUSTER

while the other pair is positioned longitudinally to provide a variable lateral field and so permit adjustment for coefficient *C*. Thus, the fields are effective in neutralizing the deviations on only the two cardinal headings appropriate to each of the coefficients.

The manner in which compensation is carried out may be understood by considering the case of an adjustment having to be made for coefficient *B*, as indicated in Fig. 3.9. When the appropriate compensator magnets are in the neutral position, as shown at (a), the fields produced are equal and opposite, and if as also shown the aircraft is heading north, then the total field is aligned with the earth's component *H* and the compass magnet system. Variation of the field strength by rotating the magnets will, therefore, have no effect. This would also be the case if the aircraft was heading south.

At (b) of Fig. 3.9, the aircraft is represented as heading east and, as before, the compensator magnets are in the neutral position. The total field of the magnets, however, is now at right angles to the earth's component and the compass magnet system, and so if the magnets are now rotated from the neutral position to the positions shown at (c), the distance between poles N_2 and S_1 is smaller. Since the intensity of a field varies in inverse proportion to the square of the distance from its source, then in this case, a strong field will exist between the poles of the magnets. The north-seeking pole of the compass magnet system will, therefore, experience a greater repulsive force, resulting in deflection of the system through an appropriate number of degrees. If the magnets are rotated so as to strengthen the field between poles N_1 and S_2, the compass magnet system will then, of course, be deflected in the opposite direction. Deflection of the compass magnet system would be obtained in a similar manner with the aircraft heading west.

The coefficient *C* compensating magnets also produce similar effects but, as a study of diagrams (d) to (f) will show, deflections of the compass magnet system are only obtainable when an aircraft is heading either north or south.

It will be apparent from the foregoing operating sequences that maximum compensation of deviation on either side of cardinal headings is obtained when the magnets are in complete alignment.

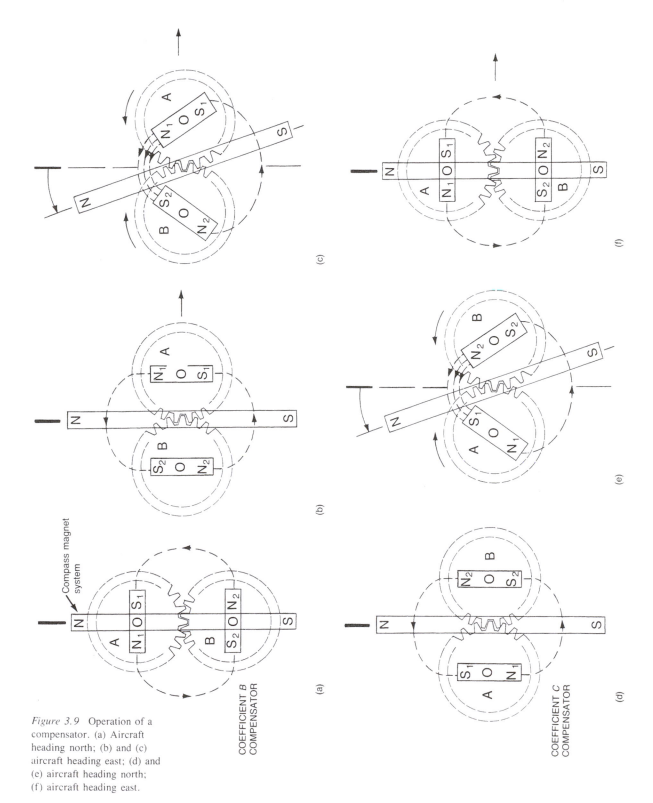

Figure 3.9 Operation of a compensator. (a) Aircraft heading north; (b) and (c) aircraft heading east; (d) and (e) aircraft heading north; (f) aircraft heading east.

COEFFICIENT *B* COMPENSATOR

COEFFICIENT *C* COMPENSATOR

Compass magnet system

95

The gears on which magnets are mounted are connected to operating heads which, depending on the type of compass, are operated either by a key or by means of a screwdriver. Indication of the neutral position of magnets is given by aligning datum marks, located typically on the ends of magnet-operating spindles and on the compensator casing.

4 Gyroscopic flight instruments

In addition to the airspeed indicator, the altimeter and the vertical speed indicator, a basic group of flight instruments also comprises instruments which provide direct indication of an aircraft's attitude. There are three such instruments, namely a *gyro horizon* (sometimes called an artificial horizon), a *direction indicator*, and a *turn-and-bank indicator*. The complete group constitutes what is termed the 'basic six' arrangement, details of which were given in Chapter 1 (see page 20).

The three additional instruments utilize a gyroscopic type of sensing element, the properties of which need to be understood before going into the construction and operating details of each instrument.

The gyroscope and its properties

As a mechanical device a gyroscope may be defined as a system containing a heavy metal wheel or *rotor*, universally mounted so that it has three degrees of freedom: (i) *spinning freedom*, about an axis perpendicular through its centre (axis of spin XX_1); (ii) *tilting freedom*, about a horizontal axis at right angles to the spin axis (axis of tilt YY_1); and (iii) *veering freedom*, about a vertical axis perpendicular to both the other axes (axis of veer ZZ_1).

The three degrees of freedom are obtained by mounting the rotor in two concentrically pivoted rings, called *inner* and *outer* rings. The whole assembly is known as the *gimbal system* of a *free* or *space gyroscope*. The gimbal system is mounted in a frame as shown in Fig. 4.1, so that in its normal operating position, all the axes are mutually at right angles to one another and intersect at the centre of gravity of the rotor.

The system will not exhibit gyroscopic properties unless the rotor is spinning; for example, if a weight is suspended on the inner ring, it will merely displace the ring about its axis YY_1 because there is no resistance to the weight. When the rotor is made to spin at high speed, however, the device then becomes a true gyroscope possessing two important fundamental properties: *gyroscopic inertia* or *rigidity*, and *precession*. Both these properties depend on the principle of conservation of angular momentum, which means that the angular momentum of a body about a given point remains constant unless

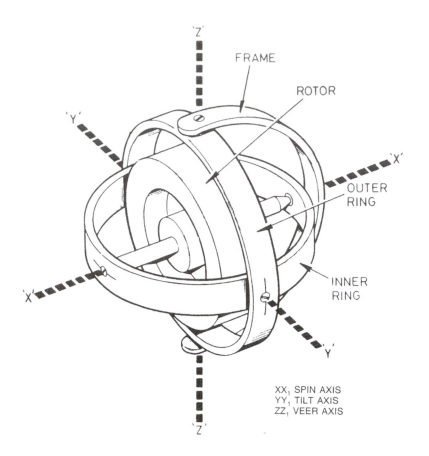

Figure 4.1 Elements of a gyroscope.

XX₁ SPIN AXIS
YY₁ TILT AXIS
ZZ₁ VEER AXIS

some force is applied to change it. *Angular momentum* is the product of the moment of inertia (I) and angular velocity (ω) of a body referred to a given point — the centre of gravity in the case of a gyroscope.

If a weight is now suspended from the inner gimbal ring with the rotor spinning it will be found that the ring will support the weight, thus demonstrating the first fundamental property of rigidity. It will also be found, however, that the complete gimbal system will start rotating about the axis ZZ_1, such rotation demonstrating the second property of precession.

These rather intriguing properties can be exhibited by any system in which a rotating mass is involved, and a simple example of such a system is the bicycle. If we lift the front wheel off the ground, spin it at high speed, and then turn the handlebars, we will feel rigidity resisting us, and precession trying to twist the handlebars from our grasp. Other familiar mechanical systems possessing gyroscopic properties are aircraft propellers, and jet engine compressor and turbine assemblies.

The two gyroscopic properties may be more closely defined as follows:

Rigidity. The property which resists any force tending to change the plane of rotor rotation. It is dependent on three factors: (i) the mass of the rotor, (ii) the speed of rotation, and (iii) the distance at which the mass acts from the centre, i.e. the *radius of gyration*.

Precession. The angular change in direction of the plane of rotation under the influence of an applied force. The change in direction takes place, not in line with the force, but always at a point 90° away in the direction of rotation. The rate of precession also depends on three factors: (i) the strength and direction of the applied force, (ii) the moment of inertia of the rotor, and (iii) the angular velocity of the rotor. The greater the force, the greater is the rate of precession, while the greater the moment of inertia and the greater the angular velocity the smaller is the rate of precession.

Precession of a rotor will continue, while the force is applied, until the plane of rotation becomes coincident with that of the force. At this point there will be no further resistance to the force and so precession will cease.

The axis about which a force is applied is termed the *input axis*, and the one about which precession takes place is termed the *output axis*.

Determining the direction of precession

The direction in which a gyroscope will precess under the influence of an applied force may be determined by means of vectors and by solving certain gyrodynamic problems, but for illustration and practical demonstration purposes, there is an easy way of determining the direction in which precession will take place, and also of finding out where a force must be applied for a required direction. It is done by representing all forces as acting directly on the rotor itself.

At (a) in Fig. 4.2, the rotor is shown spinning in a clockwise direction and with a force *F* applied upwards on the inner ring. In transmitting this force to the rim of the rotor, as will be noted from (b), it will act in a horizontal direction. Let us assume for a moment that the rotor is broken into segments and concern ourselves with two of them at opposite sides of the rim as shown at (c). Each segment has motion *m* in the direction of rotor rotation, so that when force *F* is applied there is a tendency for each segment to move in the direction of the force. This motion is resisted by rigidity, but the segments will turn about the axis ZZ_1 so that their direction of motion is along the resultant of motion *m* and force *F*. The other segments will be affected in the same way; therefore, in being combined to form the solid mass of a rotor it will precess at an angular velocity proportional to the applied force (see diagrams (d) and (e)).

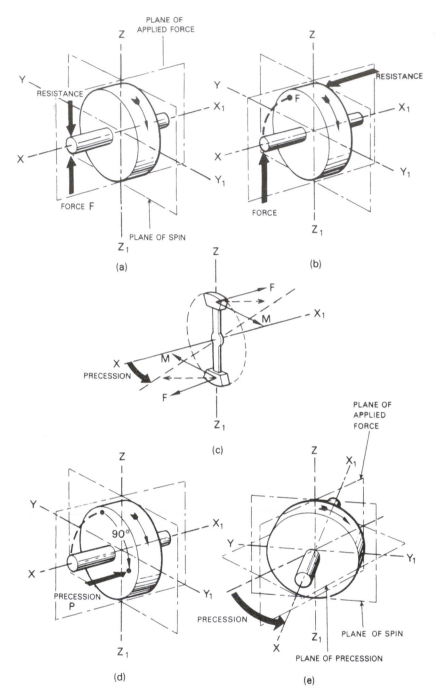

Figure 4.2 Gyroscopic precession (1). (*a*) Gyro resists force; (*b*) transmission of force; (*c*) effect on rotor segments; (*d*) generation of precession; (*e*) effect of precession.

In the example illustrated in Fig. 4.3(a), a force F is applied to the outer ring; this is the same as transmitting the force to the rotor rim at the point shown at (b). As in the previous case this results in the direction of motion changing to the resultant of motion m and force F. This time, however, the rotor precesses about the axis YY_1 as indicated at (d) and (e).

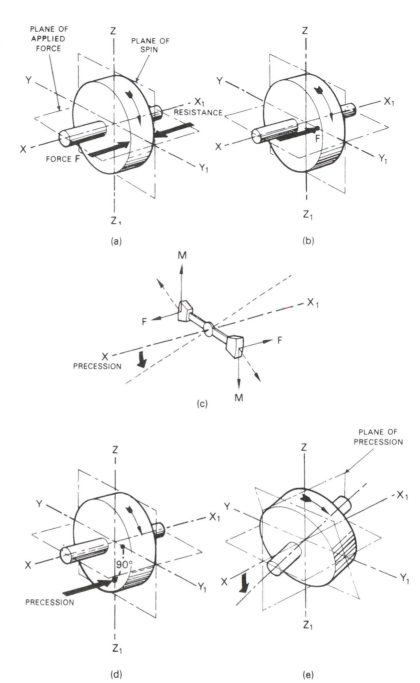

Figure 4.3 Gyroscopic precession (2). (*a*) Gyro resists force; (*b*) transmission of force; (*c*) effect on rotor segments; (*d*) generation of precession; (*e*) effect of precession.

References established by gyroscopes

For use in aircraft, gyroscopes must establish two essential reference datums: one for the detection of pitch and roll attitude changes, and the other for the detection of changes about the vertical axis, i.e. a directional reference. These datums are established by using vertical and horizontal spin-axis gyroscopes respectively as shown in Fig. 4.4. Both types utilize their fundamental properties in the following

101

Figure 4.4 References
established by gyroscopes.

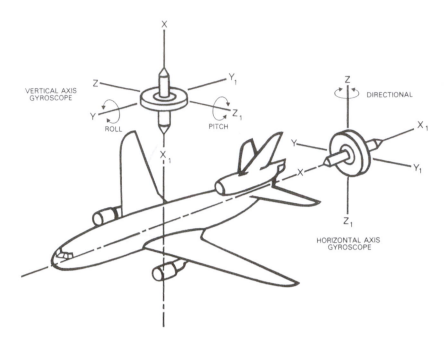

manner: *rigidity* provides a stabilized reference unaffected by movement of the supporting body, and *precession* controls the effects of apparent and real drift thus maintaining stabilized datums.

It will also be noted from Fig. 4.4 that the pitch, roll, and directional attitudes of an aircraft are determined by its displacement with respect to each appropriate gyroscope. For this reason, therefore, the gyroscopes are referred to as *displacement*-type gyroscopes. Each one has three degrees of freedom and, consequently, three mutual axes, but for the purpose of attitude sensing, the spin axis is discounted since no useful attitude reference is provided when displacements take place about the spin axis alone. Thus, in the practical case, the two types of gyroscope are further classified as *two-axis displacement* gyroscopes.

Limitations of a free gyroscope

In flight, the attitudes of an aircraft must be referenced with respect to the earth's surface, and this being so requires that a free or space gyroscope, thus far considered, be corrected for drift with respect to the earth's rotation, called *apparent drift*, and for wander as a result of carrying a gyroscope over the earth's surface, called *transport wander*.

Apparent drift

The earth rotates about its axis at the rate of 15°/hour, and in association with gyrodynamics, this is termed *earth rate* (ω_e). When a

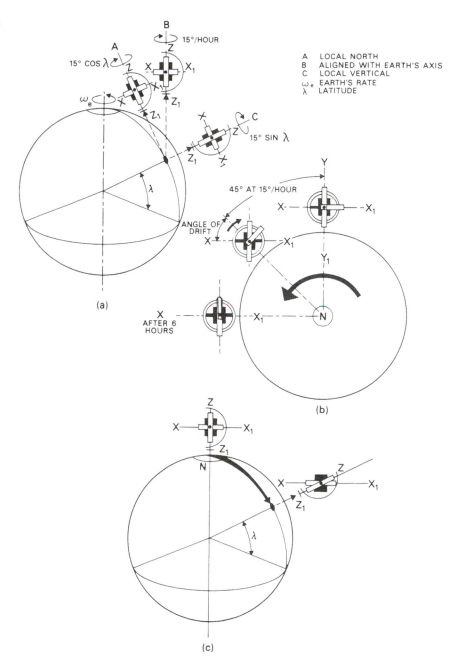

Figure 4.5 Drift and transport wander.

A   LOCAL NORTH
B   ALIGNED WITH EARTH'S AXIS
C   LOCAL VERTICAL
ω_e EARTH'S RATE
λ LATITUDE

(a)

(b)

(c)

free gyroscope is positioned at any point on the earth's surface, it will sense, depending on the latitude at which it is positioned, and on the orientation of its spin and input axes, various components of ω_e as an angular input. Thus, to an observer on the earth having no sense of the earth's rotation, the gyroscope would appear to veer or drift. This may be seen from Fig. 4.5(a) which illustrates a horizontal-axis gyroscope at a latitude λ. At 'A', the input axis is aligned with the local N–S component of ω_e; therefore, to an observer at latitude λ

103

the gimbal system would appear to drift clockwise (opposite to the earth's rotation) in a horizontal plane relative to the frame, and at a rate equal to $15°\cos\lambda$. When the input axis is aligned with that of the earth ('B'), drift would also be apparent, but at a rate equal to ω_e, i.e. $15°$/hour. If the input axis is now aligned with the local vertical component of ω_e ('C' in the diagram) the apparent drift would be equal to $15°\sin\lambda$.

In order to further illustrate drift, we may consider diagram (b) of Fig. 4.5, which is a plan view of a free horizontal-axis gyroscope positioned at the North pole with its input axis (ZZ_1) aligned with that of the earth. After three hours the earth will have rotated through $45°$, and the gyroscope will appear to have drifted through the same amount but in the opposite direction. After six hours the earth's rotation and apparent drift will be $90°$, and so on through a complete 24-hour period.

If the same gyroscope were to be positioned so that its input axis ZZ_1 was aligned with the E−W component of ω_e at any point, its spin axis would then be vertical; in other words, it becomes a vertical-axis gyroscope. Since the plane of rotation is coincident with that of the earth, there will be no apparent drift.

Real drift

Real drift results from imperfections in a gyroscope such as bearing friction and gimbal system unbalance. Such imperfections cause unwanted precession which can only be minimized by applying precision engineering techniques to the design and construction.

Transport wander

Let us again consider a horizontal-axis gyroscope which is set up initially at the North pole, with its input axis aligned with that of the earth. In this position it will exhibit an apparent drift equal to ω_e. Assume now that it is carried to a lower latitude, and with its input axis aligned with the local vertical component of ω_e. During the period of transport it will have appeared to an observer on the earth that the spin axis has tilted in a vertical plane, until at the new latitude it appears to be in the position shown at (c) of Fig. 4.5. This apparent tilt, or *transport wander*, would also be observed if, during transport, the input axis were aligned with either a local N−S component, or a local E−W component of ω_e.

Transport wander will, of course, appear simultaneously with drift, and so for a complete rotation of the earth, the gyroscope as a whole would appear to make a conical movement. The angular velocity or *transport rate* of this movement will be decreased or increased depending on whether the E−W component of an aircraft's speed is towards east or west. The N−S component of the speed will increase the maximum divergence of the gyroscope axis from the vertical, the amount of divergence depending on whether the aircraft's speed has a N or S component and also on whether the gyroscope is situated in the northern or southern hemisphere.

The relationship between ω_e, transport wander, and input axis alignment are summarized in the following table:

	Input axis alignment		
	Local north	Local east	Local vertical
Earth rate	$\omega_e \cos \lambda$	nil	$\omega_e \sin \lambda$
Transport wander	$\dfrac{U}{R}$	$\dfrac{V}{R}$	$\dfrac{U}{R} \tan \lambda$

ω_e = earth's angular velocity; λ = latitude; R = earth's radius; V = N−S component of transport velocity; U = E−W component of transport velocity.

If the input axis of a gyroscope were to be positioned such that its spin axis was vertical, then during transport it would only exhibit transport wander.

Control of drift and transport wander

Before a free gyroscope can be of practical use, drift and transport wander must be controlled so that the plane of spin of the rotor is maintained relative to the earth; in other words, it requires conversion to what is termed an *earth gyroscope*.

Drift, as already pointed out, relates only to horizontal-axis gyroscopes, and it can be controlled either by (i) calculating corrections using the earth rate formulae given in the preceding table and applying them as appropriate, e.g. to the readings of a direction indicator; (ii) applying fixed torques which unbalance the gyroscope and cause it to precess at a rate equal and opposite to ω_e; (iii) applying torques having a similar effect to that stated in (ii) but which can be varied according to the latitude in which the gyroscope is being used.

The control of transport wander is normally achieved by using gravity-sensing devices which automatically detect tilting of the gyroscope's spin axis, and applying the appropriate corrective torques.

The operation of some typical control methods will be described later under the headings of the appropriate flight instruments.

Displacement gyroscope limitations

Depending on the orientation of its gimbal system, a displacement gyroscope can be subject to certain operating limitations; one is referred to as *gimbal lock* and the other as *gimbal error*.

Gimbal lock

This occurs when the gimbal orientation is such that the spin axis becomes coincident with one or other of the axes of freedom which serve as attitude displacement references. Let us consider, for example, the case of the spin axis of a vertical-axis gyroscope shown in Fig. 4.4 becoming coincident with the ZZ_1 axis. This means, in effect, that the gyroscope would 'lose' its spin axis, and since the plane of spin would be at 90° to the ZZ_1 axis but in the same plane as displacements in roll, then the stable roll attitude reference would also be lost. If, in this 'locked' condition of the gimbal system, the gyroscope as a whole were to be turned, then the forces acting on the gimbal system would cause the system to precess or topple.

Gimbal error

This is an error which is also related to gimbal system orientation, and it occurs whenever the gyroscope as a whole is displaced with its gimbal rings not mutually at right angles to each other. The error is particularly relevant to horizontal-axis gyroscopes when used in direction indicating instruments (see page 127).

Methods of operating gyroscopic flight instruments

There are two principal methods of driving the rotors of gyroscopic flight instruments: *pneumatic* and *electric*.

Pneumatic

The pneumatic method is adopted in a number of small types of aircraft, and may be either vacuum or pressure. A typical vacuum system is shown schematically in Fig. 4.6; it consists of an engine-driven pump that is connected through pipelines to the appropriate flight instruments. A vacuum indicator, a relief valve, and a central air filter are also provided. In operation the pump creates a vacuum that is regulated by the valve at a value between 3.5 and 4.5 in Hg. Some types of turn-and-bank indicator may operate at a lower value

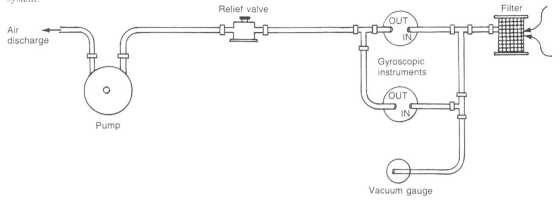

Figure 4.6 Vacuum-operated system.

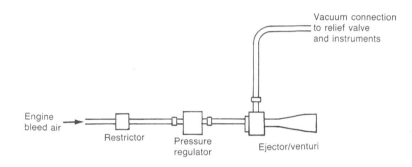

and this is obtained by including an additional relief valve in the main supply line.

Each instrument case has two connections: one is made to the pump and the other is made internally to a spinning jet system that is open to the surrounding atmosphere via the central air filter. When vacuum is applied, the pressure within the cases of the instruments is reduced to allow surrounding air to enter and emerge through the spinning jets. The jets are positioned adjacent to a series of recesses (commonly called 'buckets') formed in the periphery of each gyroscope rotor, so that as the airstreams impinge on the 'buckets', the rotors are rotated at high speed.

An example of a relief valve is shown in Fig. 4.7. During system operation the valve remains closed by compression of the spring, the tension of which is pre-adjusted to obtain the required vacuum so that air pressure acting on the outside of the valve is balanced against spring tension. If for some reason the adjusted value should be exceeded, the outside air pressure would overcome spring tension, thus opening the valve to allow outside air to flow into the system until the balanced condition was once again restored.

In some small types of turbine-engine aircraft that have pneumatically-operated instruments installed, the vacuum is created

107

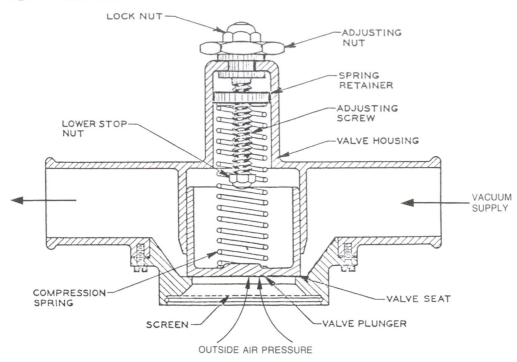

Figure 4.7 Relief valve.

by bleeding air from the engine compressor and passing it through an ejector/venturi (see Fig. 4.6).

A pressure-operated system is, as far as principal components are concerned, not unlike a vacuum system but, as will be noted by comparing Figs 4.6 and 4.8, a changeover of system inlet and outlet connections is necessary.

Electric

In electrically-operated instruments, the gyroscopes are special adaptations of ac or dc motors that are designed to be driven from the appropriate power supply systems of an aircraft. In current applications, ac motors are adopted in gyro horizons, while dc motors are more common to turn-and-bank indicators. Gyroscopes used for the purpose of direction indicating can also be motor-driven, but they normally form part of a magnetic heading reference system, or of the more widely adopted flight director systems. These systems will be covered in later chapters.

Gyro horizon principle

A *gyro horizon* indicates the pitch and roll attitude of an aircraft relative to its vertical axis, and so for this purpose it employs a

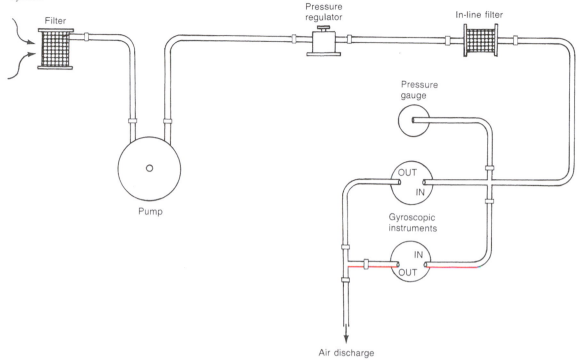

Figure 4.8 Pressure-operated system.

Filter

Pressure regulator

In-line filter

Pressure gauge

OUT

IN

Gyroscopic instruments

IN

OUT

Pump

Air discharge

displacement gyroscope whose spin axis is vertical. Indications of attitude are presented by the relative positions of two elements, one symbolizing the aircraft itself, the other in the form of a bar stabilized by the gyroscope and symbolizing the natural horizon. Supplementary indications of roll are presented by the position of a stabilized pointer and a fixed roll angle scale. Two methods of presentation are shown in Fig. 4.9.

The gimbal system (see Fig. 4.10) is arranged so that the inner ring forms the rotor casing and is pivoted parallel to an aircraft's lateral axis YY_1; the outer ring is pivoted at the front and rear ends of the instrument case, parallel to the longitudinal axis ZZ_1. The element symbolizing the aircraft may be either rigidly fixed to the case, or it may be externally adjustable for setting a particular pitch trim reference.

In operation the gimbal system is stabilized so that in level flight the three axes are mutually at right angles. When there is a change in an aircraft's attitude, it goes into a climb, say, the instrument case and outer ring will move about the axis YY_1 of the stabilized inner ring. The horizon bar is pivoted at the side and to the rear of the outer ring, and engages an actuating pin fixed to the inner ring, thus forming a magnifying lever system. The pin passes through a curved slot in the outer ring. In a climb attitude the bar pivot carries the rear

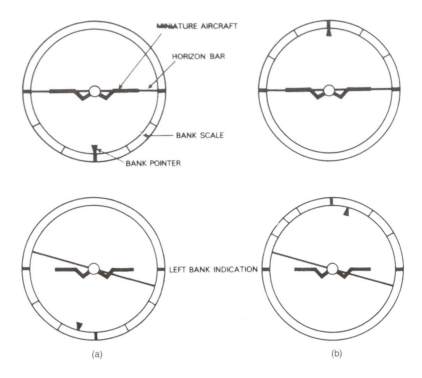

Figure 4.9 Gyro horizon presentations. (*a*) Bottom bank scale; (*b*) top bank scale.

MINIATURE AIRCRAFT

HORIZON BAR

BANK SCALE

BANK POINTER

LEFT BANK INDICATION

(a)

(b)

end of the bar upwards so that it pivots about the stabilized actuating pin. The front end of the bar is therefore moved downwards through a greater angle than that of the outer ring, and since the movement is relative to the symbolic aircraft element, the bar will indicate a climb attitude.

Changes in the lateral attitude of an aircraft, i.e. rolling, displace the instrument case about the axis ZZ_1, and the whole stabilized gimbal system. Hence, lateral attitude changes are indicated by movement of the symbolic aircraft element relative to the horizon bar, and also by relative movement between the roll angle scale and pointer.

Freedom of gimbal system movement about the roll and pitch axes is 360° and 85° respectively, the latter being restricted by means of a 'resilient stop'. The reason for this restriction is to prevent gimbal lock (see page 106).

Pneumatic type of gyro horizon

A typical instrument of the vacuum-driven type is shown in Fig. 4.11. The rotor is pivoted in ball-bearings within the inner ring/casing which is, in turn, pivoted in outer ring bearings. The upper bearing of the rotor is spring-loaded to compensate for the effects of differential expansion between the rotor shaft and casing under varying temperature conditions. A background plate which

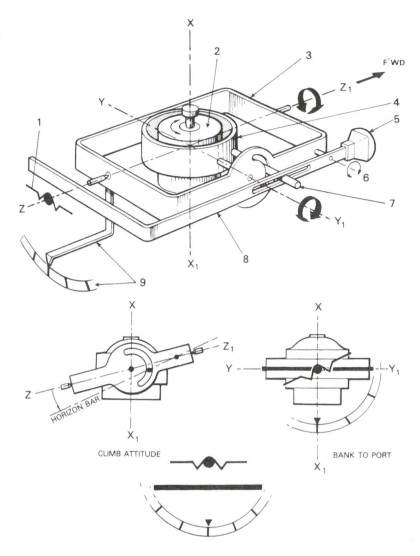

Figure 4.10 Principle of gyro horizon. 1 Symbolic aircraft, 2 rotor, 3 outer ring, 4 inner ring, 5 balance weight, 6 pivot point, 7 actuating pin, 8 horizon bar, 9 roll pointer and scale.

CLIMB ATTITUDE

BANK TO PORT

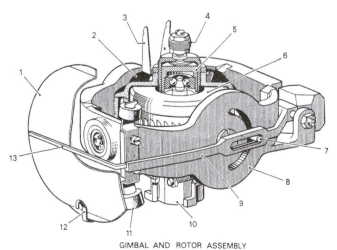

Figure 4.11 Pneumatic type of gyro horizon. 1 Sky plate, 2 inner gimbal ring, 3 resilient stop, 4 balance nut, 5 temperature compensator, 6 rotor, 7 actuating pin, 8 outer gimbal ring, 9 actuator arm, 10 pendulous vane unit, 11 buffer stops, 12 bank pointer, 13 horizontal bar.

GIMBAL AND ROTOR ASSEMBLY

111

symbolizes the sky is fixed to the front end of the outer ring and carries the roll pointer which registers against the roll angle scale.

A vacuum supply connection is provided at the rear of the instrument case, together with a filtered air inlet. The latter is positioned over the outer ring rear-bearing support and pivot which are drilled to communicate with a channel in the outer ring. This channel terminates in diametrically-opposed spinning jets within the rotor casing, the underside of which has a number of outlet holes in it.

When the vacuum system is in operation, the air pressure within the instrument case becomes lower than that of the surrounding air, which is then able to pass through the filtered inlet and to the spinning jets. The air issuing from the jets impinges on the rotor buckets, thus imparting even driving forces to spin the rotor at approximately 15 000 rev/min in an anti-clockwise direction as viewed from above. After spinning the rotor, the air passes through a pendulous vane unit (see page 114) attached to the underside of the rotor casing, and is finally drawn off by the vacuum source.

Electric gyro horizon

This instrument is made up of the same basic elements as a pneumatic type, with the exception that the gyroscope is an ac squirrel-cage induction motor which operates from a 115 V, 400 Hz, three-phase supply source.

One of the essential requirements of any gyroscope is to have the mass of the rotor concentrated as near to the periphery as possible, thus ensuring maximum inertia. This presents no difficulty where solid metal rotors are concerned, but when adopting electric motors as gyroscopes some rearrangement of their basic design is necessary in order to achieve the desired effect. An induction motor normally has its rotor revolving inside its stator, but to make one small enough to be accommodated within the confines of an instrument would mean too small a rotor mass and inertia. This is overcome by designing the rotor and its bearings so that it rotates on the outside of the stator; thus, for the same required size of motor the rotor mass is concentrated further from the centre, thereby increasing the radius of gyration and inertia.

The motor assembly is carried in a housing which forms the inner ring, this in turn being supported in the outer ring bearings. The horizon bar assembly is pivoted and actuated in a manner similar to that already described on page 109. The ac power supply is fed to the motor stator via slip rings, wire brushes and finger contact assemblies, thereby allowing for all gimbal ring movements.

When power is applied, a rotating magnetic field is set up in the stator; the field, in turn, inducing a current in the squirrel-cage rotor.

The effect of this current is to produce magnetic fields which interact with the stator's rotating field causing the rotor to turn at a speed of approximately 20 000–23 000 rev/min. A solenoid-operated 'power off' warning flag is also provided.

Standby attitude indicators

Many aircraft currently in service employ flight director systems, or more sophisticated electronic flight instrument systems, all of which comprise indicators having the capability of displaying not only attitude data, but also the data from other navigational systems. In such cases, therefore, the role of a conventional gyro horizon is relegated to that of secondary or standby, for use as a reference in the event of any failure that might occur in the attitude display sections of the aforementioned primary systems.

An example of one type of gyro horizon designed for use as a standby attitude indicator is shown in Fig. 4.12. Its gyroscope is powered by 115 V, three-phase ac supplied by a static inverter which, in turn, is powered by 28 V dc from the battery busbar of an aircraft. Power from such a source is always available, thereby ensuring continuity of indicator operation. In place of the more conventional stabilized horizon bar method of displaying attitude, a stabilized spherical element is adopted as the reference. The upper half of the element is coloured blue to display climb attitudes, and is divided, by an horizon line, from the lower half which is in black and displays descending attitudes. Each half is graduated in 10° increments, the upper one up to 80°, and the lower up to 60°. Roll or bank angles are indicated in the conventional manner.

A pitch trim adjustment and a fast-erection facility are provided, both being controlled by a knob in the lower right-hand corner of the indicator bezel. When the knob is rotated in its normally 'in'

Figure 4.12 Standby attitude indicator.

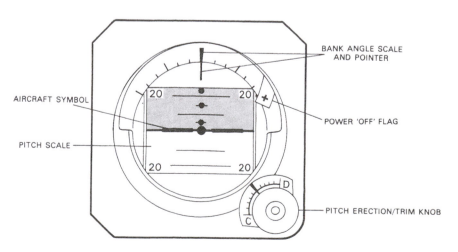

position, the aircraft symbol may be positioned through $\pm 5°$, thereby establishing a variable pitch trim reference. Pulling the knob out and holding it actuates a fast-erection circuit (see also page 119).

Erection systems for gyro horizons

These systems are designed for the purpose of erecting the gyroscope to, and maintaining it in, its vertical spin-axis position during operation. The systems adopted depend on the particular design of gyro horizon, but they are all of the gravity-sensing type and in general fall into two main categories: mechanical and electrical.

Pendulous vane unit
This is a mechanical system adopted for the gyro horizon described on page 110. It is fastened to the underside of the rotor housing and as indicated in Fig. 4.13(a) it consists of four knife-edged, pendulously-suspended vanes clamped in pairs on two intersecting shafts and passing through the unit body. One shaft is parallel to the axis YY_1 and the other to the axis ZZ_1. In the sides of the body there are four elongated ports (A, B, C and D), one under each vane.

After having spun the rotor, air is exhausted through the ports, emerging as four streams and in the directions indicated. The reaction of the air as it flows through the ports applies a force to the unit

Figure 4.13 Pendulous vane unit. (*a*) Construction; (*b*) precession due to air reaction; (*c*) gyro in vertical position; (*d*) gyro tilted.

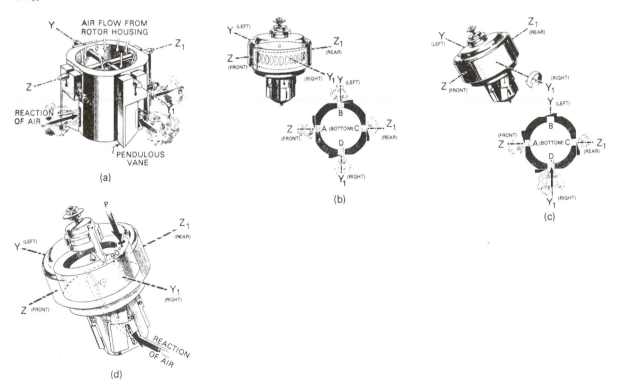

body. The vanes, under the influence of gravity, always hang in the vertical position and it is this feature that is utilized to govern the airflow from the ports and to control the reaction forces applied to the gyroscope.

When the gyroscope is in its normal vertical position as at (b) the knife edges of the vanes bisect each of the ports, making all four openings equal. All four air reactions are therefore equal and the resultant forces about each axis are in balance.

If now the gyroscope is displaced, so that, for example, its top is tilted towards the front of the instrument as at (c), the pair of vanes on the axis YY_1 remain vertical, thus opening the port D and closing the port B. The increased reaction of the air from D results in a force being applied to the body in the direction of the arrow, about axis XX_1. This force is equivalent to one applied on the underside of the rotor and to the left, or at the top of the rotor at point F as shown at (d). Precession back to the vertical position will therefore take place at point P, and the vanes will again bisect the ports to equalize the air reactions.

Ball-type erection unit

This mechanical system is applied to some designs of electric gyro horizon; it utilizes the precessional forces resulting from gravity on a number of steel balls displaced within a rotating holder suspended from the gyroscope housing as shown in Fig. 4.14. The balls are free

Figure 4.14 Ball-type erection unit. (*a*) Gyro vertical. (*b*) tilted away from front of instrument; (*c*) precession to vertical.

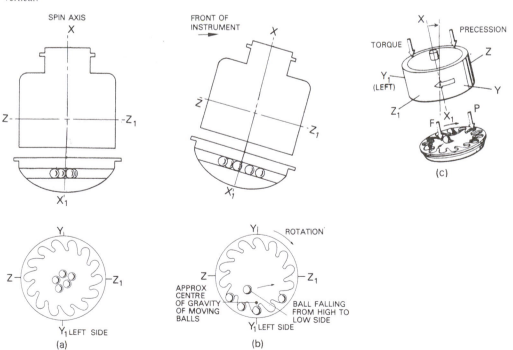

to roll across a radiused erecting disc and into and out of a number of specially profiled hooks in a plate fixed around the inner edge of the holder. The spacing of the hooks is chosen so as to regulate the release of the balls when the gyroscope tilts, and to shift their mass to the proper point on the erecting disc to apply the force required for precession. Rotation of the holder takes place through reduction gearing from the gyroscope's rotor shaft; the speed of the holder is approximately 25 rev/min.

When the gyroscope is in its normal operating position, as shown at (a), the balls change position as the holder rotates but their mass remains concentrated at the centre of the disc. Under this condition, gravity exerts its greatest pull at the centre of the mass, and therefore all forces about the principal axes of the gyroscope are in balance.

At (b) the gyroscope's vertical axis is shown displaced about pitch axis YY_1 away from the front of the instrument. The displacement of the ball holder causes the balls to roll towards the hooks, which at that instant are on the low side; therefore the force due to gravity is now shifted to this side. Since the hooked plate is rotating (clockwise viewed from above), the balls and the point at which the force is acting will be carried round to the left-hand side of the holder. In this position the balls remain hooked and their mass remains concentrated to allow a force to be exerted at the left-hand side of the holder as indicated at (c). This force may also be considered as acting directly on the left-hand bearing of the gyroscope housing and outer ring. Transferring this point of applied force to the rotor rim, precession will then take place about axis YY_1 to counteract the displacement.

As the erector mechanism continues to rotate, the balls will be carried round to the high side of the holder, but one by one they will roll into the hooks at the lower side. Thus, their mass is once again concentrated at this side, allowing the force and precession to be maintained as they are carried around to the left-hand side. This action continues with diminishing movement of the balls as the gyroscope erects to its normal vertical position, at which the balls are at the centre of the disc and the force due to gravity is again concentrated at the centre of the mass.

Displacement of the gyroscope in other directions about its lateral or longitudinal axes will result in similar actions to those described.

Torque motor and levelling switch system
This system is used in a number of electrically-operated gyro horizons and the remote vertical gyroscope units associated with flight director systems. It consists of two torque control motors operated independently by liquid levelling switches, which are mounted, one parallel to the lateral axis, and the other parallel to the longitudinal axis. The disposition of motors and switches is illustrated diagrammatically in Fig. 4.15.

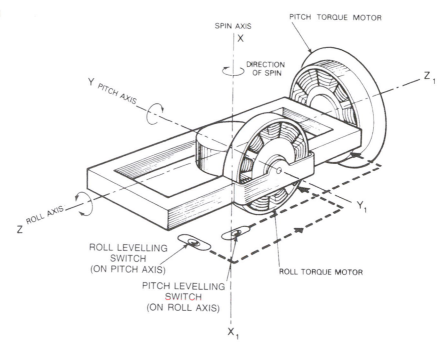

Figure 4.15 Torque motor and levelling switch erection system.

The laterally-mounted switch detects roll displacement and is connected to its torque motor so that a corrective force is applied around the pitch axis. Pitch displacements are detected by the longitudinally-mounted switch, which is connected to its torque motor so that corrective forces are applied around the roll axis. Each switch is in the form of a sealed glass tube containing three electrodes and a small quantity of either mercury or an electrolytic solution.

Each torque motor consists of a stator and a squirrel-cage rotor. The roll torque motor has its stator fixed to the outer ring of the gyroscope, and its rotor fixed to the inner ring. The stator of the pitch motor is fixed to the instrument frame, and its rotor fixed to the outer ring.

The electrical interconnection of the components that comprise each system is indicated in Fig. 4.16(a). In the case of mercury levelling switches, the mercury will lie at the centre of the tubes and, being in contact with the centre electrode, will supply a voltage to the reference windings of their respective torque motors, only when the gyro is running and in its normal operating position. The two outer electrodes are connected one to each section (designated 'A' and 'B') of their respective torque motor control windings; thus, in the normal operating position of the gyro the control winding circuit is open.

When the gyro is displaced about one of its axes, the appropriate levelling switch will also be displaced so that the mercury bridges the gap between the centre electrode and one or other outer electrode. This completes a circuit to either the 'A' or 'B' section of the

Figure 4.16 Circuit diagram of erection system.

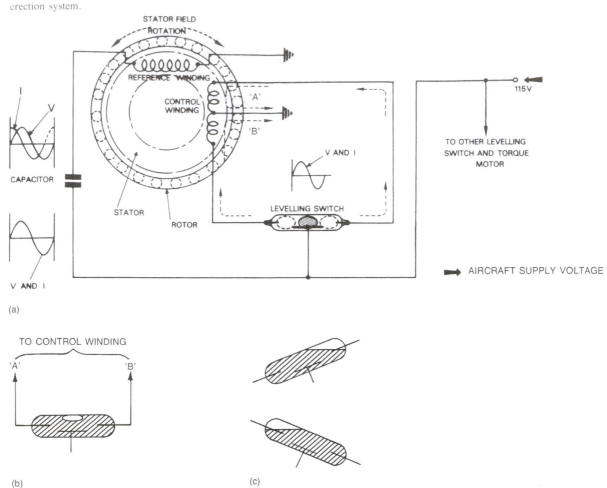

(a)

(b)

(c)

respective torque motor control winding, depending on the direction of gyro displacement.

In order for the torque motor to apply the necessary corrective torque to the gimbal system, the magnetic field of the motor stator must be made to rotate. As will be noted from Fig. 4.16, the voltage to the reference winding is applied via a capacitor, and so, as in any ac circuit containing capacitance, the phase of the current is shifted so as to lead the voltage by 90°.

The control winding circuit has no capacitance, and so the voltage and current flowing through it are in phase; therefore, and because the control and reference windings are both supplied from the same power source, reference winding current must also lead control winding current by 90°. This out-of-phase arrangement (called *phase quadrature*) applies also to the magnetic field set up by each winding.

118

Thus, with current and a magnetic field flowing through the appropriate half of the control winding resulting from a displacement of the gyro, a resultant field is produced which rotates within the torque motor stator in either a clockwise or anti-clockwise direction. As the field rotates, it cuts the conductors of the rotor and induces a current in them; this in turn produces a magnetic field that interacts with the stator field and creates a tendency for the rotor to rotate with the stator field. This tendency is opposed because of the rigidity of the gyro, and consequently a reactive torque is set up in the torque motor and is exerted on the associated gimbal ring to precess the gyro and levelling switch to their normal operating position.

In the case of levelling systems utilizing electrolytic solution-type switches, the supply of current to the control winding sections of a torque motor is controlled in a different manner to that of mercury levelling switch systems. The reason for this is, as will be noted from Fig. 4.16(b), that the electrodes are always immersed in the electrolytic solution, and the circuits to the control winding of a torque motor are always closed. In the normal stabilized vertical position of the gyro, the switch electrodes are in equal amounts of electrolyte and so the currents flowing in each section of a torque motor control winding are equally opposed. Since the electromagnetic effects on the rotors are also equally opposed then no torques will be applied to the gimbal system.

When a switch is displaced it causes a change in the amount of surface area of electrolyte in contact with the electrodes and, in turn, an imbalance in the electrical resistance of the control winding circuit. This may be noted from diagram (c): at the 'low end' electrode there is a greater amount of electrolyte and, in accordance with basic electrical principles, this means low resistance and so more current will flow in that half of the control winding connected to that electrode. A corresponding rotating field is therefore produced to set up a reactive torque in the torque motor for precessing the gyro and levelling switch to the normal stabilized position and in the same manner as that described earlier.

Fast-erection systems

These systems are used in some types of electrically-operated gyro horizons for the purpose of bringing their gyros to the vertical position as quickly as possible from large angles of tilt, particularly during starting.

A control device is therefore provided which, in a typical system, activates a set of contacts to introduce a higher voltage and current flow through the control phase windings of the erection torque control motors. The resulting higher torque applied thereby increases the precession rate. To prevent overheating of the stator coils a time limit (typically 15 seconds) is imposed on system operation. Certain types of gyro horizon utilize a system whereby the gimbal system is mechanically caged when the operating knob is operated.

Erection rate

This is the term used to define the time taken, in degrees per minute, for a vertical gyroscope to take up its normal operating position under the control of its particular type of erection system. Ideally, the rate should be as fast as possible under all conditions, but in practice such factors as speed, turning and acceleration of an aircraft, and the earth's rotation, all have their effect and must be taken into account.

Thus, erection systems are designed so that, for small angular displacements of a gyroscope from the vertical, the erecting couple is proportional to the displacement, while for larger displacements it is made constant. It is also arranged that the couple gives equal erection rates for any rotor axis displacement in any direction in order to reduce the possibility of a slow cumulative error during manoeuvring of an aircraft. Normal rates provided by some typical erection systems are $8°/min$ for pneumatic-type gyro horizons, and $3-5°/min$ for those that are electrically operated.

Errors due to acceleration and turning

Since gyro horizon erection devices are of the pendulous type, it is possible for them to be displaced by the forces acting during the acceleration and turning of an aircraft, and unless provision is made to counteract this the gyroscope spin axis would be precessed to a false vertical position, thereby presenting a false attitude indication. For example, let us consider the effects of a rapid acceleration in the flight direction, firstly on the vane type of erection device, and secondly on the levelling switch and torque motor type.

As shown in Fig. 4.17(a) the acceleration force will deflect the two athwartships-mounted vanes to the rear, thus opening the right-hand port. The greater reaction of the air flowing through this port applies a force to the underside of the rotor causing it to precess forward about the axis YY_1. The horizon bar is thus displaced downwards, presenting a false climb indication.

In the case of levelling switches (Fig. 4.17(b)), an acceleration force will deflect the liquid in the one related to pitch to the rear of its tube. A circuit will thus be completed to the control winding of the pitch torque motor which, in the manner described on page 117, will precess the gyroscope forward and will therefore also produce a false climb indication.

In both cases the precession is due to a natural response of the gyroscope, and the pendulous vanes and/or liquid always return to their neutral positions, but for as long as the disturbing forces remain, such positions apply only to a false vertical. When the forces

Figure 4.17 Acceleration error. (*a*) Vane-type erection system; (*b*) levelling switch and torque motor erection system.

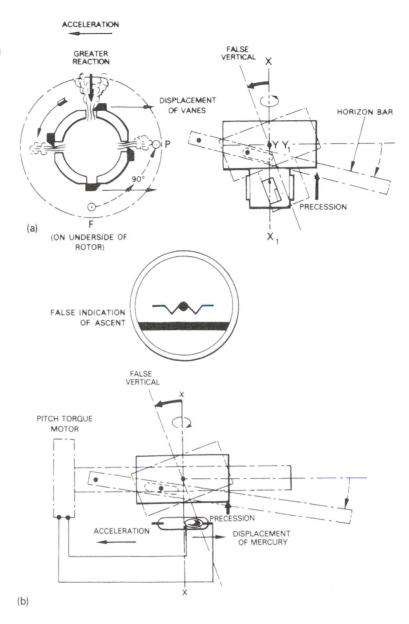

are removed the false climb indication will remain initially and then gradually diminish under the influence of precession, restoring the gyroscope to its normally true vertical position.

It should be apparent from the foregoing that, during periods of deceleration, a gyro horizon will present a false indication of descent.

When an aircraft turns, false indications about both the pitch and roll axes can occur, due to what are termed 'gimballing effects' brought about by forces acting on both sets of pendulous vanes or both levelling switches, as appropriate. There are, in fact, two errors due to turning: *erection errors* and *pendulosity errors*.

121

Erection errors As an aircraft enters a turn, the gyroscope's spin axis will initially remain in the vertical position and so an accurate indication of the roll or bank angle will be presented. In this position, however, the longitudinally-mounted pendulous vanes, or the roll levelling switch, are acted upon by centrifugal force. This force will be applied to the gyroscope in such a direction that it will tend to precess towards the perpendicular along which the resultant of centrifugal and gravity forces are acting. Thus, the gyroscope erects to a false vertical and introduces an error in roll indication. Such errors may be compensated by one of the following methods: (i) inclination of the gyroscope's spin axis; (ii) erection cut-out; and (iii) pitch-bank compensation.

Inclined spin axis This method is based on the idea that, if the top of the axis can describe a circle about itself during a turn, then only a single constant error will result. In its application, the method is mechanical in form and varies with the type of gyro horizon, but in all cases the result is to impart a constant tilt of the axis from the vertical (typically 1.6° or 2.5°). In pneumatic types of gyro horizon, the athwartships-mounted pendulous vanes are balanced so that the gyroscope is precessed to the tilted position; in certain electric gyro horizons, the pitch levelling switch is fixed in a tilted position so that the gyroscope is precessed away from the true vertical in order to overcome what it detects as a pitch error. The linkages between gyroscope and horizon bar are so arranged that during level flight the horizon bar will indicate this condition.

Erection cut-out An example of this method as applicable to electrically-operated gyro horizons incorporates additional liquid-level switches positioned in pairs on the pitch and roll axes. The switches are connected to the torque motors in such a way that under the influence of forces they isolate the control winding circuits from the main erection switches. Operation of the system may be understood from Fig. 4.18 which shows the arrangement applicable to pitch erection cut-out. The pairs of switches are set at an angle to each other in order to differentiate between acceleration and deceleration forces.

Assuming that an acceleration occurs, the electrolyte in the pitch erection switch will be displaced in the opposite direction, and as we have already learned, the change in electrical resistance of the electrolyte will produce an unbalanced condition in the torque motor control winding circuit, and a torque tending to precess the gyro; in this case, to a false vertical position. At the same time, however, the acceleration force displaces the electrolyte in cut-out switch 'A' to complete a circuit to a solid-state switch which then operates to open the ground connection of the control winding; torque motor operation

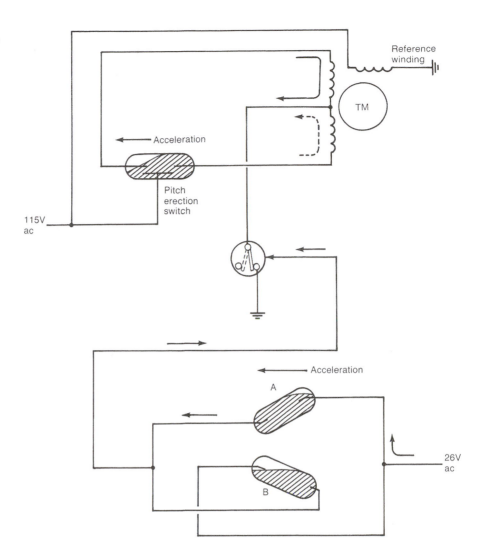

Figure 4.18 Erection cut-out.

is thereby prevented. Switch 'B' performs a similar function but under the influence of a decleration force.

Roll erection cut-out is accomplished by another identically-arranged pair of switches connected to the roll torque motor, except of course that they are angled to differentiate between the centrifugal forces corresponding to left and right turns.

Pitch-roll (bank) erection This method is a combined one (incorporated in some gyro horizons utilizing mercury-type levelling switches) in which the roll levelling switch is disconnected during a turn by the pitch levelling switch. It is intended to correct the varying pitch and roll errors and operates only when the rate of turn causes a centrifugal acceleration exceeding 0.18 g, which is equivalent to a 10° tilt of the roll erection switch. As shown schematically in Fig. 4.19, two additional switches, connected in a double-pole changeover

123

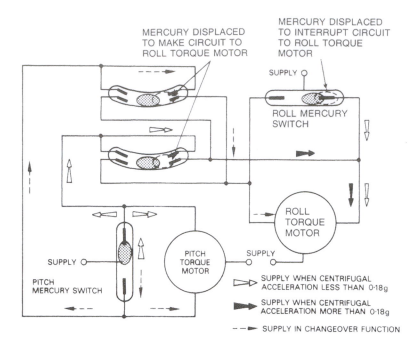

Figure 4.19 Pitch-roll (bank) erection.

MERCURY DISPLACED
TO MAKE CIRCUIT TO
ROLL TORQUE MOTOR

MERCURY DISPLACED
TO INTERRUPT CIRCUIT
TO ROLL TORQUE
MOTOR

SUPPLY

ROLL MERCURY
SWITCH

ROLL
TORQUE
MOTOR

PITCH
TORQUE
MOTOR

SUPPLY

SUPPLY

PITCH
MERCURY SWITCH

SUPPLY WHEN CENTRIFUGAL
ACCELERATION LESS THAN 0·18g

SUPPLY WHEN CENTRIFUGAL
ACCELERATION MORE THAN 0·18g

SUPPLY IN CHANGEOVER FUNCTION

arrangement, are provided and are interconnected with the normal erection systems.

Let us consider first a turn to the left and one creating a centrifugal acceleration less than 0.18 g. In such a turn, the mercury in the roll levelling switch will be displaced to the right and will bridge the gap between the supply and right-hand electrodes, thus completing a circuit to the roll torque motor. This is the same as if the gyroscope axis had been tilted to the right at the commencement of the turn; the roll torque motor will therefore precess the gyroscope back to a false vertical, and left of the true one. At the same time, the axis tilts forward due to gimballing effect, and the mercury in the ptich switch, being unaffected by centrifugal acceleration, moves forward and completes a circuit to the pitch torque motor, which precesses the gyroscope rearwards. The two additional switches, which are also mounted about the roll axis, do not come into operation since the mercury in them is not displaced sufficiently to contact the right-hand electrodes. Thus, with acceleration less than 0.18 g there is no compensation.

When acceleration is in excess of 0.18 g, the mercury in the roll levelling switch is displaced to the end of the tube and so disconnects the normal supply to its torque motor, i.e. it acts as an erection cut-out. The pitch switch, however, still responds to a forward tilt and remains connected to its torque motor, and, as will be noted from the diagram, it also connects a supply to the lower of the two additional switches. Since the mercury in these switches is also displaced by the acceleration, a circuit is completed from the lower switch to the roll torque motor, which precesses the gyroscope axis to the right to reduce the roll error. At the same time, the pitch switch completes a circuit to its torque motor, which then precesses the gyroscope axis rearward, so reducing the pitch error. Thus, during turns a constant control is applied about both the pitch and roll axes by the pitch levelling switch.

The changeover function of the additional switches depends on the direction of gyroscope tilt in pitch. This is indicated by the broken arrows in Fig. 4.19; the gyroscope and the pitch levelling switch now being tilted rearward, the latter connects a supply to the upper additional switch so that the direction of the supply to the roll torque motor is changed, and the gyroscope is precessed to the left. The change in direction of the supply to the roll torque motor is also dependent on the direction of turn, as a study of Fig. 4.19 will show.

Since the forces produced depend on an aircraft's speed and rate of turn, then all erection errors will vary accordingly, thus making it difficult to compensate for them under all conditions. It is usual, therefore, particularly for instruments employing the inclined axis method of compensation, to base compensation on standard values, e.g. a rate 1 turn of 180°/min at a speed of 200 mph.

Direction indicator

This indicator was the first gyroscopic instrument to be introduced as a heading indicator, and although for most aircraft currently in service it has been superseded by remote-indicating compass systems and flight director systems, there are still applications of it in its pneumatically-operated form. The instrument employs a horizontal-axis gyroscope and, being non-magnetic, is used in conjunction with a magnetic compass; it defines the short-term heading changes during turns, while the magnetic compass provides a reliable long-term heading reference as in sustained straight and level flight. In addition, of course, the direction indicator overcomes the effects of magnetic dip, and of turning and acceleration error inherent in the magnetic compass (see Chapter 3).

In its basic form, the outer ring of the gyroscope carries a circular card, graduated in degrees, and referenced against a lubber line fixed to the gyroscope frame. When the rotor is spinning, the gimbal system and card are stabilized so that, by turning the frame, the number of degrees through which it is turning may be read on the card.

The manner in which this simple principle is applied to practical indicators varies between types, but we may consider the vacuum-driven version illustrated in Fig. 4.20(a) and (b), which is used in the basic instrumentation of some types of small aircraft.

The rotor is enclosed in a case, or shroud, and supported in an

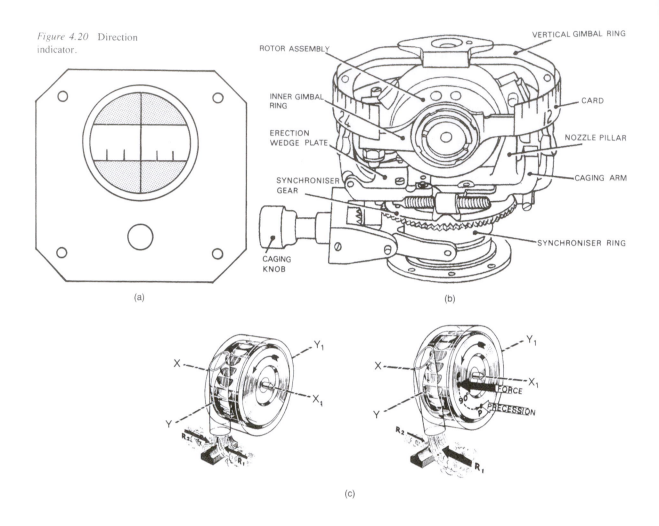

Figure 4.20 Direction indicator.

(a)

(b)

ROTOR ASSEMBLY

INNER GIMBAL RING

ERECTION WEDGE PLATE

SYNCHRONISER GEAR

CAGING KNOB

VERTICAL GIMBAL RING

CARD

NOZZLE PILLAR

CAGING ARM

SYNCHRONISER RING

(c)

inner ring which is mounted in an outer ring, the bearings of which are located at the top and bottom of the indicator case. The front of the case contains a cut-out through which the card is visible, and also the lubber line reference.

When the vacuum system is in operation, the reduced pressure created within the case allows surrounding air to enter through a filtered inlet and to pass through channels in the gimbal rings to emerge finally through jets. The air issuing from the jets impinges on the rotor 'buckets', causing the rotor to rotate at speeds between 12 000 and 18 000 rev/min.

A caging and setting knob is provided at the front of the case to set the indicator on the same heading as that of the magnetic compass. When this knob is pushed in, an arm is lifted thereby locking the inner ring at right angles to the outer ring, and at the same time meshing gearing between the knob and the outer ring. Thus, a heading can be set by rotating the knob and the whole gimbal system. The reason for caging the inner ring is to prevent it from precessing when the outer ring is rotated, and to ensure that, on uncaging, their axes are mutually at right angles.

Control of drift

Drift, as we have already learned (see page 102), is a fundamental characteristic of a horizontal-axis type of gyroscope, and so for practical direction indicating purposes, earth rate error, transport wander, and real drift must be controlled. This is generally effected by gimbal ring balancing and by erection devices.

Gimbal ring balancing

The method of controlling earth rate error is deliberately to unbalance the inner ring so that a constant force and precession are applied to the gimbal system. The imbalance is effected by a nut fastened to the inner ring, and adjusted during initial calibration to apply sufficient outer ring precession to cancel out the drift at the latitutde in which it is calibrated. For all practical purposes, this adjustment is quite effective up to 60° of latitude on the earth's surface.

Erection devices

These form part of the rotor air-drive system and are so arranged that they sense misalignment of the rotor axis in terms of an unequal air reaction. In the indicator already described, this is accomplished by exhausting air over a wedge-shaped plate secured to the outer ring as shown in Fig. 4.20(c).

In the normal horizontal position of the rotor axis, the air flowing from the outlet of the casing is equally divided, and the reaction of the air applies equal and opposite forces to the faces of the wedge. When the rotor is tilted, the air outlet is no longer bisected by the wedge; thus, the reaction forces are unbalanced, and if the greater force is visualized as being applied to the rotor rim, then it and the inner ring will be precessed until the forces are again equal and opposite.

Gimbal errors

A definition of gimbal error has already been given (see page 106). In the case of a direction indicator, errors are dependent upon: (i) the angle of climb, descent, or roll; (ii) the angle between the rotor axis and longitudinal axis of an aircraft. Fig. 4.21 illustrates the gimbal system geometry when an aircraft is in particular attitudes.

At (a) an aircraft is represented as flying straight and level on an easterly heading, and as the gimbal system geometry is such that the

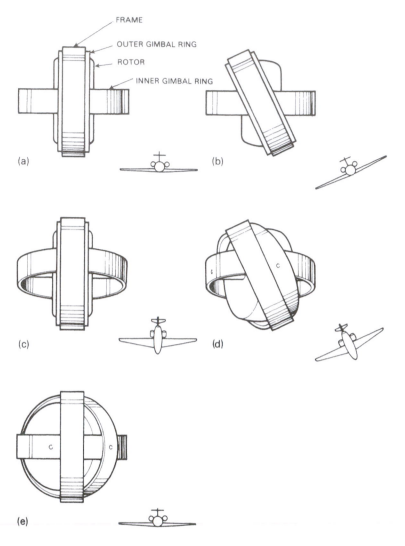

Figure 4.21 Gimbal errors.

FRAME
OUTER GIMBAL RING
ROTOR
INNER GIMBAL RING

(a)

(b)

(c)

(d)

(e)

rotor axis lies N–S, the three axes of the system are mutually at right angles, and the heading will be indicated without error. The same would also be true if an aircraft were flying on a westerly heading.

If an aircraft rolls to the left or right on either an easterly or westerly heading, or executes a left or right turn, the outer gimbal ring will be carried about the axis of the stabilized inner ring (diagram (b)). In this condition the cardinal headings, or changes of heading during turns, would also be indicated without error.

At (c) an aircraft is assumed to be descending so that, in addition to the outer gimbal ring being tilted forward about the rotor axis, the inner ring also rotates, both rings maintaining the same relationship to each other. Again, there is no gimbal error; this would also apply in the case of a climbing attitude.

When an aircraft carries out a manoeuvre which combines changes in roll and pitch attitudes, e.g. the banked descent shown at (d), the outer ring is made to rotate about its own axis, thus introducing a gimbal error causing the indicator to show a change of heading.

If an aircraft is flying on an intercardinal heading, the rotor axis will be at some angle to the aircraft's longitudinal axis, as at (e), and gimballing errors will occur during turns, rolling in straight and level flight, pitch attitude changes or combinations of these.

When the heading is such that the aircraft's longitudinal axis is aligned with that of the gyroscope rotor, rolling of the aircraft on a constant heading will not produce gimballing error because the gimbal system also rotates about the rotor axis. If, however, rolling is combined with a pitch attitude change, the effect is the same as the combined manoeuvre noted earlier (diagram (d)).

Whenever the angular relationship between the gimbal rings is disturbed during a manoeuvre, an indicator's erection device will be attempting to re-erect the rotor into a new plane of rotation and will cause false erection, the magnitude of which depends on how long the erecting force is allowed to operate, i.e. on the duration of the manoeuvre. The magnitude of the force itself will depend on the angle of the rotor to the device. Thus, on completion of a manoeuvre it is possible to have an error due to false erection, and during a manoeuvre an error can be caused which is a combination of both gimballing effect and false erection.

Turn-and-bank indicator

This indicator contains two independent mechanisms: a gyroscopically-controlled pointer mechanism for the detection and indication of the rate at which an aircraft turns, and a mechanism for the detection and indication of bank and/or slip. The dial presentation of a typical indicator is shown in Fig. 4.22(a).

Rate gyroscope

For the detection of rates of turn, a rate gyroscope is used and is arranged in the manner shown at (b) in Fig. 4.22. It differs in two respects from the displacement gyroscopes thus far described: it has only one gimbal ring, and it has a calibrated spring connected between the gimbal ring and casing to restrain movement about the longitudinal axis YY_1, i.e. it is a single-axis gyroscope.

When the indicator is in its normal operating position the rotor spin axis, due to the spring restraint, will always be horizontal and the turn pointer will be at the zero datum mark. With the rotor spinning, its rigidity will further ensure that the zero position is maintained.

Let us assume that the indicator is turned to the left about a vertical input axis. The rigidity of the rotor will resist the turning

Figure 4.22 Turn-and-bank/
slip indicator.

(a)

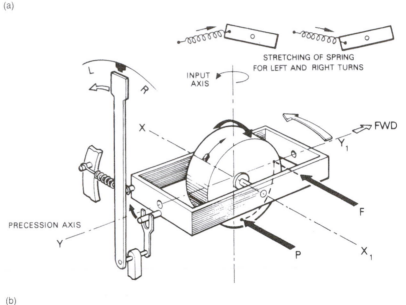

(b)

movement, which it detects as an equivalent force being applied to its rim at point F. The gimbal ring and rotor will therefore be tilted about the longitudinal axis as a result of precession at point P.

As the gimbal ring tilts, it stretches the calibrated spring until the force it exerts prevents further deflection of the gimbal ring. Since precession of a rate gyroscope is equal to its angular momentum and the rate of turn, then the spring force is a measure of the rate of turn. The actual movement of the gimbal ring from the zero position can, therefore, be taken as the required measure of turn rate.

In practice, the gimbal ring deflection is generally not more than $6°$, the reason for this being to reduce the error due to the rate of turn component not being at right angles to the spin axis during gimbal ring deflection.

The rate of turn pointer is actuated by the gimbal ring and a magnifying system which moves the pointer in the correct sense over a scale calibrated in what are termed 'standard rates'. Although they are not always marked on a scale, they are classified by the numbers 1 to 4 and correspond to turn rates of 180°, 360°, 540° and 720° per minute respectively. The marks at either side of zero of the indicator scale shown in Fig. 4.22 correspond to a Rate 1 turn.

A system for damping out oscillations of the gyroscope is also incorporated and is adjusted so that the turn pointer will respond to fast rate of turn changes and at the same time respond to a definite turn rate instantly.

It should be noted that a rate gyroscope requires no erecting device or correction for random precession, for the simple reason that it is always centred by the control spring. For this reason also, it is unnecessary for the rotor to rotate at high speed, a typical speed range being 4000−4500 rev/min. The most important factor in connection with speed is that it must be maintained constant, since precession of the rotor is directly proportional to its speed.

Bank indication

In addition to the primary indication of turn rate, it is also necessary to have an indication that an aircraft is correctly banked for the particular turn. A secondary indicating mechanism is therefore provided which depends for its operation on the effect of gravitational and centrifugal forces. A method commonly used for bank indication is one utilizing a ball in a curved liquid-filled glass tube as illustrated in Fig. 4.23.

Figure 4.23 Ball-type bank indicating element. (*a*) Level flight; (*b*) correctly banked; (*c*) underbanked (skidding out of turn); (*d*) overbanked (slipping into the turn).

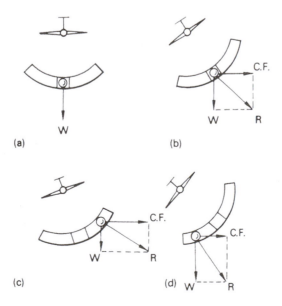

In normal level flight (diagram (a)) the ball is held at the centre of the tube by the force of gravity. Let us assume now that the aircraft turns to the left at a certain airspeed and bank angle as at diagram (b). The indicator case and the tube move with the aircraft, of course, and because of the turn, centrifugal force in addition to that of gravity acts upon the ball and tends to displace it outwards from the centre of the tube. However, when the turn is executed at the correct bank angle and matched with airspeed, then there is a balanced condition between the two forces and so the resultant force holds the ball at the centre of the tube as shown. If the airspeed were to be increased during the turn, then the bank angle and centrifugal force would also be increased, but so long as the bank angle is correct for the appropriate conditions, the new resultant force will still hold the ball at the centre of the tube.

If the bank angle for a particular rate of turn is not correct, say underbanked as in diagram (c), then the aircraft will tend to skid out of the turn. Centrifugal force will predominate under such conditions and will displace the ball from its central position. When the turn is overbanked, as at (d), the aircraft will tend to slip into the turn and so the force due to gravity will now have the predominant effect on the ball. It will thus be displaced from centre in the opposite direction to that of an underbanked turn.

Typical indicator

The mechanism of a typical pneumatic type of indicator is shown in Fig. 4.24. Air enters through a filtered inlet situated at the rear of the case and passes through a jet from which it is directed onto the rotor buckets. The direction of spin is in the direction of flight. Adjustment of gyroscope sensitivity is provided by a screw attached

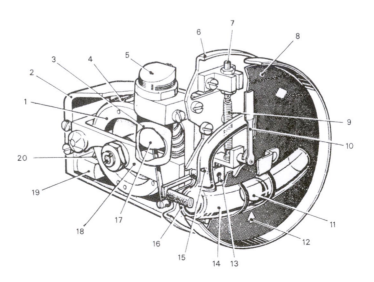

Figure 4.24 Mechanism of a pneumatic-type turn-and-bank indicator.

to one end of the rate control spring. A stop is provided to limit gimbal ring movement to an angle which causes slightly more than full-scale deflection (left or right) of the rate of turn pointer.

A feature common to all indicators is damping of gimbal ring movement to provide 'dead beat' indications. In this particular type, the damping device is in the form of a piston, linked to the gimbal ring, and moving in a cylinder or dashpot. As the piston moves in the cylinder, air passes through a small bleed hole, the size of which can be adjusted to provide the required degree of damping.

The slip indicator is of the ball and liquid-filled tube type, the operation of which has already been described.

Turn coordinator

This instrument is a development of the turn-and-bank indicator, and is adopted in lieu of this in a number of small types of aircraft. The primary difference, other than the display presentation, is in the setting of the precession or output axis of the rate gyroscope. This is set at about 30° with respect to an aircraft's longitudinal axis, thus making the gyroscope sensitive to rolling or banking *as well as* turning. Since a turn is initiated by banking, then the gyroscope will precess, and thereby move the aircraft symbol to indicate the direction of the bank, enabling a pilot to anticipate the resulting turn. The turn is then controlled to the required rate as indicated by the alignment of the symbol with the graduations on the outer scale. In the example illustrated in Fig. 4.25 the graduations correspond to a

Figure 4.25 Turn coordinator.

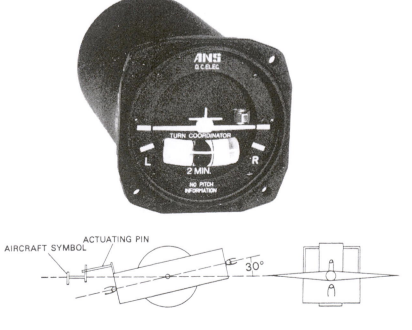

rate 2 (2-min) turn. Coordination of the turn is indicated by the ball-type indicator remaining centred in the normal way (see page 131). In some indicators, a pendulous type of indicator may be adopted for this purpose.

The gyroscope is a dc motor operating at approximately 6000 rev/min; in some cases, a constant-frequency ac motor may be used. The annotation 'no pitch information' on the indicator scale is given to avoid confusion in pitch control which might result from the similarity of the presentation to that of a gyro horizon.

Damping of the gyroscope may be effected by using a silicone fluid or, as in the indicator illustrated, by a graphite plunger sliding in a glass tube.

5 Synchronous data-transmission systems

The instruments that have been described thus far are very basic in concept, in that the display of appropriate data is achieved by mechanical-type elements which require either a direct connection to remotely-located sensing elements, as in the case of pneumatic airspeed indicators, altimeters and vertical speed indicators, or which provide indications directly from their sensing elements, as in compasses and gyroscopic instruments. While such instruments can still satisfy a primary role requirement in the instrumentation of small types of aircraft, their application to large aircraft is restricted by increase in distances between the locations of sensing elements and flight deck panels. In order, therefore, to overcome some of the problems that can arise, e.g. having to run lengthy pipelines between sensors and instruments, an electrical method of transmitting changes in measured quantities is adopted. This is not new, of course, having originated in the days when it became necessary to improve the measuring accuracy of those instruments associated with the operating parameters of engines and, in particular, their use in multi-installation arrangements.

As part of the technique of 'remote indicating', synchronous data-transmission systems, or *synchro systems* as they are generically known, were introduced. They consist of transmitting and receiving elements which, in varying circuit configurations, are now utilized not only in certain engine instruments, but also in analogue-type air data computers, remote-indicating compasses, and in flight director systems for heading and aircraft attitude sensing.

Categories of synchro systems

Synchro systems are divided into four main categories as follows:

1. *Torque*   This is the simplest form of synchro, in which torque is derived solely from the input to its transmitting element; no amplification of this torque takes place. Moderate torque only is developed at the output shaft of the receiving element, and for this reason the system is used for data-indicating purposes, e.g. oil or

fuel pressure, and for the indication of the position of mechanical controlling devices, e.g. airflow control valves.

2. *Control*   This type of synchro normally forms part of a servomechanism to provide the requisite signals which, after amplification, are used for the control of a drive motor.

3. *Resolver*   This is used where precise angular measurements are required. It converts voltages, which represent the *cartesian coordinates* of a point, into a shaft position and a voltage which together represent the *polar coordinates* of the point. They can also be used for conversion from polar to cartesian coordinates. Typical applications are in analogue computers, remote-indicating compasses, and flight director systems.

4. *Differential*   This type of synchro is used where it is necessary to detect and transmit error signals representative of two angular positions, and in such a manner that the difference or the sum of the angles can be indicated. They can be utilized in conjunction with either torque, control or resolver synchro systems.

The foregoing synchros are designated by standardized abbreviations and symbols as in Table 5.1.

Torque synchro system

This consists of a TX element and a TR element interconnected as shown in Fig. 5.1. Both elements are electrically similar, each consisting of a rotor carrying a single winding around a laminated iron core, mounted co-axially within a stator core carrying windings that are spaced 120° apart. The principal physical differences between the two elements are that the TR is usually fitted with a mechanical damper to reduce oscillation, and that the TX rotor is mechanically rotated whereas the TR rotor is rotated by a magnetic field produced by the TX.

Both rotors are supplied from a source of single-phase ac power via sliprings and fine wire brushes, and the corresponding stator connections are joined by transmission lines to form a closed circuit. When the axis of a rotor is aligned with that of the S_1 winding of a stator, a synchro is in what is termed the *electrical zero* position. This serves as a datum when testing a synchro for accuracy.

When power is applied to the system, current flows in both rotor windings and sets up an alternating magnetic flux. Since the rotor windings and stator windings of the TX and TR correspond, in effect, to the primary and secondary windings of a transformer, then the flux induces an alternating voltage in the stator winding coils. The induced voltages are at maximum, and in phase with the rotor voltages, in the 'electrical zero' position, and are zero when the rotors are at 90° to this position. At 180°, the induced voltages are again maximum but

Table 5.1

Synchro type	Abbreviation	Circuit	Symbol

Torque:
  Transmitter
  Receiver
	TX
	TR		

Control: Transmitter	CX		
Receiver	CT		
Resolver	RS		

Differential (in a torque system):
  Transmitter
  Receiver
	TDX
	TDR		

Differential (in a control system):
  Transmitter
  Receiver
	CDX
	CDR		

out of phase with the rotor voltages. At 270°, the induced voltages will again be zero. Due to the 120° spacing of the stator coils, the voltages appearing between the connection points S_1, S_2 and S_3 will be the sums or differences between the voltages induced in the stator coils, depending on whether such voltages are in phase or 180° out of phase.

When the TX and TR rotors are in the same angular positions, e.g. the electrical zero position, the voltages induced in the stators are equal and opposite, and so no current will flow in the stator coils. If, however, the rotors occupy different angular positions, e.g. the TX motor is moved through an angle of 30° while the TR rotor is at 'electrical zero', then an unbalanced condition between voltages induced in the stator coils arises.

This imbalance causes current to flow through the closed circuit between TX and TR stators. The magnitude and phase of the currents are in proportion to that of the induced voltages. The currents are greatest in the coil sections of stator windings when the voltage

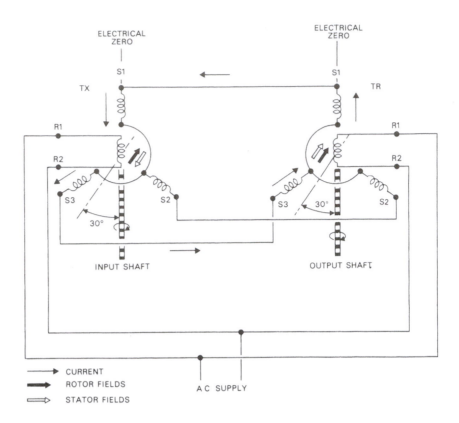

Figure 5.1 Torque synchro system.

ELECTRICAL ZERO

ELECTRICAL ZERO

S1

TX

S1

TR

R1

R1

R2

R2

S3

S2

S3

S2

30°

30°

INPUT SHAFT

OUTPUT SHAFT

CURRENT
ROTOR FIELDS
STATOR FIELDS

A C SUPPLY

imbalance is greatest; thus, with the TX rotor at 30°, the imbalance is greatest in the section comprising coils S_1 and S_3. The greatest current therefore flows through these coils to the corresponding ones in the TR stator.

The current flow through the TX stator produces a resultant magnetic field, and as with normal transformer action, this field and the one produced by current flowing in the rotor winding must always be in balance; the directions of the fields are, therefore, in opposition.

The current flowing through the TR stator also produces a resultant magnetic field, but as the direction of current flow is opposite to that through the TX stator, then the direction of the field will also be opposite. The interaction of this field with that of the rotor will develop a torque and thereby turn the rotor from 'electrical zero' to the same position as that of the TX. As the rotor turns, the imbalance of induced voltages decreases, and in turn the currents produced by them also decrease. When the TR rotor synchronizes with the 30° position of the TX rotor, its field will be in alignment with the resultant field, its voltages will be balanced, and current no longer flows between the stators. The system is then said to be at the *'null' position*.

The foregoing action takes place as a result of positioning the TX rotor at any other angle; it is apparent, therefore, that if this rotor were to be continuously rotated, then rotating magnetic fields would be set up to provide synchronized and continuous rotation of the TR rotor.

During operation, and because of the current flow in the stator coils of the TX, a torque is also set up tending to turn its rotor out of alignment. This torque, however, is overcome by the loads exerted by the prime mover that actuates the TX.

In the event that the rotor and stator connections of a TX and TR system are changed over from the normal symmetrical arrangement indicated in Fig. 5.1, then different operating results will be produced. The TR rotor can still move synchronously with the rotor, but it can do so from a different reference position, or in the reverse direction. The possible rearrangements are illustrated in Fig. 5.2.

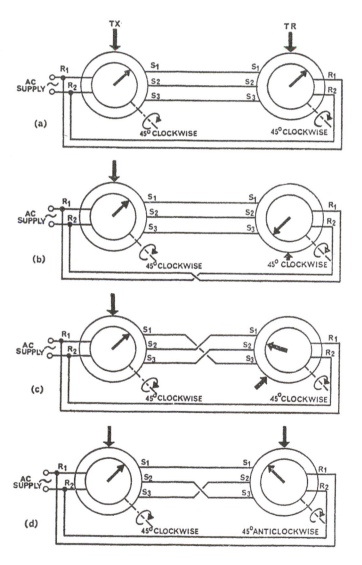

Figure 5.2 Interchange of TX synchro connections. (*a*) Symmetrical connections — data coincide; (*b*) rotor connections reversed — output datum advanced 180°; (*c*) cyclic shift of stator connections — output datum advanced 240°; (*d*) two stator leads interchanged — output rotor reverses direction of rotation.

Control synchro system

A control synchro system differs from the one just described in that its function is to produce an error voltage signal in the receiver element, as opposed to the production of a rotor torque. Typical uses of the system are in servomechanisms such as: altimeters and Mach/airspeed indicators that operate in conjunction with air data computers, and in the indicators of flight director systems.

The interconnection of the two elements of the system are shown in Fig. 5.3. The transmitter is designated as CX, and the receiver as CT, signifiying *control transformer*. The CX is similar to a TX, and from the diagram it will be noted that the single-phase ac power is connected to its rotor only. The CT rotor is not energized since it acts merely as an inductive winding for detecting the magnitude and phase of error voltage signals which are supplied to an amplifier. Other differences in a CT element are as follows:

(a) the rotor winding is on a cylindrical core, to ensure that the rotor is not subjected to any torque when the magnetic field of the CT stator is displaced.
(b) it operates as a single-phase transformer with the stator windings acting as the primary, and the rotor winding as the secondary.
(c) the stator winding coils are of high impedance to limit the alternating currents through them.
(d) it is at electrical zero when the rotor is at 90° as shown.

When the CX rotor is energized, then, as in the case of a TX, voltages are induced in the stator windings to produce resultant magnetic fields when the rotor is at electrical zero (see diagram (a)). The induced voltages are applied to the CT stator coils and the alternating flux produced induces a voltage in the rotor. The magnitude of this voltage depends on the relative position of the rotor; with the rotor at its 'electrical zero' position of 90° as shown, the induced voltage is zero.

If the CX rotor is now rotated clockwise from its electrical zero position as in diagram (b), the resultant flux in the CT stator will be displaced from its datum point by the same angle, and relative to the rotor at that instant. An error voltage is therefore induced in the rotor, and with the connections as shown, the magnitude of the voltage increases from zero, and is also in phase with the voltage applied to the CX rotor. For an anti-clockwise rotation of the CX rotor from electrical zero (as in diagram (c)), the error voltage induced in the CT rotor again increases in magnitude, but this time it is in anti-phase with the voltage applied to the rotor.

As commonly used in servomechanisms, the error voltage signal from a CT rotor is supplied to an amplifier which, in turn, supplies its output to the *control phase* of an ac servomotor. The other phase

Figure 5.3 Control synchro system.

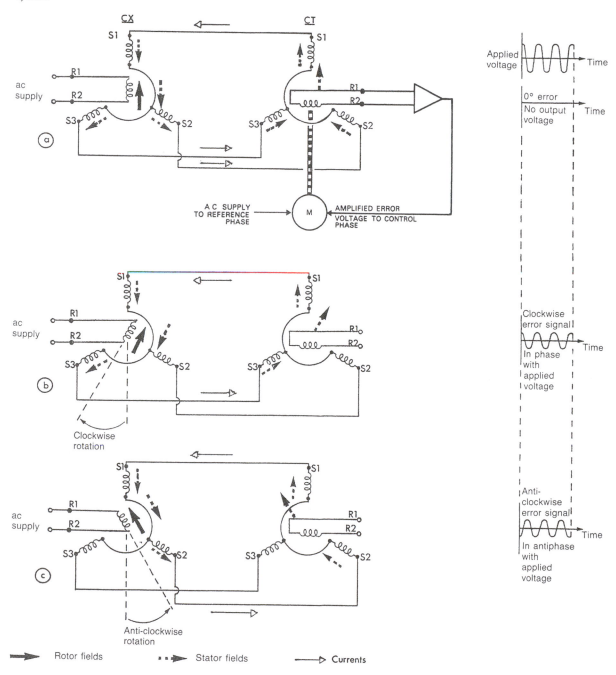

(the *reference phase*) of this motor is continuously supplied with ac. Since the control phase of a two-phase motor can either lead or lag the reference phase voltage, then the phase of the error voltage will determine the direction in which the motor will rotate, and its magnitude will determine its speed of rotation.

The servomotor drives the mechanism being controlled, and in addition it turns the CT rotor in the direction appropriate to that in which the CX rotor has been turned, thereby reducing the error voltage. At zero error voltage, the synchros are at 'null' and the mechanism being controlled is at the new datum established by the positioning of the CX rotor.

In some control synchro servomechanisms, the servomotor also drives a tacho-generator which produces a feedback signal that is supplied to the amplifier for the purpose of controlling the rate at which the servomotor rotates.

Differential synchros

In some applications it is necessary to detect and transmit error signals representative of two angular positions, and in such a manner that the receiver element of a synchro system will indicate the algebraic difference or the sum of the two angles. This is achieved by introducing a *differential* synchro into either a torque or control synchro system, and then using it as a transmitter. Unlike TX or CX synchros, the rotor of a differential synchro also has three star-connected windings; the rotor core is of cylindrical shape. When utilized in *torque* or *control* synchro systems, differential synchros are designated as TDX and CDX respectively.

Figure 5.4 shows the three synchros comprising a TDX system, their interconnection in this case being set up for detecting the *algebraic difference* between two inputs. One input shaft controls the angular position of the TX rotor, and the second input shaft controls the angular position of the TDX rotor. Clockwise rotations of the rotors are taken as positive, and anti-clockwise rotations as negative. The TDX rotor windings are connected to the TR stator windings. The TX and TDX rotors are at their electrical zero positions, and so the resultant magnetic fields are as shown. The effects of the voltages induced in the TR stator will, therefore, produce resultant fields such that its rotor will also be at electrical zero.

If now the TX rotor is rotated clockwise through 60° while the TDX rotor remains at electrical zero as in diagram (a), the signals generated in the stator of TX will be transmitted, unmodified, to the TR stator windings and so the resultant fields will rotate its rotor clockwise through 60°.

In diagram (b) the TX rotor is shown at the electrical zero position, while the TDX rotor is rotated clockwise through 15°. The fields of both synchros remain at electrical zero because their position is determined solely by the orientation of the TX rotor. However, a 15° clockwise rotation of the TDX rotor without a change in the position of its field is equivalent to moving the rotor field 15° anti-clockwise whilst leaving the rotor at electrical zero. This relative angular

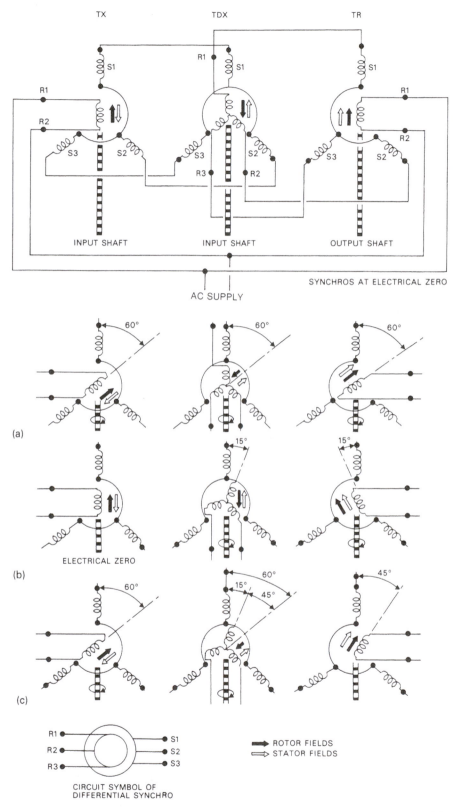

Figure 5.4 TDX synchro system.

TX   TDX   TR

S1   R1   S1   S1

R1   R1

R2   R2

S3   S2   S3   S2   S3   S2

R3   R2

INPUT SHAFT   INPUT SHAFT   OUTPUT SHAFT

SYNCHROS AT ELECTRICAL ZERO

AC SUPPLY

60°   60°   60°

(a)

15°   15°

ELECTRICAL ZERO

(b)

60°   60°   45°
              15°   45°

(c)

R1 ● — ● S1   ➡ ROTOR FIELDS
R2 ● — ● S2   ⇨ STATOR FIELDS
R3 ● — ● S3

CIRCUIT SYMBOL OF
DIFFERENTIAL SYNCHRO

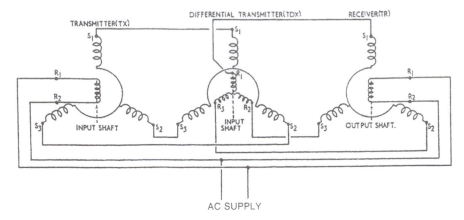

Figure 5.5 Algebraic addition.

change is duplicated in the TR stator and so its rotor will align itself with the field, i.e. for a 15° clockwise rotation of the TDX rotor, the TX rotor will rotate 15° anti-clockwise.

When the *algebraic addition* of two inputs is required, the TX and TDX stator connections S_2 and S_3, and also the TDX rotor connections R_2 and R_3 to the stator connections S_2 and S_3 of the TR, are interchanged as shown in Fig. 5.5.

Figure 5.6 illustrates the effects produced by interchanging the connections of a TDX system.

An alternative method of integrating two inputs into a single output is to utilize a differential synchro as a receiver (TDR) in conjunction with two TXs (one for each input) as shown in Fig. 5.7. The interconnections are arranged to obtain the algebraic difference between the two inputs.

As in the case of the basic TR synchro, the rotor of a TDR always aligns itself so that its magnetic field coincides with that of the receiver stator. Assume now that the rotor of TX (A) is displaced 60° clockwise and the rotor of TX (B) is displaced 15° from their electrical zero positions. This displacement of the rotor of TR (A) displaces the stator magnetic field of TDR 60° clockwise from the electrical zero position. The magnetic field of the TDR rotor is derived from TX (B), thus a 15° displacement of its stator magnetic field also causes a 15° displacement of the field of the TDR rotor. The consequent rotation of the rotor to align its magnetic field with that of its stator is thus 45° clockwise, thereby indicating the difference (60° − 15°) in the angular displacements of the two TX rotors.

If the connections R_2 and R_3 of the TDR rotor and the connections S_2 and S_3 are interchanged, the angular rotation of the rotor of TX (B) from the electrical zero position gives rise to an equal and opposite displacement of the TDR rotor field. The angular movement of this rotor to align its magnetic field with that of its stator is now

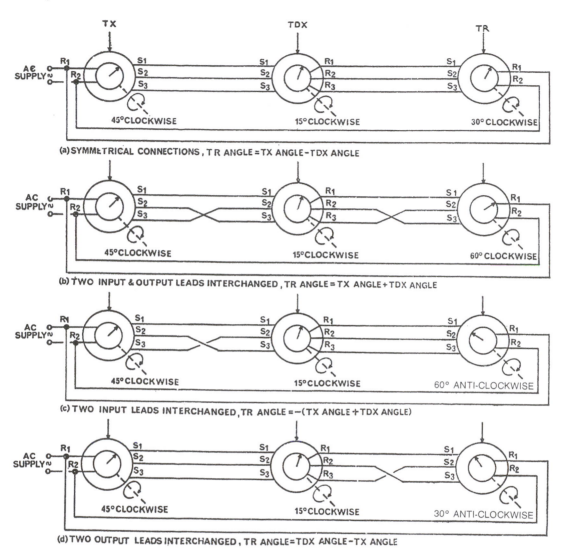

Figure 5.6 Interchange of TDX system connections.

(a) SYMMETRICAL CONNECTIONS, T R ANGLE = TX ANGLE − TDX ANGLE

(b) TWO INPUT & OUTPUT LEADS INTERCHANGED, TR ANGLE = TX ANGLE + TDX ANGLE

(c) TWO INPUT LEADS INTERCHANGED, TR ANGLE = −(TX ANGLE + TDX ANGLE)

(d) TWO OUTPUT LEADS INTERCHANGED, TR ANGLE = TDX ANGLE − TX ANGLE

75° clockwise, thus indicating the algebraic sum of the two inputs to the TDR.

In the same way that differential synchros can be used in torque synchro systems, so they can be used in systems utilizing control synchros; the basic arrangement is shown in Fig. 5.8.

Resolver synchros

As was pointed out at the beginning of this chapter, resolver synchros are employed to convert voltages which represent the cartesian coordinates of a point into a shaft position and a voltage which

145

Figure 5.7 TDR arrangement.

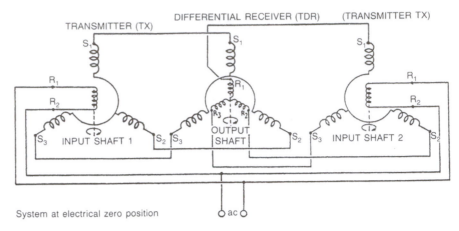

System at electrical zero position

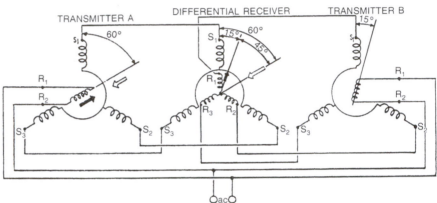

TX rotor displacements from electrical zero position

➡ ROTOR FIELDS
⇨ STATOR FIELDS

together represent the polar coordinates of a point. Let us now see what is meant by these terms and how they are related to a voltage.

If a vector representing an alternating voltage is drawn, then, as indicated in Fig. 5.9, it can be defined in terms of its length (designated r) and also of the angle θ it makes with a horizontal axis X; these are referred to as the *polar coordinates*. This same vector can also be defined in terms of x and y, where:

$$x = r \cos \theta \quad \text{and} \quad y = r \sin \theta$$

These two expressions are the *cartesian coordinates*.

Unlike torque or control synchros, a resolver has four stator windings and four rotor windings arranged as shown in Fig. 5.10(a). Stator windings S_1 and S_2 are in series and have a common axis which is at right angles to that formed by S_3 and S_4 in series. A similar arrangement applies to the rotor windings. For the purpose of simplification, arrangements are usually shown as at the right of the diagram.

Figure 5.8 CDX synchro
system.

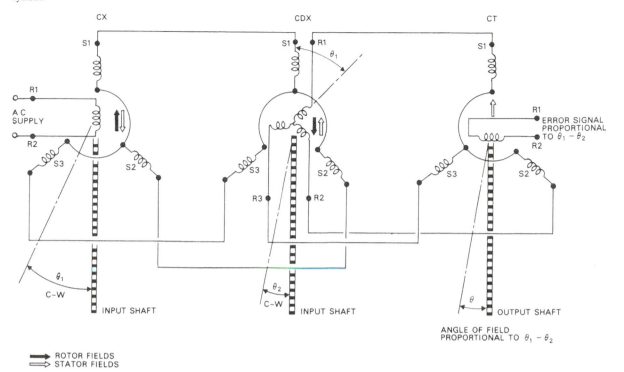

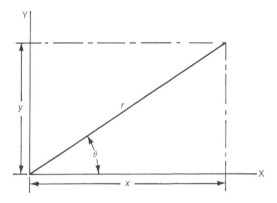

Figure 5.9 Polar and cartesian
coordinates.

Figure 5.10(b) illustrates a common application of a resolver
whereby polar coordinates are converted to cartesian. An alternating
voltage is applied to one of the rotor windings and represents the
length r of the vector shown in Fig. 5.9. The other winding is
unused and is normally shorted-out to improve accuracy and to limit
spurious response.

The flux produced by the rotor current links with the stator
windings, and voltages are induced in them depending on the relative

147

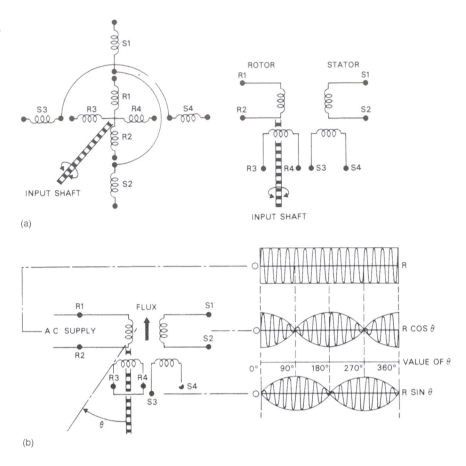

Figure 5.10 Resolver synchro.

(a)

(b)

position of the rotor. In the position shown in Fig. 5.10, maximum voltage is induced in the stator coils aligned with the rotor windings in use, i.e. S_1 and S_2 aligned with R_1 and R_2. No voltage is induced in S_3 and S_4 which are at right angles to the rotor flux.

Movement of the rotor at a constant speed will therefore induce sinusoidal voltages across the two stator coils; these variations will be equal to $r \cos \theta$ in S_1 and S_2 and to $r \sin \theta$ in S_3 and S_4. The sum of these two defines, in *cartesian coordinates*, the input voltage and rotor shaft rotation, which together are the *polar coordinates* (r and θ of Fig. 5.9).

Figure 5.11 shows the arrangement in which cartesian coordinates are converted to polar. An alternating voltage $V_x = r \cos \theta$ is applied to the cosine winding S_1 and S_2, and a voltage $V_y = r \sin \theta$ is applied to the sine winding S_3 and S_4. An alternating flux of amplitude and direction dependent on these voltages and representing the cartesian coordinates is, therefore, produced inside the stator.

One of the rotor windings R_1 and R_2 is connected to an amplifier and a servomotor which drives the output load, and also the rotor in such a direction as to return the rotor to a 'null' position, i.e. at 90°

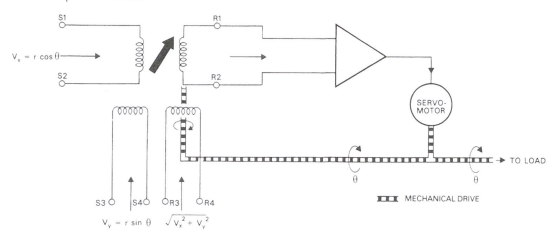

Figure 5.11 Conversion of cartesian to polar coordinates.

to the stator flux; the motor then stops. Rotor winding R_3 and R_4 must now lie parallel to the stator field and has induced in it a voltage proportional to the amplitude of the alternating flux; in other words, proportional to the length of vector r (Fig. 5.9) and equal to $\sqrt{(V_x^2 + V_y^2)}$. The shaft position then represents the angle θ. Thus, the input defined in cartesian coordinates has been converted to an output in terms of polar coordinates.

Synchrotel

A synchrotel is generally used as a low-torque CT and is interconnected with a CX synchro, but unlike the CT synchro applied to the more conventional type of control system, it serves as a *signal transmitting element*. The construction of a synchrotel, and its interconnections, are shown schematically in Fig. 5.12. It employs a stator core carrying three windings at 120° apart, but, as will be noted, the rotor section differs from the conventional form of construction in that: (i) the rotor, which is made of aluminium, is hollow and of oblique section; (ii) the rotor rotates in an air gap formed between a fixed cylindrical core, the stator core, and the single-phase rotor winding which is also fixed. The rotor shaft is supported in jewelled bearings and concentric with the cylindrical core, and is mechanically connected to the element whose position or displacement is to be measured.

In a typical application, e.g. measurement of a fluid pressure, the synchrotel rotor is connected to the appropriate pressure-sensing element. The CT is located within a panel-mounted indicator, together with a servo-amplifier and a servomotor. The CT rotor is energized by a 26 Volts, 400 Hz single-phase ac supply which induces voltages in its stator; since the stator is connected to the

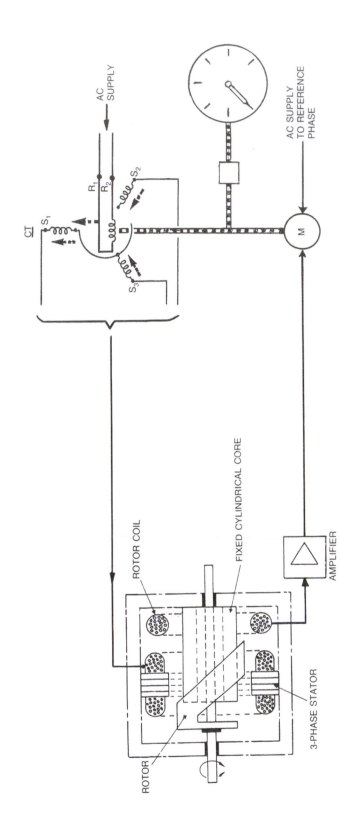

Figure 5.12 Synchrotel.

synchrotel stator then a resultant radial alternating flux is established across it.

For any particular pressure applied to the sensing element there will be a corresponding position of the synchrotel rotor and, due to its oblique shape, sections of it will be cut by the radial stator flux. Currents are thus produced in the rotor and, since it is pivoted around the cylindrical core, an axial component of flux will be created in the core. The core flux will also induce an alternating voltage in the fixed rotor winding, and the amplitude of this voltage will be a sinusoidal function of the relative positions of the rotor and stator flux. This voltage is fed, via the servo-amplifier, to the control phase of the two-phase servomotor which drives the CT rotor round in its stator, thereby causing a change in synchrotel stator flux, to the point where no voltage is induced in the rotor winding, i.e. the CT is driven to the 'null' position. This position corresponds to the pressure measured by the sensing unit at that instant, and since the servomotor also drives the indicator pointer, then this will also be positioned to register the pressure on the indicator scale.

Synchros and electronic display systems

The circuits of synchro systems, irrespective of their type and data measurement application, are of the linear or analogue type, i.e. their output signals vary continuously as a given function of the input. This does not, however, preclude their use in conjunction with electronic display systems whose circuits process all incoming data in a binary digit and coded signal format. Operating details of the devices used for converting analogue output signals from synchros into digital format will be given in later chapters.

6 Digital computers and data transfer

Instruments and certain associated integrated systems utilizing signals from digital computers have been in use for a very long time, and they have operated alongside, and been integrated with, those systems dependent upon outputs from the earlier types of analog computer. For example, the digital computer of an inertial navigation system (INS) can be supplied with altitude and airspeed data signals from an analog type of air data computer (ADC). In such applications, signal interfacing devices referred to as analog-to-digital (A/D) and digital-to-analog (D/A) converters are also used. The A/D conversion is also necessary in applications whereby analog sensors are retained for the supply of signal outputs to digital computers.

In the mid- to late-'70s, the operational requirements for aircraft became more demanding and, in consequence, design concepts underwent drastic changes which were to pave the way for greater automation of in-flight management of aircraft and their systems. The integration of systems which, of necessity, were to assume higher levels of importance demanded greater capacity for processing of their data outputs, and a faster means of data transfer; the application of mixed computer technology was, therefore, no longer acceptable. Thus 'new technology' types of aircraft have come into commercial service, each having a large number of digital computers operating within what is still an analog environment, and distributing their data in binary-coded format via a 'data highway' bus system.

Digital computer fundamentals

A digital computer is essentially a device which uses circuits that respond to, and produces signals of, two values: namely, logic high or binary 1, and logic low or binary 0. It is capable of performing operations on data represented as a series of discrete impulses arranged in the binary-coded or 'bit' format. Figure 6.1 illustrates what is generally termed the organization (sometimes architecture) of the principal *hardware* elements of a computer. The central processor unit (CPU) executes the individual machine instructions which make up the computer *program*.

The binary-coded format of the program consists of an *operation code* which tells the computer what operation it is to start with next,

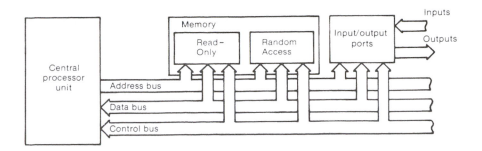

Figure 6.1 Computer organization.

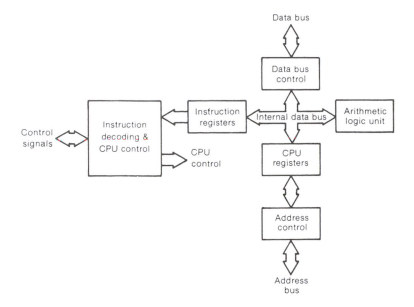

Figure 6.2 Typical CPU organization.

and an *operand* which is the data to be operated on. The program, together with procedures and associated documentation, form what is termed the *software*.

The CPU contains a number of registers or temporary storage units which can each store a single *byte* or *word*, an arithmetic logic unit (ALU) which performs the binary arithmetic and logic functions associated with data manipulation, and a timing and control section for co-ordinating CPU internal operation so that fetching and execution of the instructions specified by a program is performed. The typical organization of the CPU is shown in Fig. 6.2.

Communication between the CPU and memory, and the input/output ports, is by means of a computer highway consisting of three separate busses: the *data bus, address bus* and *control bus*. The term 'bus' signifies a group of conductors carrying one 'bit' per conductor, and is represented on diagrams by a broad arrow identified by function.

The data bus carries the data associated with a memory or input/output transfer, and the number of lines constituting the bus is the same as the number of bits (e.g. eight) in the CPU's word length.

153

The bus is usually bi-directional, i.e. the CPU can write data to be read by a memory, or it can read data from the bus presented by the memory. Thus, data transfer between the two can be effected over a single set of data lines. All information transferred under program control travels on the data bus via the CPU.

The address bus specifies the memory locations or input/output ports involved in a transfer. The number of bits constituting this bus has no direct relationship to the data bus word-length and depends on the operation being performed; for example, at the beginning of an instruction cycle the CPU must supply the address of the next instruction in sequence to be fetched from the memory. During the execution of the instruction, data may then be required to move between the CPU and either the memory or an input/output port. If this is the case, then the data memory address or input/output address must be placed on the address bus by the CPU. Typically, the bus contains 16 lines, and so gives the CPU the capability of addressing up to 2^{16} or 65 536 individual locations.

The control bus is also bi-directional since, being made up of individual control lines for CPU memory and input/output control, it synchronizes the transfer of 'read-out' and/or 'write-in' data along the data bus.

Memory

This consists of a number of storage locations for instruction words whose bit patterns define specific functions to be performed, and for data words to be used for carrying out the operations specified by the instruction words. Each memory word is given a numbered location or *address* which is itself a binary word. There are two types of memory:

Random-Access Memory (RAM) in which stored data at any location can be changed by 'writing in' new data at that location. It can therefore also be called a read/write memory.

Read-Only Memory (ROM) in which the binary information it contains is permanently stored in it. The data, which can be accessed in random fashion, are written in at the time of manufacture, and so the specific program cannot usually be changed afterwards.

In the organization of some digital computers, the two types of memory are used together.

Memories are also classified as: *volatile*, i.e. one which loses its stored data when the power supply is switched off, or *non-volatile*, i.e. one that retains stored data even though the power supply is off.

Capacity and addressable locations

The capacity of a memory relates to 'bit storage' and is quoted in *kilobits* (K); the prefix 'kilo' does not stand for 1000 in the usual sense, but for 2^{10} or 1024. Thus, $8K = 8 \times 1024 = 8192$ bits capacity.

The number of *addressable locations* in a memory is dependent on its number of input/output data lines, and is derived from the bit storage capacity divided by the number of data lines. This is because each address location generally contains as many bits as it can pass through the data bus. If, for example, a 1K memory has only one data line, it will have 1024 separate addressable locations, but with four data lines it can only be addressed at 256 locations. The number of lines are decided by design and specified in the appropriate manufacturer's data sheets.

Input/Output (I/O) ports

These form the interface between a computer and the sources of input data and subsequent output data, and are generally under the control of the CPU. Special I/O instructions are used to transfer data into and out of the computer.

More sophisticated I/O units can recognize signals from extra peripheral devices called *interrupts* that can change the operating sequence of the program. In some cases, units permit direct communications between the memory and an external peripheral device without interference from the CPU; such a function is called *direct memory access (DMA)*.

Computer languages

In the same way that we humans communicate with each other through language, so a digital computer must use a language of one sort or another to carry out its functions. There is, however, a big difference between us and the computer in that when we are, say, given an instruction to do something, the understanding of our own language enables us to understand the instruction directly, and apart from acting upon it, no other conversion is required. This is not true for a computer, because when we want to give it an instruction a conversion from our language into the binary-coded language must first be carried out. The digital code is called the *machine code*, and if instructions for the computer can be programmed directly in this code, the program is written in *machine language*, and overall it is called a *machine language program*.

The task of converting to machine code is usually delegated to the computer, which follows what is called an *assembler program* telling the computer what to do. The choice of instruction can be made by selecting a mnemonic which is an abbreviation of what the instruction does. This programming with mnemonic instructions is called *assembly language programming* because, after the sequence is written, it is fed into the assembler program which makes the conversion to machine code and assembles it into the memory in the proper order.

Such mnemonic codes are still unlike the human language, and so a higher-level language-programming concept can be adopted in which instructions are written in a problem-oriented or procedure-oriented notation with each statement corresponding to several machine code instructions. The conversion of the statements to machine code is done by a more involved computer program called a *compiler*. The easier the programming is made by bringing the machine language closer to the human language, the more complex is the computer program needed to convert the machine language statements in machine code. Once the conversion is available, however, it can be used over and over again as necessary.

Data conversion

The majority of data to be processed by digital computers are, in the first instance, in analog form, and so in order for the computers to carry out their interpreting function, the data must be converted to a binary-coded format. In many cases it is also necessary for data to be converted from digital to analog format. The conversion devices used for such purposes are of the integrated logic circuit type, and respectively they perform encoding and decoding functions as shown graphically in Fig. 6.3.

The ideal A/D converter has a 'staircase' transfer characteristic, with the analog input quantized into a number of levels corresponding to the number of 'bits' resolution. The true analog value corresponding to a given output code is centred between two decision levels. For the ideal D/A converter, there is a one-to-one correspondence between input and output.

Data transfer

The transfer of data between the individual computer systems of an aircraft is a necessary feature of in-flight operation. For example, under automatically-controlled flight conditions, an automatic flight control system operates in conjunction with an INS, ADC, FDS (flight director system) and radio navigation systems, and all these involve an exchange of data to provide appropriate commands to the control system.

When conventional techniques are used to interconnect all the units comprising individual systems, the extent of the cabling required is

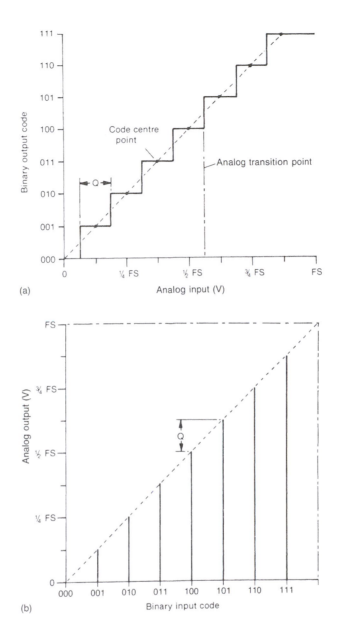

Figure 6.3 Data conversion.
(*a*) Analog−digital;
(*b*) digital−analog.

considerable, particularly as individual wires must transfer signals dedicated to each of the parameters being monitored. In maximizing the utilization of digital computer-based systems, therefore, it became necessary to adopt an alternative method by which the exchange of information could be effected by a network of single data busses, known as a *data highway*, within an aircraft. In other words, this is an adaptation of the data highway concept that is utilized within digital computers themselves.

Each data bus consists of shielded and twisted pairs of wires, and the voltage difference between them encodes a binary 0 or binary 1.

All outgoing encoded data from the computers are identified by an additional binary-coded word called a *label*. The label takes up the first eight bits of each word and is octal-coded, i.e. coded to the base 8.

The designation of labels to particular functions is arranged by an aircraft manufacturer in relation to each of the specific systems installed in the type of aircraft concerned, and in accordance with standard specifications. A specification accepted as an air transport industry standards reference for the transfer of digital data is known as ARINC 429 (ARINC is the abbreviated name of a US organization 'Aeronautical Radio Incorporated'). As separate bus systems are predictable for the different classes of aircraft systems, ARINC 429 includes some duplication of labels where it is known that the use of a common label on the same bus for two different purposes will occur. For example, label 315_8 defines 'wind shear' for navigational purposes, but for flight control systems the same label defines 'stabilizer position'.

Systems providing data outputs (referred to as transmitters) each have their own data bus connecting them to the 'receiver' systems in need of the data, as shown in Fig. 6.4. The shielding of the wires comprising each data bus is connected to ground and, in particular, at each branch to receivers. The maximum number of receivers that can be connected to the same bus line is 20.

The digital computers of the different aircraft systems process data in the form of specific messages or parallel binary words. The messages are converted and transmitted in serial form, the reason for this being that weight of transmission lines is reduced, and also reliability is improved. The serial messages are then adapted into high- and low-voltage levels, and transmitted along the data bus lines in the form of strings of pulses. These comprise the word strings of a message and correspond to those appropriate to all of the systems detailed in the ARINC 429 specification. Each word is formed of 32 bits, each bit being either a binary 1 or a binary 0. As noted earlier, the first eight bits comprise the label which identifies the source of the message; the remaining bits are assigned to data, parity, sign and status or validity. Two examples of a message are also shown in Fig. 6.4; one is labelled 'DME distance' and the other 'Radio Altitude'.

In each case, bits 9 and 10 are assigned to what is termed a Series Destination Identifier (SDI); this applies when specific words need to be directed to a specific system of a multi-system installation, or when the source system needs to be recognized from the word content. In the examples indicated the systems are, of course, the DME and Radio Altimeter respectively.

The bits 11 to 29 are those assigned to the actual data being transmitted which, in the case of our examples, are distance in nautical miles, and radio altitude in feet. The groups of binary 1s and

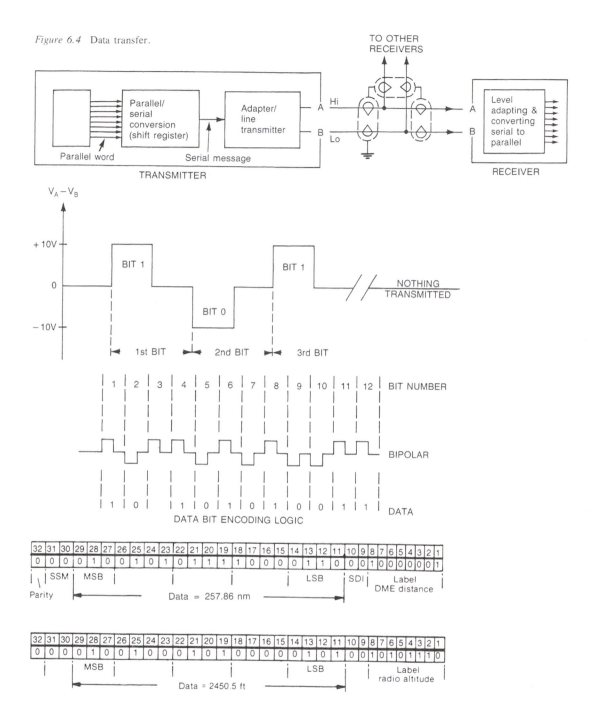

Figure 6.4 Data transfer.

0s (in bits 11 to 29) correspond to equivalent decimal numbers, which indicate that the DME system computer is transmitting encoded data corresponding to a distance of 257.86 nautical miles, while the Radio Altimeter computer is transmitting data corresponding to an altitude of 2450.5 feet.

Bits 30 and 31 are assigned to what is termed the Sign/Status

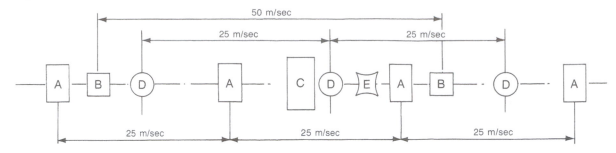

Figure 6.5 Refreshment rate.

Matrix (SSM), which refers to plus, minus, north, south, left, right, etc. of binary-coded decimal numeric data. They also refer to the validity of data, and failure warning.

The detection of errors in codes and their correction is a very important aspect in the transmission of digital data, and for this purpose a *parity check* method is provided whereby a computer can test whether bits in a binary word have been accidentally changed during transmission. The test is done by automatic summation of the bits comprising a word to determine whether the total number is odd or even, and by calculating what is termed a *parity bit*: this forms the last bit of a word, i.e. bit 32. If, for example, there is an odd number of binary 1s among the first 31 bits, the parity bit is set to '1' to make the word of 'even parity'. 'Odd parity' can also be used where the parity bit is set to binary 0 to make the total number of binary 1s odd. This latter form of parity is adopted in the ARINC 429 specification.

Data are transmitted in batches at a specified repetition or refreshment rate along the appropriate busses, and at either high speed (100 kilobits/sec) or low speed (12−14.5 kilobits/sec) according to the frequency at which interfacing systems require an update of information. This is shown in Fig. 6.5.

A standards specification of comparatively recent origin is the ARINC 629. It relates to a data bus system called Digital Autonomous Terminal Access Communications DATAC, conceived by Boeing for use in the B777. Unlike ARINC 429 it is a two-way bus requiring fewer wires and having a very much faster data transfer rate.

160

7 Air data computers

The term 'air data', as we learned from Chapter 2, relates to the sensing and transmission of pitot and static pressures to indicators which, on the basis of physical laws, are specifically designed to measure such pressures in terms of airspeed, altitude and rate of altitude change. In addition to these three indicators, however, there are many other systems whose operation depends on an air data input. The utilization of such systems in an aircraft does, in turn, depend on its size and operational category.

Although it would not be impossible to connect these systems to pressure probes and/or vents by pipelines, then, as may be imagined, the amount of 'plumbing' required would have to be considerably increased. Apart from causing additional weight problems, there could be others associated with maintenance. In order therefore to minimize these problems the principle is adopted whereby the pressures are transmitted to a centralized *air data computer* (ADC) unit, which then converts the data into electrical signals and transmits these through cables or data busses to the dependent indicators and systems. Another advantage of an ADC is that circuits may be integrated with their principal data modules in such a way that corrections for pressure error (PE), barometric pressure changes, and compressibility effects can be automatically applied; in addition, provision can also be made for the calculation of true airspeed (TAS) from air temperature data inputs. The modulator arrangement of an ADC, its associated indicators, and details of systems that utilize air data inputs are shown in Fig. 7.1

An ADC may either be of the analogue type, or of the type which processes and transmits data in digital signal format. The latter type is now more widely used, but as analog computers are still adopted in some types of aircraft we can, at this stage, and by way of introduction to ADC operating principles overall, consider a typical analog arrangement.

Analog ADC

The arrangement of the basic modules of this type of computer and their interfacing is shown in Fig. 7.2. Each module constitutes what is termed a servomechanism, and is comprised of certain mechanical elements, and synchros which perform the various functions already described in Chapter 5. The output signals from each module are

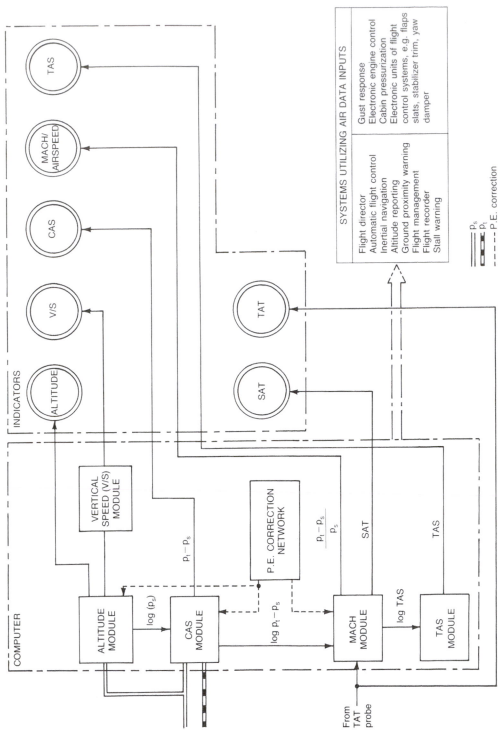

Figure 7.1 Modular arrangement and data transmission of an ADC.

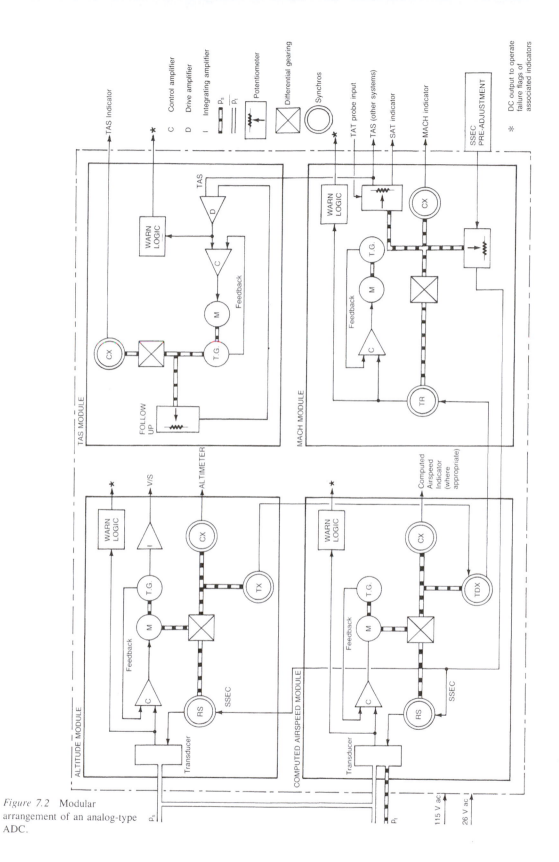

Figure 7.2 Modular arrangement of an analog-type ADC.

163

transmitted to their relevant indicators which, as we shall see later in this chapter, are of the servo-operated type.

Transducers

The pipelines from an aircraft's pitot pressure probe and static vent system are connected via a manifold in the computer mounting to pressure transducers in the computed airspeed and altitude modules. The transducers are of the electro-mechanical type, the constructional features of which vary dependent on those adopted by any one manufacturer. One example we may consider is known as a *force-balance transducer* which, as can be seen from the schematic diagram in Fig. 7.3, consists of a capsule-type pressure sensor that actuates an 'E' and 'I' bar pick-off element.

The 'E' bar has an ac-powered primary input winding on its centre limb, and a secondary output winding on each of its outer limbs; these windings are connected to an amplifier. The 'I' bar is mechanically connected to the capsule, the displacements of which pivot the bar such that the gaps between its ends and the outer limbs of the 'E' bar are increased or decreased. The 'I' bar is also interconnected with a servomotor via a torsion bar, gear train and a cam follower; the servomotor also forms part of the synchro system appropriate to the computed airspeed and altitude modules.

In the static condition, i.e. a capsule is not subjected to a pressure change, the gaps between the ends of the 'I' bar and outer limbs of

Figure 7.3 Force-balance transducer.

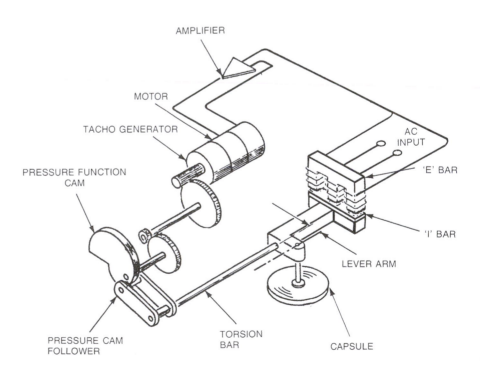

the 'E' bar are equal. When ac is applied to the primary winding then magnetic fields will be produced which, in the static condition, will be equal and opposite; thus, no signals will be induced in the secondary coils.

When a pressure change occurs, the capsule responds accordingly and the force it produces displaces the 'I' bar so that one air gap increases and the other decreases. The resulting changes in the magnetic fields cause out-of-balance signals to be induced in the secondary coils. After amplification, the signals are applied to the control phase of the servomotor, which then drives the cam follower, and torsion bar, to produce opposing torsional effects which start balancing the force exerted by the capsule, to 'back-off' the signals induced in the secondary coils. When a constant pressure condition is attained, equilibrium between capsule force and torsion is established, and no further amplified signals are supplied to the servomotor.

In some types of ADC, the pressure transducers take the form of a solid-state circuit device which utilizes what is termed the piezoelectric* effect, i.e. the generation of electrical signals by certain crystalline materials when subjected to pressure. The device consists of quartz disks with a metallic pattern deposited on them, and arranged in a thin stack such that they serve as a flexible diaphragm. Thus, when subjected to pressure changes, the resultant flexing sets up an electrical polarization in the disks so that electrical charges are produced. The polarity of the charges depends on the direction of flexing, in other words, on whether the pressure applied is increasing or decreasing. All ouptut signals are supplied to the appropriate type of transmission link adopted for the airspeed and altitude modules.

Let us now refer once again to Fig. 7.2, in order to see how pressure transducer output signals are processed and transmitted by the various modules for the purpose of operating their associated indicators.

Module operation

In the case of the *computed airspeed* module, the servomotor, in response to the amplified output signals from the transducer, drives the rotor of a CX synchro whose stator is connected to a CT synchro within the indicator.

The servomotor also drives, via differential gearing, the rotor of an RS that forms part of a static source error correction (SSEC) network which, as shown in Fig. 7.2, originates in the Mach module of the computer. The circuit of this network is pre-adjusted so that the signal input to the RS is a correction factor signal corresponding to the position error (PE) of the aircraft (see also page 34) in which

* From the Greek *piezein*, meaning 'to press'.

the ADC is installed. The output signals from the RS are supplied to the pressure transducer circuit so that its output, which is a measure of the pressure difference $p_t - p_s$, is in turn also corrected. Thus, the servomotor and CX synchro rotor position are controlled to produce an output compensated for PE as a function of Mach number.

In operation, the servomotor also drives a tachogenerator which supplies rate feedback signals to the control amplifier to reduce the input error voltage signals, and thereby prevent the motor from 'overshooting' its controlled positions.

The *altitude* module is comprised of a servomechanism arrangement, whose only difference from the one just described is that it operates in response to signals which are a measure of the pressure p_s. In addition to supplying signals to a servo-operated altimeter, the module also determines rates of altitude change, i.e. vertical speed (V/S), and produces the corresponding signals. Since the rate of change involves a time factor, the measurement of V/S is accomplished by supplying the rate signals produced by the servomotor-driven tachogenerator to an integrating amplifier. This is a device that performs the mathematical operation of integration so that its output is substantially the integral with respect to time of the input to the device. After integration, the signals are amplified and supplied to a servo-operated VSI and/or to V/S mode select modules which form part of the pitch channels of automatic flight control and flight director systems.

An indication of speed in terms of Mach number can, as we learned from Chapter 2 (see page 46), be derived by measuring it in terms of the pressure ratio $p_t - p_s/p_s$. In the case of basic pneumatically-operated indicators this, as we also learned, necessitates that altitude and speed measuring elements be used in combination with a mechanism that will perform the required dividing function. Fundamentally, this arrangement also applies to the *Mach module* of an ADC, but in adopting synchronous transmission and servomechanism methods of accurately measuring the three parameters involved, it is incumbent to use an equally accurate method of performing the dividing function. In the example of ADC shown in Fig. 7.2, the dividing is done by means of a differential synchro in combination with a torque synchro system.

The differential synchro (TDX) is part of the computed airspeed module servomechanism, the TX synchro is in that of the altitude module, while the TR synchro is part of the Mach module servomechanism.

When the altitude and computed airspeed modules are in operation, the TX synchro rotor will be driven to some angular position within its stator corresponding to the pressure sensed by the altitude module transducer. The signals induced in the stator will be of a related value, and these are transmitted to the TDX. In response to the signals produced by the transducer of the computed airspeed module, the TDX rotor will also be at some corresponding angular position within its stator. Since the angular positions of the TX and TDX rotors are different, then, by virtue of the connection arrangements between the two synchros, the output signals from the TDX are the difference between those produced by its rotor and the TX synchro, and in terms of the required pressure ratio.

The signals are transmitted to the control amplifier in the Mach module via the TR synchro. The servomechanism arrangement of this module is the same as that of the computed airspeed and altitude modules. The CX synchro transmits signals in terms of Mach number to a digital counter which may be individually mounted on a panel or combined with a Mach/airspeed indicator. The 'nulling out' of signals under constant speed and altitude conditions is obtained by driving the TR synchro rotor from the servomotor.

For the measurement of true airspeed (TAS) it is necessary to utilize signals that are a measure of total air temperature (TAT). These signals are generated by externally mounted sensing probes (see page 62) and, in addition to an independent indicator, they are also transmitted to the *TAS module* of the ADC via a potentiometric network in the Mach module as shown in Fig. 7.2. This network serves as a function generator in that it produces TAS output signals that correspond to the values of a specified function of independent variable inputs, in this case TAT and Mach speed. The output signals are supplied to drive and control amplifiers for the operation of a servomechanism consisting of a motor and CX synchro, the output from which is supplied to an independent TAS indicator. The servomotor also drives a 'follow-up' device which provides a signal to the drive amplifier for the purpose of balancing out incoming TAS signals.

Failure warning

Each module of the ADC incorporates a warning logic circuit network which activates a warning flag in the associated indicators in the event of loss of the respective data signals. Annunciator lights corresponding to each module are provided on the end panel of the computer, and are also illuminated in the event of failures. Once a warning circuit has been triggered it remains latched.

Indicators

The indicators that are used in conjunction with an ADC of the analog type just described also contain servomechanisms, and when connected to the computer they each form a complete servo loop with the respective modules of the computer. These indicators may, in some applications, be of the combined pneumatic and servo type, as for example the airspeed indicator shown in Fig. 2.16 of Chapter 2, or they may be entirely servo-operated.

Airspeed indicators

In the case of the indicator referred to above, its indicated (IAS) and maximum operating speed pointers are operated by pressure-sensing capsules within the indicator, while a servomechanism is used for driving a digital counter for the display of computed airspeed. The servomechanism, which is illustrated in Fig. 7.4, operates in response to the signals supplied to its CT synchro by the relevant module of the ADC (see also Fig. 7.2).

A failure monitor circuit is also incorporated in the indicator and comes into operation in the event of loss of power, or data signal input from the ADC, and also if excessive 'nulling' occurs in the digital counter servo loop. The circuit controls a solenoid-operated flag such that it obscures the digital counter display. A check on flag operation can be carried out by moving a computed airspeed switch (see also Fig. 2.21) to its 'off' position, thereby isolating the excitation circuit of the CT synchro.

This indicator is also used in conjunction with an autothrottle system, the purpose of which is to adjust the power settings of engines in order to acquire, and then maintain, a commanded airspeed. The system is also integrated with an aircraft's automatic flight control system (AFCS). Airspeed commands may be selected either from the AFCS mode select panel, or by a command set knob in the airspeed indicator.

Figure 7.4 Servo-operated airspeed indicator.

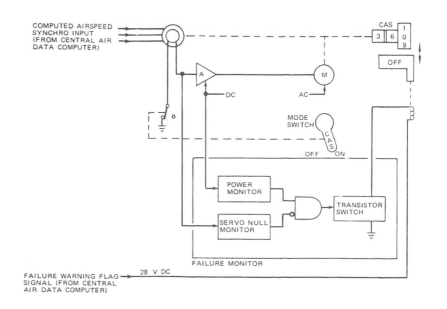

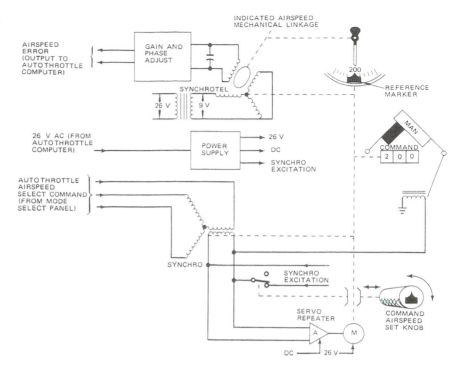

Figure 7.5 Command airspeed circuit arrangement.

The command airspeed circuit arrangement within the indicator is shown in Fig. 7.5, and from this it will be noted that it consists of a CT synchro system, and a synchrotel mechanically connected with the command speed set knob, a reference marker and a command speed indicator.

Under normal operating conditions of the autothrottle system and the AFCS, command airspeeds are set on a digital counter display on the AFCS mode select panel. This setting also positions a CX synchro rotor so that it can transmit equivalent signals to the indicator servomechanism for the purpose of positioning the speed reference marker and command speed indicator. In order that it may do so, however, the clutch in the drive must, of course, be disengaged by pulling out the command set knob; at the same time, a switch in the CT synchro excitation circuit is held in the closed position.

The servomotor is also mechanically coupled to the synchrotel transmitter, which differs from that described in Chapter 5 in that its stator can also be rotated. The rotor is mechanically positioned within the stator by the indicated airspeed pointer mechanism. The relative positions of the two therefore produce an error signal ouptut representing the difference between indicated airspeed and commanded airspeed at any one instant. This output is then supplied to the autothrottle system computer which then causes the power output of the engines to be automatically adjusted to attain the commanded speed.

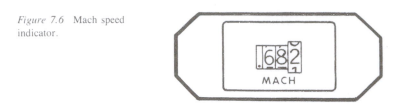

Figure 7.6 Mach speed indicator.

Figure 7.7 Servo-operated Mach/airspeed indicator.

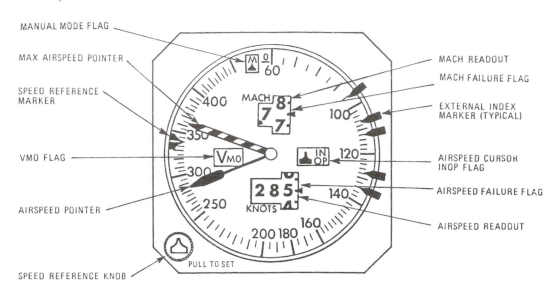

MANUAL MODE FLAG

MAX AIRSPEED POINTER

SPEED REFERENCE MARKER

VMO FLAG

AIRSPEED POINTER

SPEED REFERENCE KNOB

MACH READOUT

MACH FAILURE FLAG

EXTERNAL INDEX MARKER (TYPICAL)

AIRSPEED CURSOR INOP FLAG

AIRSPEED FAILURE FLAG

AIRSPEED READOUT

PULL TO SET

If it is required to set a command airspeed on the indicator itself, the set knob must be pushed in to engage the servomechanism drive clutch. This action also opens the switch in the excitation circuit of the CT synchro, thereby isolating it from the servomotor. Rotation of the set knob now provides for manual positioning of the reference marker and synchrotel stator and, therefore, manual control of the output signals to the autothrottle system computer. The command speed digital counter is also rotated, but its display is obscured by a yellow 'MAN' flag, the solenoid control circuit of which is also isolated when the set knob is pushed in. In the event that a command airspeed exceeding a certain value (in this example 250 knots) is set, a black flag is triggered to obscure the counter display.

Figure 7.6 illustrates a display presentation of a Mach speed indicator that is used in conjunction with the indicated/computed airspeed indicator just described. The digital counter is servo-operated by a CT synchro supplied with input signals from the Mach module of the ADC.

The display presentation of a pure servo-operated indicator (referred to as a Mach/airspeed indicator) is shown in Fig. 7.7; it may be used in conjunction with an ADC of either the analog or digital type. Computed airspeed is displayed in knots by a distinctly-shaped pointer and by a digital counter. The indication of speed in terms of Mach number is shown by a digital counter display. A striped pointer, which is also servo-driven, provides an indication of V_{mo} and M_{mo} (see page 42).

The speed reference knob and marker perform the same function as that of the indicated/computed airspeed indicator described earlier. The other markers are 'memory bugs' that are pre-set to indicate certain operating speeds appropriate to the type of aircraft, e.g. take-off speeds, flap extension speed.

Five warning and indicating flags are provided as follows: 1. airspeed flag to indicate a failure in the airspeed circuit within the indicator or ADC; 2. Mach flag to indicate failure in the Mach circuits; 3. V_{mo} flag to indicate failure in the V_{mo} and M_{mo} circuits; 4. 'INOP' flag that comes into view to indicate that the speed reference marker is inoperative; and 5. 'M' flag that operates in conjunction with the 'INOP' flag to indicate manual setting of the speed reference marker.

The internal circuit arrangement of the indicator is shown in Fig. 7.8. Power requirements are 26 V ac for synchro operation, and this is distributed within the indicator via a power supply module. The module also supplies 28 V dc for the operation of servomotors, amplifiers, flag monitor circuits, etc.

Signals corresponding to computed airspeed are supplied from the relevant module in the ADC to a CT synchro, the error signal output from which is amplified to drive the servomotor connected to the airspeed pointer and digital counter. At the same time, it drives the synchro rotor to 'null out' the error signal. The servomotor also drives, through 2:1 gearing, a potentiometer which supplies a dc signal to an 'anti-ambiguity' circuit connected to the servo amplifier. The purpose of the circuit is to ensure that the airspeed pointer is not driven to a position 180° out with respect to 'null'.

The airspeed servomotor also drives a synchro transmission loop, the purpose of which is to transmit computed airspeed to an autothrottle system.

The airspeed pointer and counter drive mechanism also incorporates a specially calibrated cam and follower to provide square-law compensation (see page 43). As the cam rotates it varies the magnification rate of the pointer movement so as to maintain linearity as speed increases.

The maximum airspeed pointer is driven by a servomotor which receives its signals from a V_{mo} and overspeed processor circuit module via an amplifier. The pointer is always driven to, and 'nulled' out at, a scale reading higher than that of the airspeed pointer, by signals from a synchro whose rotor is also driven by the servomotor. If airspeed is increased to the maximum value, the

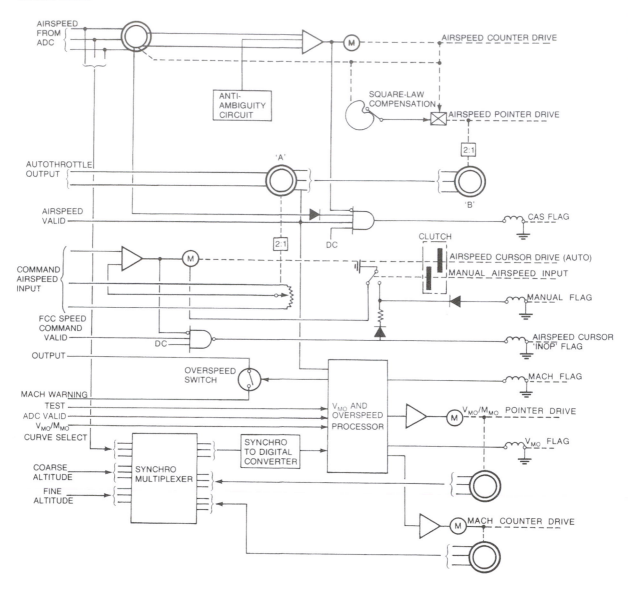

Figure 7.8 Mach/airspeed indicator circuit.

airspeed pointer will be driven to coincide with the maximum airspeed pointer position, and the higher airspeed signal will be detected by the V_{mo} and overspeed processor. This produces an output signal that triggers a solid-state overspeed switch, causing it to activate a Mach/airspeed aural warning system (see page 70).

In order to drive the Mach number counter of the indicator, synchro output signals corresponding to computed airspeed and altitude are supplied from the respective modules of the ADC. As can be seen from Fig. 7.8, the signals are supplied to a synchro multiplexer, and then after conversion from synchro to digital they are fed to the V_{mo} and overspeed processor; after amplification they drive the Mach counter via its servomotor. The motor also drives a synchro whose output is fed back to the multiplexer to 'null' the signals corresponding to Mach number, when constant speed values are obtained.

The setting of command airspeeds and associated signals for autothrottle system operation is done in a similar manner to that described earlier. For *automatic operation*, i.e. settings made on an AFCS mode select panel, or, in some cases, on a display unit of a Performance Data Computer (PDC) system, the speed reference knob of the indicator remains in its normal pushed-in position. In this position a clutch is disengaged, and a switch in the servomotor circuit is closed to provide a path to ground as shown in Fig. 7.8. When the commanded airspeed is set a command signal is supplied via an amplifier to the speed reference marker servomotor so that it now rotates the marker to the commanded speed. The servomotor also drives the rotor of a synchro (indicated 'A') which then becomes de-synchronized with respect to a second synchro (indicated 'B'). Thus, an error voltage signal corresponding to the difference between computed and commanded airspeeds is transmitted to the autothrottle system. As the airspeed changes in response to the commanded engine power change, the airspeed pointer and counter are driven so as to indicate the speed change. At the same time, the rotor of synchro 'B' is rotated in order to reduce the error signal voltage produced by synchro 'A'. When the null position is reached, no further output is supplied to the autothrottle system and the airspeed pointer and counter are then at the commanded airspeed. A dc potentiometer is also driven by the servomotor to provide position feedback.

For *manual operation*, the speed reference knob is pulled out to engage a gear type of clutch, and, as can be seen from Fig. 7.8, the switch in the marker drive motor circuit now rotates the marker; through a 2:1 ratio gear it also rotates the rotor of synchro 'A' to establish an error signal for transmission to the autothrottle system in the same manner as that resulting from automatic operation.

A logic circuit is provided in the speed reference system, its purpose being to monitor the system (while the speed reference knob is pushed in) for loss of power, nulling of the synchro/servo system, and validity of the input signals from an AFCS mode select panel or PDC display unit. If an invalid reference display should occur, the 'INOP' flag appears as shown in Fig. 7.8. This flag also appears when the reference knob is pulled out for manual operation (the monitor circuit is disabled in this case) together with the 'MANUAL' flag.

The remaining flags, i.e. V_{mo}, 'MACH' and 'A/S', appear under the conditions referred to earlier.

The indication of *true airspeed* (TAS) is provided by a digital counter type of indicator, the servomotor of which is supplied with signals from the TAS module of the ADC. A failure monitor circuit is incorporated in the indicator for the operation of a yellow 'OFF' flag.

Altimeters

The display presentation of one example of *pneumatic/servo-operated* altimeter, and the basic arrangement of its mechanism, are shown in Fig. 7.9. The pneumatic section consists of two capsules which in responding to changes in static pressure admitted to the indicator case drive the pointer and digital counter in a manner similar to that of a conventional pneumatic altimeter. The pointer and counter are also driven by signals supplied to a CT synchro from the altitude module of the ADC, and since these signals are of higher resolution and accuracy, pointer and counter operation is predominantly controlled through the servo drive. The pneumatic section, therefore, performs a standby role so that it can provide altitude indications in the event of failures in the synchronous transmission loop. A control knob, located at the front of the instrument, is provided for use in such cases, and when moved from the 'CADC' position to 'STBY', it

Figure 7.9 Pneumatic/servo-operated altimeter.

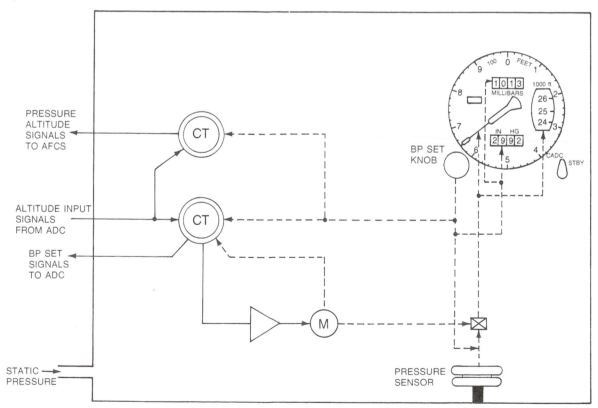

isolates the CT synchro signal circuit and also electrically activates a red 'STBY' flag. The flag is also automatically operated by a failure monitor circuit similar to that incorporated in Mach/airspeed indicators.

As in the case of pure pneumatically-operated altimeters, indicated altitudes are corrected to standard pressure data by means of a barometric pressure setting knob and counter mechanism. In addition, however, the mechanism also positions the CT synchro stator with respect to its rotor, thereby modifying the input signal from the ADC altitude module. The resulting error signal voltage induced in the rotor therefore drives the servomotor and, via the differential gearing, the altitude pointer and counter are driven to the required pressure altitude value. The servomotor also drives the CT synchro stator for 'nulling out' the error voltage signal. The barometric pressure setting knob also positions the stator of a second CT synchro provided for the purpose of supplying equivalent pressure altitude signals to the altitude selection facility and pitch control computer of an AFCS.

The internal arrangement and display presentation of a *servo-operated* altimeter is shown in Fig. 7.10. Although used principally with ADCs of the digital type, it may in some cases be interfaced with some analog types of ADC.

Altitude signals designated as coarse and fine are transmitted from

Figure 7.10 Servo-operated altimeter.

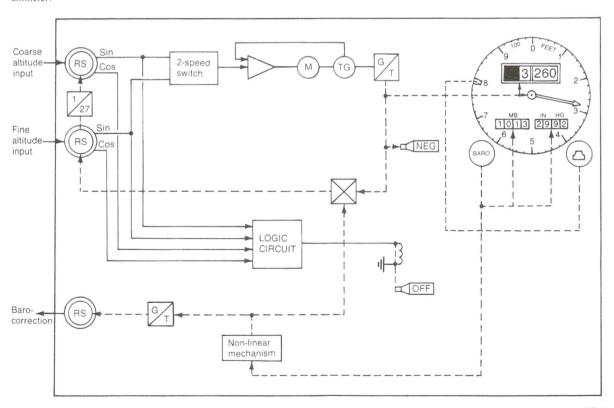

the ADC altitude module to the stator windings of corresponding resolver synchros. The rotors are mechanically interconnected by a 1:27 ratio gear train, and their sine windings are connected to a solid-state switch referred to as a speed switch. The purpose of the switch is to control the servomotor operation so that fine altitude signals are supplied to its amplifier at altitudes below 1000 ft, while at altitudes above this value, coarse altitude signals are supplied. The servomotor drives the altitude pointer and counter through a clutch and gear train, and directly drives a tachogenerator which provides rate feedback signals to the amplifier. The reduction and 'nulling out' of altitude error signals is effected by driving the rotors of the resolvers through a differential gear.

The cosine windings of the resolvers are connected to a logic circuit that monitors the presence of the following: 1. coarse and fine servo 'nulls' (sine windings); 2. coarse and fine excitation (cosine windings); 3. indicator power supplies; and 4. valid altitude data. If either of these is unreliable, a solenoid-operated 'OFF' warning flag is activated to obscure the digital counter display of altitude.

Barometric pressure setting is done in a manner similar to that of the altimeter described earlier, except that the setting knob rotates the stators of two resolver synchros for establishing the error voltage signals necessary to obtain the required pressure altitude indications. The purpose of the 'non-linear' mechanism shown in the diagram is to compensate automatically for the non-linear relationship between barometric pressure and altitude, so that for any setting of the pressure counters, the corresponding altitude will be indicated. The pressure setting knob also changes the rotor position of a third resolver, the purpose of which is to supply pressure-corrected signals to such systems as AFCS and altitude alerting.

The second knob in the bottom right-hand corner of the altimeter permits the setting of a reference marker to align with an altitude indication corresponding to a specific operating condition. The purpose of the servo-driven 'NEG' flag is to obscure the digital counter display at altitudes below sea-level.

Static air temperature indication

The most basic method of obtaining an indication of SAT is to use charts of pre-calculated values of ram rise related to sensing probe recovery factors (see page 62) and Mach number and then subtract the values from the readings of the TAT indicator. Such conversion charts are provided by manufacturers and normally form part of an aircraft's operations or flight manual. It is, however, more advantageous to provide an automatic method of conversion so that corrections, in the form of electrical signals, can be applied to the

TAT signal output to derive SAT, and then utilize the corrected signals to operate a separate indicator.

In the case of the analog ADC described in this chapter, the conversion and correction is effected by a circuit network whose electrical characteristics are matched to those of the TAT sensing probe. As shown in Fig. 7.2, the network is incorporated within the Mach module of the computer to accept TAT probe output signals, as well as a drive input from the servo loop in order to vary the SAT signals as a function of Mach number.

The circuit arrangement of one example of SAT indicator is shown in Fig. 7.11. It utilizes a drum type of counter, the left-hand and right-hand sections of which display temperatures in the plus and minus parts, respectively, of the range. The centre drum displays the sign of the temperature being indicated; the drums not in use are automatically masked. The computed SAT is supplied as a dc analog voltage to a chopper circuit, and is compared with a voltage supplied as a reference via a re-balanced potentiometer. The chopper circuit produces a 400 Hz ac error signal representative of the difference between the two inputs, which is then amplified to drive the servomotor and counters. The motor also repositions the re-balancing potentiometer to 'null out' the error signal. In the event of loss of dc or ac power, or an excessive 'null' voltage in the re-balance/feedback system, an 'OFF' flag is triggered by a failure monitor circuit to obscure the counter display.

Digital ADC

The modular arrangement and data signal flow of a typical computer are shown in schematic form in Fig. 7.12. It processes the same basic parameters as the one already described, but with the major difference that all the signals corresponding to the variables measured are converted and transmitted in digital format. The pitot and static pressure sensors are of the piezoelectric crystal type (see also page 165) and their frequency-modulated signals are supplied to the altitude, computed airspeed, and Mach calculation circuit modules via a frequency-to-digital converter. The analog inputs from the synchros of angle of attack (alpha) sensors, and altimeter barometric pressure setting controls, are converted by means of synchro-to-digital converters. Outputs from all modules of the computer are supplied to an ARINC 429 transmitter connected to four data busses from which all interfacing systems requiring air data are then supplied. The purpose of the discretes coder module is to monitor signals relating to the status and integrity of particular circuits, e.g. the heater circuits of TAT probes, pitot probes, and angle of attack sensors, and to initiate appropriate warnings. In order for the computer automatically to take into account the pressure error of the air data system of a

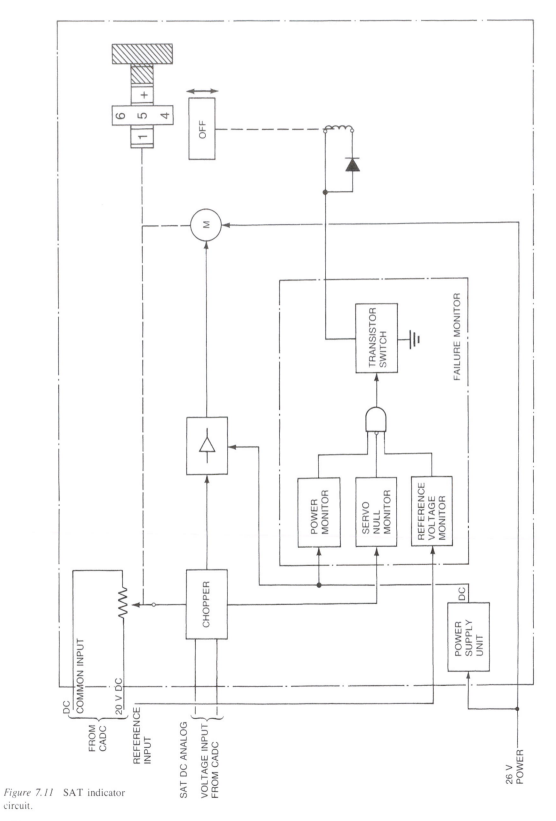

Figure 7.11 SAT indicator circuit.

178

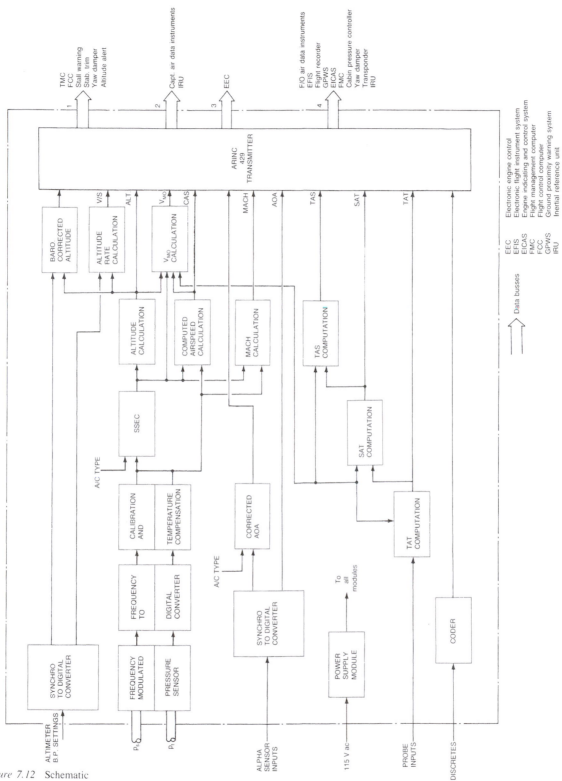

Figure 7.12 Schematic arrangement of a typical digital ADC.

179

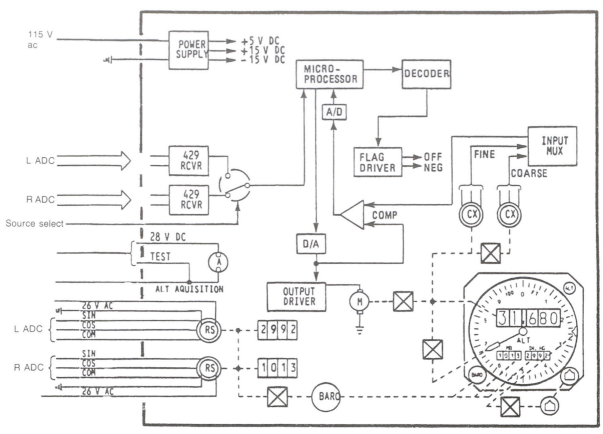

Figure 7.13 Digital/servo-operated altimeter.

particular type of aircraft, and also its stall characteristics, it is 'matched' by programming the SSEC and angle of attack modules with the relevant data.

The indicators associated with a digital ADC are of the pure servo-operated type, and as an example of their operation generally, we may consider the altimeter circuit shown in Fig. 7.13. Data may be supplied from either of two ADCs as selected by the triggering of a solid-state switch; the right ADC is shown. Under changing altitude conditions the corresponding signals pass to a microprocessor, and from this unit they are transmitted to a D/A converter which then provides the drive signals for operating the servomotor, and the altitude pointer and counter mechanisms. At the same time the motor drives two CX synchros which supply coarse and fine analog inputs to an input multiplexer. The output signals from this unit are then compared with those from the D/A converter, and the difference between them (as a result of altitude change) is fed back into the microprocessor via an A/D converter. The signals will 'null out' when a constant altitude condition has been attained. The setting of barometric pressures is done in the usual manner, i.e. by means of a set knob and digital counters, and, as will be noted, the stators of two resolver synchros are also repositioned. These produce sine- and cosine-related signals which are fed back to the corresponding synchro-to-digital converter in the computer (see Fig. 7.12). The change in the converter output signal is supplied to the altimeter, via the relevant data bus, so that its servomotor will drive the pointer and counter mechanism to indicate the attitude change corresponding to the barometric pressure setting. If input signal failure or a negative altitude condition should occur, the microprocessor activates decoder and flag driver circuits which then cause the appropriate flag to appear across the altitude counter display.

8 Magnetic heading reference systems

A magnetic heading reference system (MHRS), sometimes called a remote-indicating compass, is basically one in which an inductive type of element detects an aircraft's heading with respect to the horizontal component of the earth's magnetic field in terms of flux and induced voltage changes, and then transmits these changes via a synchronous/servo and stabilized reference system to a heading indicator. Thus, in concept an MHRS is a combination of the functions of a direct-reading magnetic compass and a direction indicator, but one in which the individual errors associated with these two instruments are considerably reduced.

In practice, there are two types of MHRS: (i) that in which the detector element monitors a directional gyroscope unit linked with a heading indicator, and (ii) one in which the detector element operates in conjunction with the platform of an inertial navigation system (INS) to supply magnetic heading and stabilized heading data respectively to a compass coupler unit linked with a heading indicator.

Detector elements

These elements (variously known as flux detectors, or flux valves), unlike those of direct-reading compasses, are of the fixed type which detect the effect of the earth's magnetic field as an electromagnetically induced voltage and control a heading indicator by means of a variable secondary output voltage signal. In general, the construction of an element is as shown in Fig. 8.1. It takes the form of a three-spoked wheel, slit through the rim between the spokes so that they, and their section of rim, act as three individual flux collectors.

Around the hub of the wheel is an exciter coil which has an electrical function corresponding to that of the primary winding of a transformer. Coils are also wound around the spokes, and these correspond in function to that of a transformer's secondary windings. The reason for adopting a triple spoke and coil arrangement will be made clear later in this chapter, but at this stage the operating principle can be understood by considering a single-turn coil placed in a magnetic field. The magnetic flux passing through the coil is a

Figure 8.1 Detector element construction.

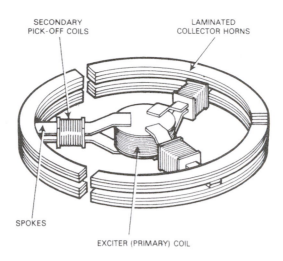

SECONDARY
PICK-OFF COILS

LAMINATED
COLLECTOR HORNS

SPOKES

EXCITER (PRIMARY) COIL

maximum when it is aligned with the direction of the field, zero when it lies at right angles to the field, and maximum but of opposite sense when the coil is turned 180° from its original position. Fig. 8.2(a) shows that for a coil placed at an angle θ to a field of strength H, the field can be resolved into two components, one along the coil equal to $H \cos \theta$, and the other at right angles to the coil equal to $H \sin \theta$. This latter component produces no effective flux through the coil so that the total flux passing through it is proportional to the *cosine* of the angle between the coil axis and the direction of the field. In graphical form this total flux may be represented as at (b).

If the coil were to be positioned in an aircraft so that it lay in the horizontal plane with its axis fixed on, or parallel to, the aircraft's longitudinal axis, then it would be affected by the earth's horizontal component and the flux passing through the coil would be proportional to the magnetic heading of the aircraft. It is therefore apparent that in this arrangement we have the basis of an MHR system able to detect the earth's magnetic field without the use of a permanent magnet. Unfortunately, this simple system would be of little practical use because, in order to determine the magnetic heading, it would be necessary to measure the magnetic flux, and there is no simple and direct means of doing this. If, however, a flux can be produced which changes with the earth's field component linked with the coil, then we can measure the voltage induced by the changing flux, and interpret the voltage changes so obtained in terms of heading changes. This is achieved by adopting the construction method shown at (c) of Fig. 8.2.

Each spoke consists of a top and bottom leg suitably insulated from each other and shaped so as to enclose the hub core around which the primary coil is wound. The material from which the spokes are made is an alloy especially chosen for its characteristic property of being easily magnetized but losing almost all of its magnetism once the

183

Figure 8.2 Detector element coil.

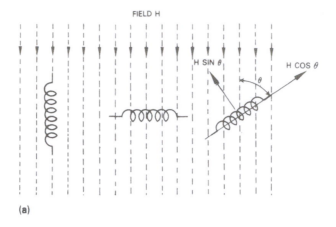

FIELD H

H SIN θ H COS θ

θ

(a)

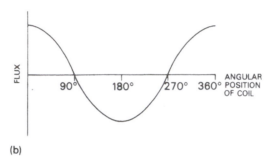

FLUX

90°   180°   270° 360° ANGULAR POSITION OF COIL

(b)

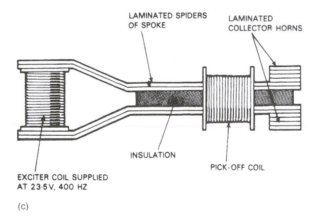

LAMINATED SPIDERS OF SPOKE

LAMINATED COLLECTOR HORNS

INSULATION

PICK-OFF COIL

EXCITER COIL SUPPLIED AT 23·5 V, 400 HZ

(c)

external magnetizing force is removed (a typical material is one known as Permalloy). With this arrangement there are two sources of flux to be considered: (i) the alternating flux in the legs due to the current flowing in the primary coil; this flux is at the same frequency as the current and is proportional to its amplitude; (ii) the static flux due to the earth's component H, the maximum value of which depends upon the magnitude of H and the cosine of the angle between H and the axis of the detecting element.

If we consider first that the axis of the element lies at right angles to H, the static flux linked with the coils will be zero. Thus, with an

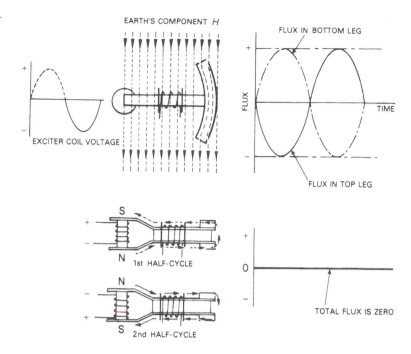

Figure 8.3 Total flux-detector spoke at right angles to the earth's field.

EARTH'S COMPONENT *H*

EXCITER COIL VOLTAGE

FLUX IN BOTTOM LEG

FLUX

TIME

FLUX IN TOP LEG

S

N
1st HALF-CYCLE

N

S
2nd HALF-CYCLE

0

TOTAL FLUX IS ZERO

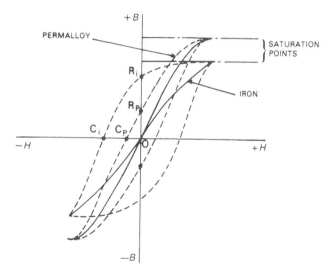

Figure 8.4 *B/H* curve and hysteresis loops.

PERMALLOY

+ *B*

SATURATION POINTS

R_i

R_P

IRON

C_i C_P

O

− *H*

+ *H*

− *B*

alternating voltage applied to the primary coil, the total flux linked with the secondary will be the sum of only the alternating fluxes in the top and bottom legs and must therefore also be zero as shown graphically in Fig. 8.3.

The transition from primary coil flux to flux in the legs of the detector element is governed by the magnetic characteristics of the material, such characteristics being determined from the *magnetization* or *B/H curve*. In Fig. 8.4, the curve for Permalloy is compared with that for iron to illustrate how easily it may be magnetized. There are several other points about Fig. 8.4 which

185

should also be noted because they illustrate the definitions of certain terms used in connection with the magnetization of materials, and at the same time show other advantages of permalloy. These are:

1. *Permeability*: which is the ratio of magnetic flux density B to field strength or *magnetizing force H*; the steepness of the curve shows that Permalloy has a high permeability.

2. *Saturation point*: the point at which the magnetization curve starts levelling off, indicating that the material is completely magnetized. Permalloy is more susceptible to magnetic induction than iron, as shown by its higher saturation point.

3. *Hysteresis* curve and loop*: these are plotted to indicate the lagging behind of the induced magnetism when, after reaching saturation, the magnetizing force is reduced to zero from both the positive and negative directions, and also to determine the ability of a material to retain magnetism. The magnetism remaining is known as *remanence* or remanent flux density, and it will be noted that Permalloy has an extremely low remanence, thus making it admirably suitable for use in detector elements.

4. *Coercivity*: this refers to the amount of negative magnetizing force (*coercive force*) necessary to completely demagnetize a material, and is represented by the distances OC_i and OC_p. Coercivity and not remanence determines the power of retaining magnetism.

In order to show the characteristics of the flux waves produced in the legs of a detector, a graphical representation in the form of that illustrated in Fig. 8.5 is adopted.

The waveshapes of the alternating primary fields are drawn across the axis B of the B/H curve, and those of the corresponding flux densities in the legs are then deduced from them by projection along the H axis. The total flux density produced in the legs is the sum of the individual curves, and with the detector element at right angles to the earth's horizontal component H then, like the static flux linked with the secondary coil, it will be zero. Since the total flux density does not change, the output voltage in the secondary coil must also be zero.

Let us now consider the effects of saturation when the detector element lies at any angle other than a right angle to the horizontal component H as indicated in Fig. 8.6(a). The alternating flux due to the primary coil changes the reluctance, i.e. the magnetic resistance, of the material, thus allowing the static flux due to component H to flow into and out of the spoke in proportion to the reluctance changes. This effect is analogous to that produced by the opening and closing of a valve, hence the name *flux valve* being applied to a detector element.

During those stages of the primary flux cycle when the reluctance

---

* From the Greek *hysteros*, meaning 'later'.

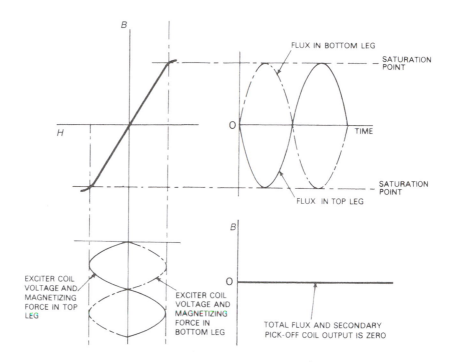

Figure 8.5 Flux wave characteristics.

is greatest, the static flux links with the secondary coil, and the effect of this is to displace the axis, or datum, about which the magnetizing force alternates. The amount of this displacement depends upon the angle between the earth's field component H and the flux detector axis. This is shown graphically at (b) of Fig. 8.6.

If we now apply a graphical representation similar to Fig. 8.5, and include the static flux of component H, the result will be as shown at (c) of Fig. 8.6. It should be particularly noted that a flattening of the peaks of the flux waves in each leg of a spoke has been produced. The reason for this is that the amplitude of the primary coil excitation current is so adjusted that, whenever the datum for the magnetizing forces is displaced, the flux material is driven into saturation. Thus a positive shift of the datum drives the material into saturation in the direction shown, and produces a flattening of the positive peaks of the fluxes in a spoke. Similarly, the negative peaks will be flattened as a result of a negative shift driving the material into saturation at the other end of the B/H curve. The total flux linked with the secondary coil is, as before, the sum of the fluxes in each leg, and is of the waveshape as also indicated in Fig. 8.6.

When the detector element is turned into other positions relative to the earth's field, then dependent on its heading the depressions of the total flux value become deeper and shallower. Thus the desired changes of flux are obtained and a voltage is induced in the secondary coil. The magnitude of this induced voltage depends upon the change of flux due to the static flux linked with the secondary

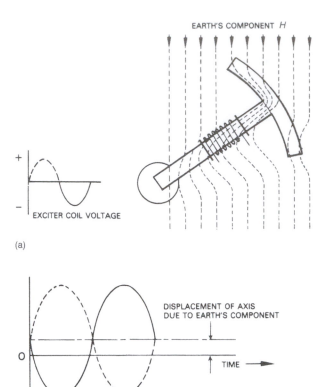

Figure 8.6 Effect of earth's component *H*. (*a*) Detector at an angle to component *H*. (*b*) displacement of axis due to static flux; (*c*) total flux and emf.

(a)

DISPLACEMENT OF AXIS
DUE TO EARTH'S COMPONENT

TIME

(b)

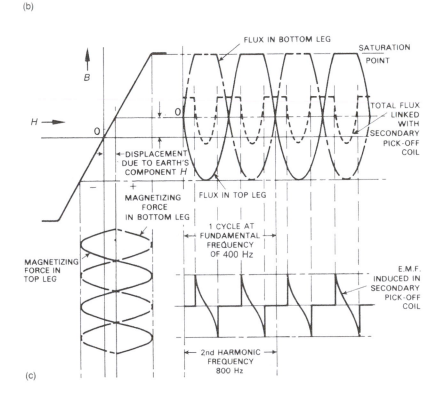

(c)

188

coil which, in turn, depends upon the value of the effective static flux. As pointed out earlier, the value of the static flux for any position of the detector element is a function of the cosine of the magnetic heading; thus the magnitude of the induced voltage must also be a measure of the heading.

One final point to be considered concerns the frequency of the output voltage and current from the secondary pick-off coil and its relationship to that in the primary excitation coil. During each half-cycle of the primary voltage, the reluctance of the material goes from minimum to maximum and back to minimum, and in flowing through the material the static flux cuts the pick-off coil twice. Therefore, in each half-cycle of primary voltage, two surges of current are induced in the pick-off coil, or for every complete cycle of the primary, two complete cycles are induced in the secondary pick-off coil.

The ac supply for primary excitation has a frequency of 400 Hz; therefore, the resultant emf induced in the secondary pick-off coil has a frequency of 800 Hz, as shown in Fig. 8.6(c), and an amplitude directly proportional to the earth's magnetic component in line with the particular spoke of the detector element.

Having thus far studied the operation of a single spoke of an element, the reasons for having three may now be examined a little more closely. If we again refer to Fig. 8.3, and also bear in mind the fact that the flux density is proportional to the cosine of the magnetic heading, it will be apparent that for one detector spoke there will be two headings corresponding to zero flux, and two corresponding to a maximum. Assuming for a moment that we were to connect an ac voltmeter to the detector, the same voltage reading would be obtained for both maximum values because the voltmeter cannot take into account the direction of the voltage. For any other value of flux there will be four headings corresponding to a single reading of the voltmeter. However, by employing a triple spoke and coil arrangement at a spacing of 120°, the paths taken by the earth's field through the spokes, and for 360° rotation, will be as shown in Fig. 8.7. Thus, varying magnitudes of flux and induced voltage can be obtained and related to all headings of the detector element without ambiguity of directional reference. The resultant of the voltages induced in the three spokes at any one time can be represented by a single vector which is parallel to the earth's component H.

Figure 8.8 is a sectional view of a practical detector element. The spokes and coil assemblies are pendulously suspended from a universal joint which allows a limited amount of freedom in pitch and roll, to enable the element to sense the maximum effect of the earth's component H. It has no freedom in azimuth. The case in which the assemblies are mounted is hermetically sealed and partially filled with fluid to damp out excessive oscillations of the assemblies. The complete unit is secured to an aircraft's structure at locations which

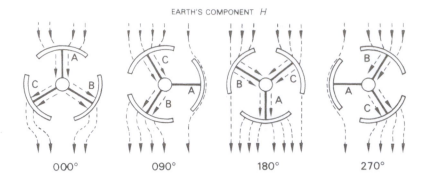

Figure 8.7 Path of earth's field through a detector.

000° 090° 180° 270°

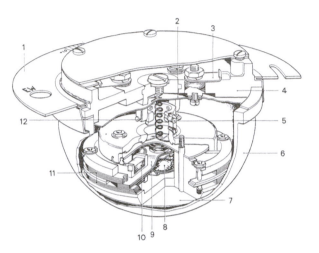

Figure 8.8 Practical detector element. 1 Mounting flange (ring seal assembly), 2 contact assembly, 3 terminal, 4 cover, 5 pivot, 6 bowl, 7 pendulous weight, 8 primary (excitation) coil, 9 spider leg, 10 secondary coil, 11 collector horns, 12 pivot.

afford maximum protection against the deviating effects of aircraft magnetism. Typical locations are in wing tips and vertical stabilizers. One of the slots in the mounting flange is calibrated a limited number of degrees on each side of a zero position corresponding to an aircraft installation datum, for adjustment of deviation coefficient *A* (see also page 93). In the example illustrated, provision is made at the top cover of the casing for electrical connections and attachment of a deviation compensating device.

In many MHRS installations, the detector elements are of the *pre-indexed* type, i.e. they can be removed and replaced without subsequent adjustments having to be made for coefficient *A* compensation. The element (see Fig. 8.9) is supplied by its manufacturer, together with a mounting plate on which it has been accurately aligned; its fixing, or indexing, screws are sealed and must not thereafter be removed. The complete unit is then secured to an alignment reference bracket which, having already been accurately aligned parallel to the longitudinal axis of an aircraft, ensures that the detector element is similarly aligned. The forward indexing screw of the bracket is also sealed and must not be removed.

From the foregoing description of a detector element, the similarity

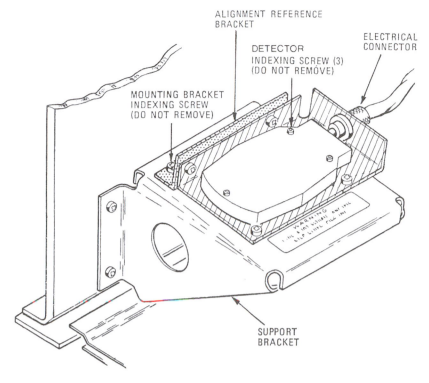

Figure 8.9 Pre-indexed detector element.

ALIGNMENT REFERENCE
BRACKET

DETECTOR
INDEXING SCREW (3)
(DO NOT REMOVE)

ELECTRICAL
CONNECTOR

MOUNTING BRACKET
INDEXING SCREW
(DO NOT REMOVE)

SUPPORT
BRACKET

between its operation and that of a basic type of synchro transmitter
will no doubt have been observed. Such an observation is not
incorrect, of course, and in fact if the detector were directly
connected to an electrically-matched receiver synchro such as a TR,
then in combination they would form a simple MHR or remote-
indicating compass system. It would not, however, be very accurate
in its indications since, by virtue of the limited pendulous suspension
arrangement of a detector element, errors can occur as a result of its
tilting under the influence of acceleration forces, e.g. during speed
changes on a constant heading, or turning of an aircraft.

In order therefore to compensate for these effects and so reduce
errors, it is necessary to incorporate within the system a means by
which the *long-term* azimuth or magnetic reference established by the
detector element is continuously stabilized and monitored. A
stabilizing technique of early origin, but nonetheless still widely used,
is that in which a horizontal-axis gyroscope unit is referenced initially
to the magnetic meridian, and then, in order to maintain this
relationship, precessional forces created by a slaving synchro/torque
motor system are applied to the gyroscope. The degree of control of
the detector element over the gyroscope, i.e. the monitoring rate, is
of considerable importance. For example, during a turn the detector
element heading is likely to be in error, and so the monitoring rate
must be such that the induced heading is that of the gyroscope. At
the same time, there must be sufficient control to correct for drift of

the gyroscope. The gyroscope, therefore, provides *short-term* azimuth references.

Monitored gyroscope system

As will be seen from Fig. 8.10, this type of MHRS is comprised of five individual units: (i) detector element; (ii) slaving/servo amplifier; (iii) directional gyroscope unit (DGU); (iv) radiomagnetic indicator (RMI); and (v) a deviation compensator.

The units (i) to (iv) are interconnected through a transmission loop consisting of control synchros (see page 140) which produce the required slaving and monitoring signals appropriate to the heading reference signals transmitted by the detector element. This element can, therefore, be considered as a special form of CX synchro, whereby the transmitter rotor field is represented by the resultant of the earth's field component H, as shown in Fig. 8.11. The secondary pick-off coils are connected to the corresponding windings of the stator of a CT receiver synchro whose function is to produce error signals which, after amplification, precess the DGU, thereby slaving it to the detected magnetic heading reference.

When the detector element is positioned as shown at (a) of Fig. 8.11, the path of the earth's field component H through the spokes will cause a maximum voltage signal to be induced in the pick-off coil A, while in coils B and C, signals of half the amplitude and of opposing phase will be induced. These signals produce fluxes in the CT synchro stator to establish a resultant field which is in alignment

Figure 8.10 Monitored gyroscope system.

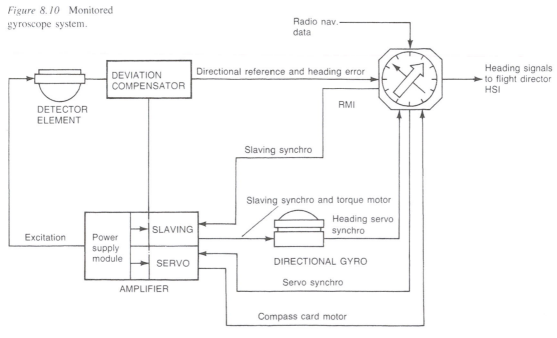

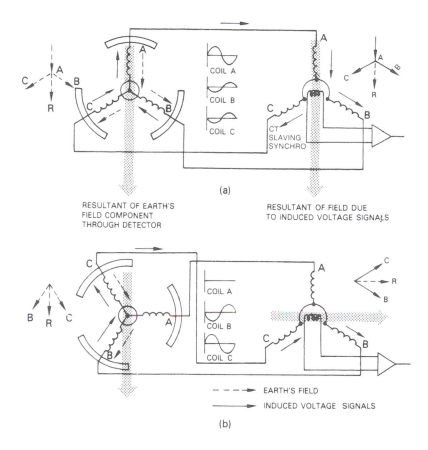

Figure 8.11 Slaving.
(*a*) Heading = 000°; (*b*)
heading = 090°.

COIL A

COIL B

COIL C

CT
SLAVING
SYNCHRO

(a)

RESULTANT OF EARTH'S
FIELD COMPONENT
THROUGH DETECTOR

RESULTANT OF FIELD DUE
TO INDUCED VOLTAGE SIGNALS

COIL A

COIL B

COIL C

– – – –▶ EARTH'S FIELD

————▶ INDUCED VOLTAGE SIGNALS

(b)

with that passing through the detector element. If, at that instant, the CT rotor is at right angles to the resultant field, no voltage will be induced in its winding. The DGU alignment will correspond to that of the resultant earth's field, and the RMI will indicate the heading 000°.

Let us now assume that the detector element is turned through 90° say; the disposition of the pick-off coils will then be as shown at (b) of Fig. 8.11. No signal voltage will be induced in coil A, but coils B and C have increased voltage signals induced in them, the signal in coil C being opposite to what it was on the previous heading. The resultant flux across the CT stator will also have rotated through 90°, and assuming for a moment that the rotor is still at the original position, then the flux will induce maximum voltage in the rotor. This heading error voltage signal is supplied to the slaving module of the amplifier in which its phase is detected, and after amplification it is supplied to a slaving torque motor which then precesses the gyroscope to the new magnetic heading reference. At the same time, the DGU operates a servo control synchro system, the function of which is to rotate the slaving synchro rotor so as to start 'nulling' out the heading error signal, and also to rotate the compass card of the RMI as the heading change takes place.

Figure 8.12 Operation of
monitored gyroscope system.

Figure 8.12 Operation of monitored gyroscope system.

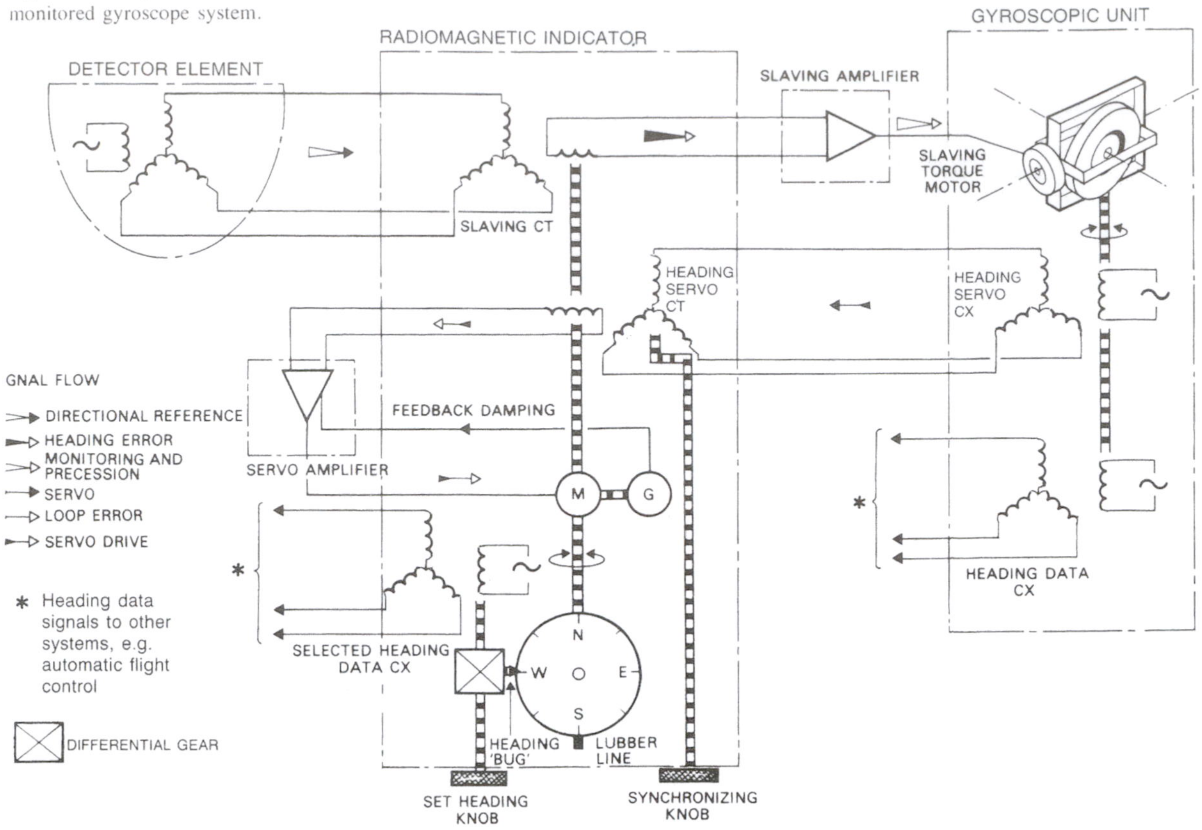

In practice, the rotation of the field in the slaving synchro, and slaving of the DGU, occur simultaneously with the turning of the detector element so that synchronism between the element and the DGU is continuously maintained.

The synchronous transmission link between the four principal units of the MHRS is shown in more detail in Fig. 8.12. The rotor of the servo CX synchro is rotated whenever the DGU is precessed, or slaved, to a magnetic heading reference, and the signals thereby induced in its stator are applied to the CT synchro in the RMI. During slaving, the rotors of both synchros will be misaligned, and a servo loop error voltage is therefore induced in the CT rotor and then applied to the servo module of the amplifier. After amplification, the voltage signal is applied to a servomotor which is mechanically coupled to the CT rotor and to the heading dial of the RMI. Thus, both the rotor and dial are rotated, the latter indicating the direction of the heading change taking place. On cessation of the heading change, the rotor reaches a 'null' position, and as there will no longer be an input to the servo module of the amplifier, the servomotor ceases to rotate and the RMI indicates the new heading. The servomotor also drives a tachogenerator which supplies feedback signals to the amplifier to damp out any oscillations of the servo system.

DGU

This unit is located at a remote point in an aircraft (typically in an electronics equpment compartment), and although it normally forms part of an MHR system, it is also designed to serve as a centralized source of heading data for use in the operation of flight director and automatic flight control systems. For example, and as shown in Fig. 8.12, an additional CX synchro is provided in the RMI for the purpose of supplying magnetic heading data to the horizontal situation indicator (HSI) of a flight director system (see page 216) and the DGU is provided with a similar synchro for supplying this data to an automatic flight control system (AFCS).

In a typical unit (see Fig. 8.13) the gyroscope is based on a two-phase induction motor operating from a 115 V ac power source. The inner gimbal ring is fitted with hemispherical covers to totally enclose the motor, and this assembly is filled with helium and hermetically sealed. The rotors of the slaving/heading CX synchro, and the additional heading data CX synchro, are mounted on the top of the

Figure 8.13 Example of a DGU.

outer gimbal ring; the stators are fixed to the frame supporting the gimbal ring. The slaving torque motor is positioned on one side of the outer gimbal ring, and on the inner gimbal ring axis, to provide precession of the outer gimbal ring appropriate to magnetic heading reference changes; the precession rate is $1°-2°/$min.

In order to control drift of the gyroscope (see page 102) its spin axis is maintained in the horizontal position by a liquid levelling switch and torque motor system which operates in a similar manner to the erection system used in electrically-operated gyro horizons (see page 116). The levelling switch is mounted on the inner gimbal ring and is connected to the control winding of the torque motor mounted on the lower part of the outer gimbal ring.

A speed-monitoring circuit system (sometimes called 'spin-down braking') is also incorporated to prevent the oscillating effect, or nutation, of the gyroscope which can occur when, at low rotational speeds, its gimbal ring axes are not mutually perpendicular. The system holds the gimbal system steady for short periods during the start-up and run-down stages of gyroscope operaton. Circuits are also provided for the monitoring of the slaving/heading synchro output signals, and the input signals from the servo module of the system's amplifier. Should the signals from any one of these monitoring circuits become invalid, a relay is energized to complete circuits controlling warning flags in the RMI, and the HSI of a flight director system.

The complete gyroscope assembly is mounted on anti-vibration mountings contained in a base which provides for attachment at the appropriate location in an aircraft, and also for the connection of the relevant electrical circuits.

Radio magnetic indicator (RMI)

This is a triple display type of indicator which derives its name from the fact that, in addition to magnetic heading data, it also displays the magnetic bearing of an aircraft with respect to ground-based transmitting stations of radio navigation systems. The systems concerned are: ADF (Automatic Direction Finding) and VOR (Very high-frequency Omnidirectional Range).

The display presentations of two examples of RMI are shown in Fig. 8.14. Magnetic heading data is displaced by the heading card which is rotated relative to a fixed lubber line in the manner already described. Magnetic bearing indications are provided by two concentrically-mounted pointers, one called a 'double-bar' pointer and the other a 'single-bar' pointer. Both are referenced against the heading card, and are positioned by synchros that are supplied with the appropriate bearing signals from the ADF and VOR navigation receivers.

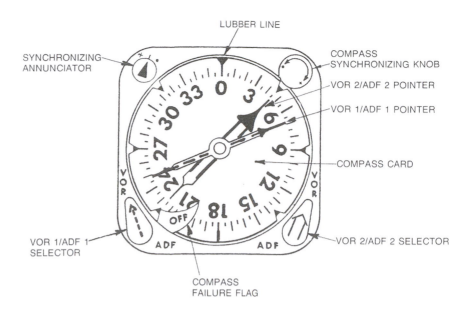

Figure 8.14 RMI displays.

SYNCHRONIZING ANNUNCIATOR

LUBBER LINE

COMPASS SYNCHRONIZING KNOB

VOR 2/ADF 2 POINTER

VOR 1/ADF 1 POINTER

COMPASS CARD

VOR 1/ADF 1 SELECTOR

VOR 2/ADF 2 SELECTOR

COMPASS FAILURE FLAG

(a)

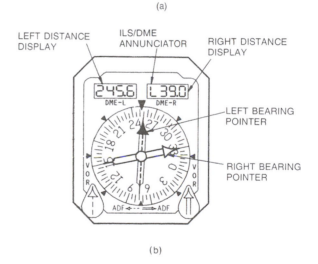

LEFT DISTANCE DISPLAY

ILS/DME ANNUNCIATOR

RIGHT DISTANCE DISPLAY

LEFT BEARING POINTER

RIGHT BEARING POINTER

(b)

There are normally two ADF and VOR systems installed in an aircraft, and by means of the selector knobs shown, they can be individually selected so that their output signals can operate the pointers in a corresponding manner. For example, when the 'VOR-1/ADF-1' and 'VOR-2/ADF-2' selector knobs are each at the 'VOR' position, the single-bar pointer and double-bar pointer synchro systems will, respectively, respond to bearing signals received from the No. 1 and No. 2 VOR system receivers. A similar response will be obtained with both selector knobs at the 'ADF' position. Bearing information can also be displayed with one selector knob set at 'ADF' and the other at 'VOR'.

Figure 8.15 Magnetic bearing
and heading display.

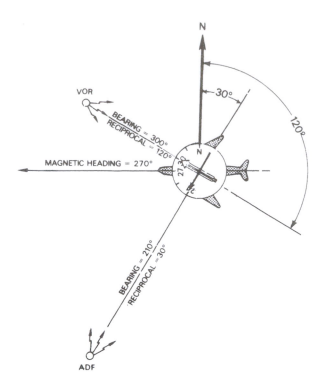

The signals transmitted to the synchros are such that the pointers always point to the stations from which the signals are received. This may be seen from Fig. 8.15, which is a representation of how the air position of an aircraft may be determined from the display of magnetic bearings and magnetic heading.

The function of the synchronizing knob and annunciator incorporated in the indicator illustrated at (a) of Fig. 8.14 will be described later.

MHRS integration with an INS

When an MHRS is to be integrated with an INS (see Fig. 8.16) there is no longer a requirement for it to have its own individual directional gyroscope unit. The reason for this is that short-term stabilizing of heading references is readily available from the inertial platform of the INS. The system also differs from the one described earlier in that the slaving/heading and servo signal transmission and control circuits are contained within a unit referred to as a *compass coupler*. The interconnection between all relevant units of a typical integrated system is shown in more detail in Fig. 8.17.

The stabilized heading reference, or platform heading signal, is derived from an azimuth CX synchro the rotor of which is positioned by the gimbal system of the inertial platform during changes in aircraft heading. The signals so produced are supplied to the rotor

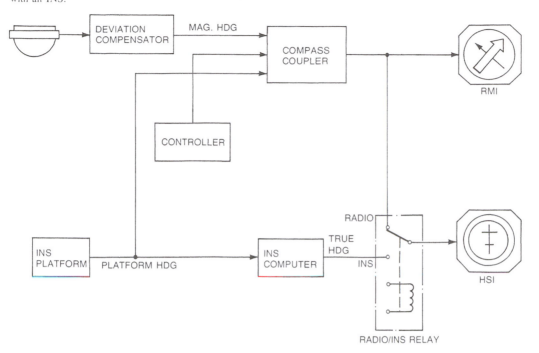

Figure 8.16 MHRS integration with an INS.

DEVIATION COMPENSATOR

MAG. HDG

COMPASS COUPLER

RMI

CONTROLLER

INS PLATFORM

PLATFORM HDG

INS COMPUTER

RADIO

TRUE HDG

INS

HSI

RADIO/INS RELAY

windings of a CDX synchro and a resultant field vector is produced in the conventional manner (see page 142). During a heading change, the detector element will also induce voltage signals corresponding to magnetic heading, and these are supplied in the normal manner to a slaving CT synchro. In this case, however, the amplified heading error signals are supplied to a stepper motor via logic circuit control modules which perform the functions of frequency comparison and voltage/frequency conversion. In addition a polarity detector circuit is provided to determine the direction of stepper motor rotation. The motor is mechanically coupled to the CDX synchro rotor and, as its name implies, it is one whose shaft rotates a step at a time as it responds to signals supplied to its windings. In the motor control circuit module shown in Fig. 8.17, the output pulses from the voltage/frequency converter are combined with the polarity detector output to energize the motor windings in pairs. The sequence, and frequency, of energizing determines the direction and rate respectively of step rotation; each step corresponds to 1.3 min of arc.

As may be seen from Fig. 8.17, the stator of the CDX synchro is connected to a CT synchro which utilizes the platform heading signals for driving a heading servomotor. The motor is mechanically coupled to the rotor of this synchro, the rotor of a CX synchro which supplies heading output signals to the RMI, and also to the magnetic heading CT synchro. Thus, during a heading change, the servomotor

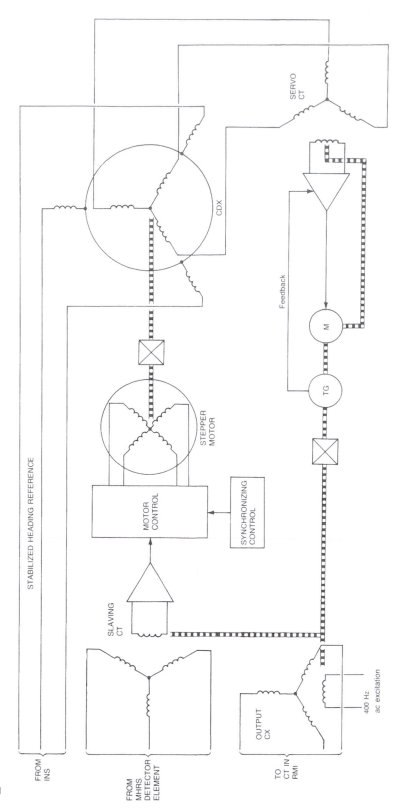

Figure 8.17 MHRS/INS signal transmission.

is driven at a rate to maintain synchronism between the platform heading and magnetic heading signals, and so no heading error signals are supplied to the stepper motor. The servomotor drives a tachogenerator which supplies rate feedback signals for speed control of the motor, and it also repositions the rotor of the servo CT synchro to ensure 'nulling out' of the heading reference signals on completion of a heading change.

In the event of de-synchronizing occurring, e.g. as a result of inertial platform drift, the signal produced by the CDX synchro would cause the heading servomotor to drive the slaving input CT synchro rotor out of 'null', thereby producing a heading error signal. This signal, after amplification, is supplied to the stepper motor which then repositions the CDX synchro rotor in a direction opposite to that of the field vector input produced by the drift. This vector is therefore 'physically' repositioned so as to produce a corresponding directional change in the stator of the CDX. The resulting reversal of the stator output signal to the heading servo system then causes the servomotor to drive the synchros until the heading error signal caused by drift has been completely 'nulled'.

Synchronizing

Whenever heading error signals are produced as a result of, say, selection of a new magnetic heading to be flown, or drift of a DGU or inertial platform, it must be ensured that in the 'nulling out' of such signals, synchronization of the slaving circuit system and RMI indications with the appropriate heading references is maintained. In order to accomplish this it is necessary, therefore, to provide additional circuitry and devices that will control and annunciate synchronized conditions and any departure therefrom.

The rate at which synchronization is carried out depends in the first instance on the magnitude of the heading error, e.g. if it is less than 2°, synchronization takes place at a slow rate of 1°–2°/min, this being the normal automatic slaving rate of typical systems. In the event that errors are greater than 2°, synchronization must take place at a faster rate, and to achieve this, the circuits also include manual and automatic control facilities.

A typical annunciator consists of a dc micro-ammeter, the centre-zero position of which indicates the synchronized state of the slaving system. Depending on the type of MHRS, the instrument may be incorporated within the RMI (see Fig. 8.14(a)) or within a control panel as shown in Fig. 8.18. If the system becomes desynchronized, the annunicator pointer will be deflected to one or other side of zero. For example, if the servo CT synchro output signal to the RMI is such that it produces an indicated heading that is less than the sensed magnetic heading, the annunciator pointer is deflected to the left of

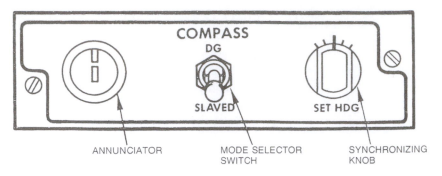

Figure 8.18 Control panel annunciator.

zero. Conversely, a deflection to the right signifies a greater indicated heading. Annunciator pointers are also deflected during turning of an aircraft as a result of the heading error signals produced by displacement of the detector element. Under synchronized conditions, the annunciator should, ideally, remain steady at its centre-zero position. During flight, however, it oscillates slowly due to pendulosity effects on the detector element, and this can in fact serve as a useful indication that slaving is taking place.

In the monitored gyroscope system described earlier (see page 192) fast synchronizing is initiated by manually operating a synchronizing knob which is also incorporated in the RMI. Referring to Fig. 8.14(a) again, it will be noted that the knob is marked with arrows and signs which correspond to the deflected positions of the annunciator pointer. Thus, if, as in the example already noted, de-synchronization produces a deflection to the left of zero, the plus sign signifies that the heading indicated by the RMI must be increased to regain synchronism. The synchronizing knob is therefore rotated in the direction of its plus sign and, in so doing, it rotates the stator of the RMI heading CT synchro (see Fig. 8.12) to induce a large error signal in its rotor. This signal, after amplification by the servo amplifier, drives the servomotor and compass card at a much faster synchronizing rate, which typically is 300°/min. At the same time, the servomotor drives the rotor to start 'nulling out' the error signal, and it also rotates the slaving CT synchro rotor to produce a slaving signal for precessing the DGU into synchronism with the magnetic heading reference established by the detector element. When the error signal is reduced to 2°, synchronizing takes place at the normal slow rate of 1°−2°/min. In dual MHR systems, the foregoing synchronizing process is also activated when switching from one system to another.

In the case of an MHRS/INS integrated system, the slaving circuit has to be maintained in synchronism with both magnetic and inertial platform heading references, and so the appropriate control circuit is connected to that of the stepper motor (see Fig. 8.17). In response to error signals produced by a de-synchronized condition, the stepper motor controls the relative positions between the stator and rotor of the CDX synchro, the resultant output of which drives the servomotor to 'null out' the error signals in the manner already described. Since the direction of motor rotation to attain synchronism is automatically determined by passing error signals through a polarity detector, it is not necessary to utilize a synchronizing knob as in the case of a monitored gyroscope system. A slow synchronizing rate of $1°-2°$/min up to $2°$ heading error is also utilized, but for larger errors the stepper motor is driven at an increased rate of $600°-800°$/min.

The fast synchronizing rate is also activated when: power is initially applied to the system; the sources of heading references are changed over as is possible in dual systems; a system is switched to the slaving mode from the DG mode, when there is a valid inertial platform reference signal, and the magnetic heading is greater than $2°$.

MHRS operating modes

MHR systems provide for the selection of two modes of operation, namely *slaved* and *DG*. The slaved mode is the one normally selected, and provides for operation in the manner already described. When operating in this mode, however, the accuracy of a system is affected by the range of latitudes over which an aircraft is flown. The reason for this is that the magnetic intensity of the earth's field component *H* varies with latitude such that beyond $70°$ north or south of the equator, it becomes an unreliable primary heading reference. Such a reference would also be obtained if, regardless of latitude, a malfunctioning of heading reference signal circuits were to occur. Thus, as in the case of direct-reading compasses and direction indicators (see page 125), an MHRS can also be selected to operate in the *DG* mode to obtain a short-term stability reference irrespective of magnetic field variations. Once selected, the heading information displayed must be frequently updated in order to maintain its integrity.

In a *monitored gyroscope* type of system, the selection of the DG mode disables the slaving control circuit in the DGU so that its gyroscope then functions as a basic direction indicator for controlling the heading servo loop coupled to the compass cards of the RMI and the HSI of the flight director system. In an MHRS/INS integrated system, the slaving control and stepper motor circuit in the compass coupler is similarly disabled, and the heading servo loop coupling the RMI and HSI compass cards is controlled solely by inertial platform heading references established by the azimuth gyroscope and synchro.

Heading selection

The method of selecting a magnetic heading to be flown varies between types of MHRS and their integration with other associated

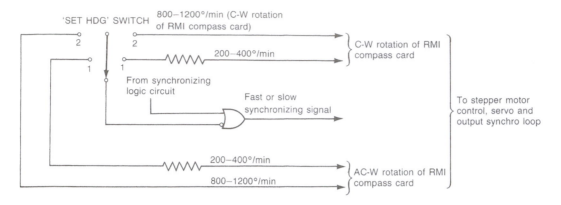

Figure 8.19 Heading selection
— MHRS/INS.

systems. In a basic *monitored gyroscope* type (see Fig. 8.12) the RMI
is provided with a 'SET HDG' knob which, on being rotated,
positions a heading 'bug' relative to the compass card; it also
positions the rotor of a heading data CX synchro whose stator is
connected to that of a CT synchro in the roll control module of an
AFCS computer. The resulting error signal corresponds in magnitude
to the selected heading, and after processing by the roll control
module the aircraft is automatically turned onto this heading. As the
DGU and RMI respond normally to the changing position of the
detector element, the heading data synchro rotor is repositioned to
'null out' the error signal supplied to the roll control module, and the
compass card is rotated to indicate the new heading with reference to
the heading 'bug'.

When a system is integrated with a flight director system, the HSI
provides the facility for selecting heading changes, and its operation
will be described in Chapter 9.

In the MHR/IN system thus far used as an example, heading
selection is accomplished by a 'SET HDG' switch (see Fig. 8.19) on
the system control panel. The switch is supplied with direct current
and is of the limited-travel rotary type which is spring-loaded back to
its centre position. It has two positions left and right of centre, and
the contacts corresponding to these positions are connected to the
stepper motor control circuit in the compass coupler. When the
switch is selected to the first right-hand position, the dc supply passes
through a resistor and causes the stepper motor to rotate at a rate of
200°−400°/min, and through the servo and output synchro loop, the
RMI compass card is rotated in a clockwise direction. In the second
right-hand position there is no resistance in the control circuit, and so
when selected the dc supply rotates the stepper motor at a faster rate
of 800°−1200°/min. A similar operation results when the first and
second left-hand positions are selected, except that the compass card

of the RMI is rotated in an anti-clockwise direction. When the switch is released, the system reverts to the normal slaved mode and slow synchronizing rate.

Heading selections can also be made when the MHRS is operating in the DG mode, except that on releasing the switch to its centre position, the heading will remain at the 'set' position until the aircraft's heading changes, at which moment the servo loop positions the synchros to the relative heading change.

Deviation compensation

Deviation is, as pointed out in Chapter 3, an error in heading indication that results from the effects of hard- and soft-iron components of aircraft magnetism (see page 87) on the detector element of a compass. Although the detector elements of MHR systems have the advantage over those of direct-reading compasses, in that they are fixed in azimuth, and can be located at specifically chosen remote points in an aircraft, they are not entirely immune from extraneous fields that may be present in their vicinity. The principal reason for this is that the element material has a high permeability and so is very receptive to magnetic flux (see page 186). Thus, any flux additional to that of the desired earth's field component will displace the H axis of the material's B/H curve to a false datum, and thereby induce heading error signals. It is therefore necessary to incorporate deviation compensation devices in an MHR system.

As far as compensation for the deviation coefficient A is concerned, it is normally the practice to utilize detector elements of the pre-indexed type as referred to on page 190. In some early designs of detector element, compensation was effected by rotating the element in its mounting by the requisite amount, and referencing it against a scale and fixed datum mark.

An electromagnetic method is normally adopted for the compensation of deviation coefficients B and C, and this is illustrated in the basic circuit diagram of Fig. 8.20. The potentiometers, which are incorporated in a remotely-located compensator unit, are connected to the pick-off coils of the detector element. When rotated with respect to calibrated scales, they inject very small dc signals into the coils, so that the fields they produce are sufficient to oppose those causing deviations. The output of the detector element is thereby modified to correct the readings of the RMI via the synchronous transmission loop. In some types of electromagnetic compensator, provision is also made for coefficient A adjustment. This is achieved by the inclusion of a differential synchro between the detector element and the servo synchro within the RMI. When the position of the differential synchro rotor is adjusted in the appropriate direction,

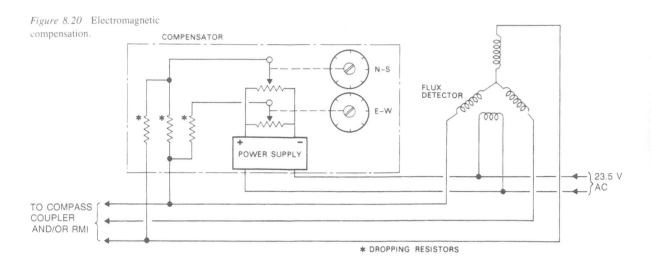

Figure 8.20 Electromagnetic compensation.

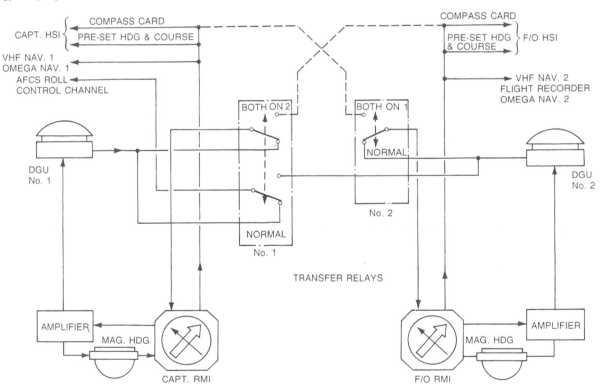

Figure 8.21 Dual monitored gyroscope system.

it offsets the deviation by changing the magnitude of the heading signals transmitted by the detector element.

Dual systems

The interconnection of dual systems of a typical monitored gyroscope type is shown in Fig. 8.21. The transfer relays are controlled by a

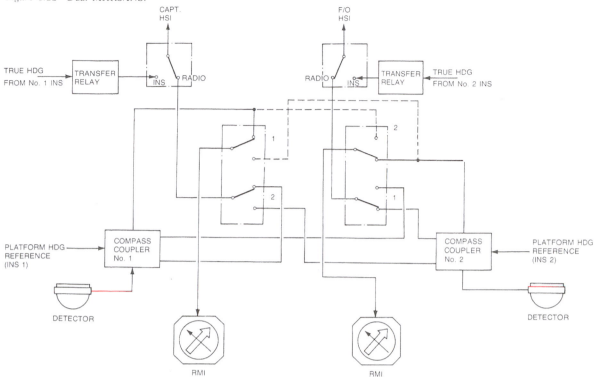

Figure 8.22 Dual MHRS/INS.

panel-mounted selector switch, and in the 'normal' position the systems operate independently of each other. In the event of failure of magnetic heading data input to one or other system, the operating system can be selected to take over by energizing the appropriate transfer relay. For example, if the input to the captain's or No. 1 system should fail, the selector switch is moved to a position placarded 'BOTH ON 2', and so, as may be seen from the diagram, the contacts of the No. 1 transfer relay change over to connect an alternate data input from the first officer's or No. 2 system to the captain's system.

The arrangement of a typical dual MHR/IN system is illustrated in Fig. 8.22; this is also drawn to represent normal independent operation. The method of transferring magnetic heading data from one MHR system to the other is similar to that described above. The system also incorporates selector switches placarded 'RADIO' and 'INS', and as will be noted they are connected independently between each INS and each flight director system's HSI. When the switches are in the 'RADIO' positions each HSI is supplied with magnetic heading data, while in the 'INS' positions these indicators are supplied with true heading data from the respective inertial platforms. Further details of these aspects of data transfer and switching will be given in Chapter 10.

9 Flight director systems

A flight director system (FDS) is one in which the display of pitch and roll attitudes and heading of an aircraft are integrated with such radio navigation systems as automatic direction finding (ADF), very high-frequency omnidirectional range (VOR), and instrument landing system (ILS) so as to perform a total directive command function. It also provides for the transmission of attitude and navigational data to an AFCS so that in combination they can operate as an effective flight guidance system.

The components comprising a typical FDS, their connections and signal interfacing with the systems providing essential navigational data are shown in Fig. 9.1.

Vertical gyroscope unit (VGU)

This unit performs the same function as a gyro horizon, i.e. it establishes a stabilized reference about the pitch and roll axes of an aircraft. Instead, however, of providing attitude displays by direct means, it is designed to operate a synchro system which produces, and transmits, attitude-related signals to a computer (sometimes referred to as a 'steering' computer) and to an amplifier unit. After processing and amplification, the signals are then transmitted to servo-operated indicating elements within a separate attitude director indicator (ADI). The synchro system also supplies attitude-related signals to the appropriate control channels of an AFCS. The gyroscope and its levelling switch and torque motor system is basically the same as that adopted in electrically-operated gyro horizons (see pages 116 and 119).

The synchro system referred to earlier senses changes in pitch and roll attitudes by means of a CX synchro positioned on each corresponding axis of the gyroscope's gimbal system. The stator of the *roll* synchro is secured to the frame of the unit, while its rotor is secured to the outer gimbal ring. The *pitch* synchro has its stator secured to the outer gimbal ring, and its rotor secured to the inner gimbal ring. The stators supply attitude error signals to corresponding CT synchros in the ADI, and also to pitch and roll circuit modules of the computer.

Computer

This unit contains all the solid-state circuit module boards, or cards, necessary for the processing of attitude reference and command

Figure 9.1 Typical FDS and signal interfacing.

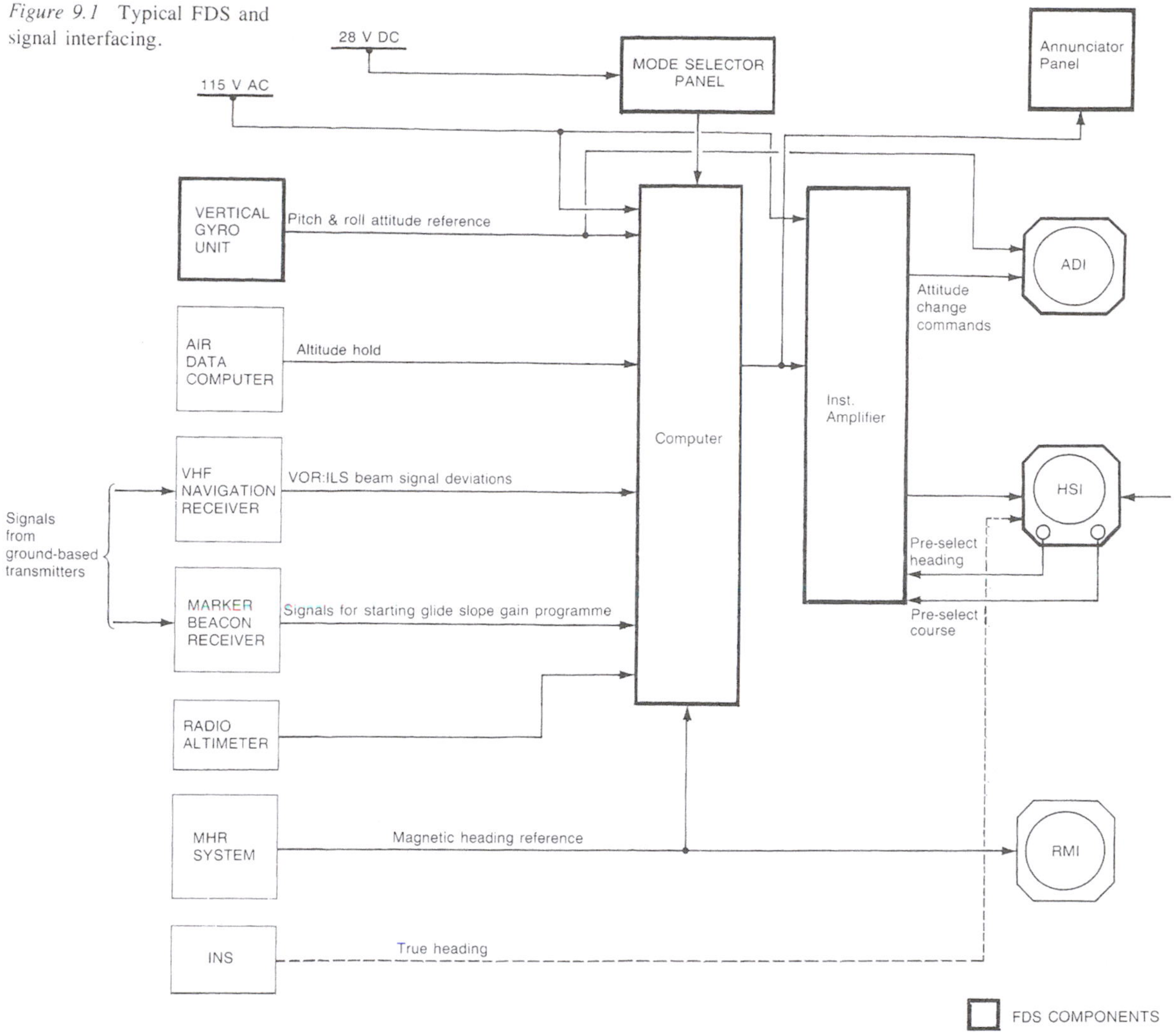

28 V DC

115 V AC

MODE SELECTOR PANEL

Annunciator Panel

VERTICAL GYRO UNIT — Pitch & roll attitude reference

AIR DATA COMPUTER — Altitude hold

Signals from ground-based transmitters

VHF NAVIGATION RECEIVER — VOR:ILS beam signal deviations

MARKER BEACON RECEIVER — Signals for starting glide slope gain programme

RADIO ALTIMETER

MHR SYSTEM — Magnetic heading reference

INS — True heading

Computer

Inst. Amplifier

ADI

Attitude change commands

HSI

Pre-select heading

Pre-select course

RMI

□ FDS COMPONENTS

signals. Logic circuit boards are also provided for the purpose of adjusting the scaling and gain values of signals appropriate to the type of FDS and aircraft in which it is installed. In many cases it is usual for a status code number to be quoted on the front panel of a computer; this number relates to the required pre-adjusted scaling and gain values which are listed in the form of charts in maintenance and overhaul manuals.

Instrument amplifier

The primary function of this unit is to convert the attitude reference and command signals supplied to it by the computer into servo-actuating power signals for driving the display elements of the FDS indicators. Like the computer, all circuits are of the solid-state type contained on plug-in type module boards or cards.

209

Attitude director indicator (ADI)

This indicator, like the gyro horizon it basically resembles, provides information on an aircraft's pitch and roll attitude. In addition, however, and as may be seen from Fig. 9.2, it provides attitude commands and information related to an aircraft's position with respect to the glide slope (GS) and localizer (LOC) beams transmitted by an ILS. A series of warning flags are also provided, and a ball-in-tube indicator provides indication of slip during turns.

The symbol representing the aircraft is fixed and is referenced against a moving 'sky/ground' background tape on which are presented an horizon line, and markings spaced at a specified number of degrees to indicate pitch-up and pitch-down attitudes. The tape is positioned around two rollers which, on being driven by a servomotor and gear train, move the tape up or down as appropriate to the pitch attitude change (see Fig. 9.3). The servomotor is activated by amplified signals from a CT synchro connected to the pitch CX synchro of the VGU, and its direction of rotation is determined by

Figure 9.2 Attitude director indicator.

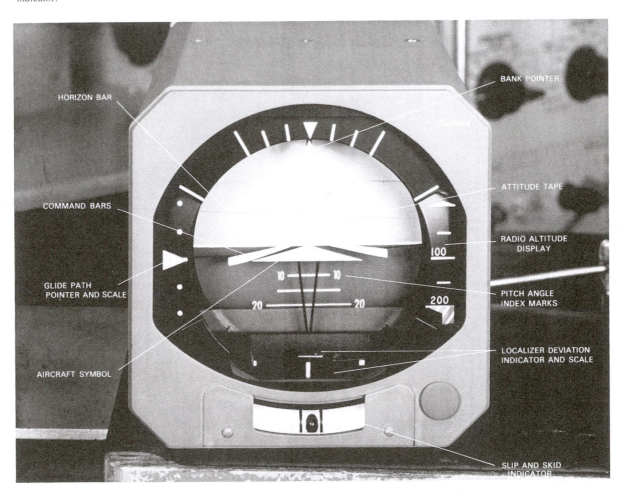

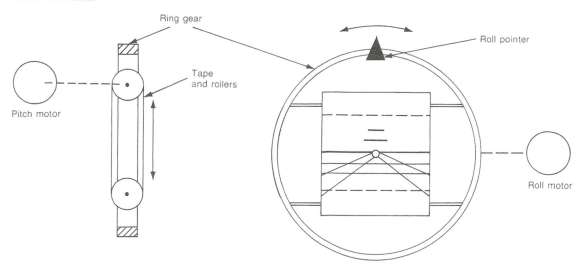

Figure 9.3 Pitch and roll attitude indication.

Ring gear

Tape and rollers

Pitch motor

Roll pointer

Roll motor

the phase relationship between the CX synchro error signal voltages and the servomotor excitation voltage.

Roll attitude is indicated on the roll attitude scale by the relative position of a bank pointer. The pointer is fixed to the ring gear which supports the pitch attitude tape rollers, and is also coupled to a servomotor via a gear train. This servomotor is activated by amplified signals from a CT synchro connected to the roll synchro of the VGU, and as in the case of the pitch servomotor its direction of rotation is determined by the phase relationship between error signal voltages and servomotor excitation voltage. A differential gear is provided in the drive system so that whenever there is a change in roll attitude, the pitch attitude tape also rotates together with the roll attitude pointer.

The GS and LOC indicating elements are respectively located at the left and bottom of the ADI's attitude display. Each element consists of a scale and a pointer that is deflected by miniature-type micro-ammeter movements that respond to the appropriate beam deviation signals supplied from an aircraft's VHF radio navigation receivers via the FDS computer. The scales of the indicators are shown in a little more detail in Fig. 9.4. The pointers represent the position of the beams relative to an aircraft, and the pointer positions in relation to the commands should be particularly noted. The dots on the GS scale represent $75\mu A$ (one dot) and $150\mu A$ (two dots) of aircraft deviation above or below the beam ($75\mu A = 0.35°$; $150\mu A = 0.7°$). The LOC scale has a single dot to the left and right of the centre position, each dot also representing $75\mu A$, but in this case $= 1°$ of aircraft deviation. The LOC pointer is distinctly shaped to represent

Figure 9.4 GS and LOC indicating elements.

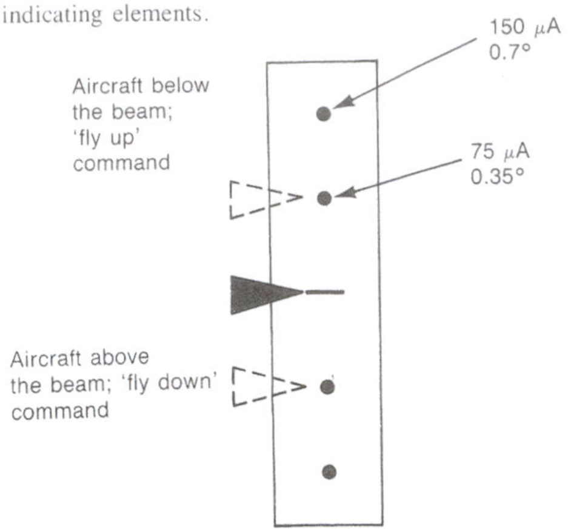

Aircraft below the beam; 'fly up' command

150 µA 0.7°

75 µA 0.35°

75 µA 1°

Aircraft above the beam; 'fly down' command

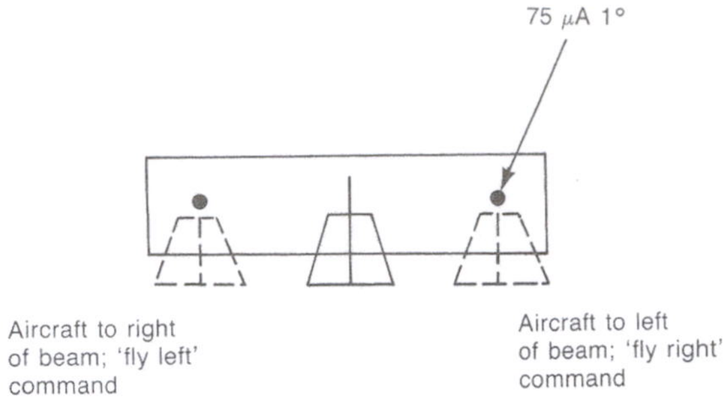

Aircraft to right of beam; 'fly left' command

Aircraft to left of beam; 'fly right' command

symbolically the converged shape of the runway as it appears during an approach. In some ADIs the pointer is also deflected upwards by signals from a radio altimeter to simulate the runway coming up to the aircraft. At touchdown this 'rising runway' symbol, as it is called, just touches the fixed aircraft symbol.

The GS pointer and scale come into view when the FDS is operating in the GS mode and when there is a valid and reliable signal from the VHF navigation receiver. At all other times it is covered by a red warning flag controlled through the signal circuit. The LOC pointer and scale are normally covered by a shutter which is retracted when valid and reliable signals are supplied during the VOR/LOC mode of operation of the FDS.

Two further red warning flags are provided, one placarded 'GYRO' and the other 'COMPUTER'. The 'GYRO' flag is controlled by a monitoring circuit within the VGU and warns of loss of power to this unit and any failures occurring in the attitude reference circuits. The 'COMPUTER' flag monitors and warns of failures in inputs to the computer itself, the instrument amplifier, and the ADI.

The command bars, as the name implies, provide the commands relating to the changes that are to be made to manoeuvre an aircraft into required pitch and/or roll attitudes. They are driven by servomotors that receive their input signals from the pitch and roll channels of the computer via the instrument amplifier. Although the two bars are not physically connected to each other, they move together up or down to represent pitch commands, and tilt to the left or right to represent roll commands (see Fig. 9.5). When the commands being generated by the computer are satisfied, the bars take up a position coincident with the top of the fixed aircraft symbol.

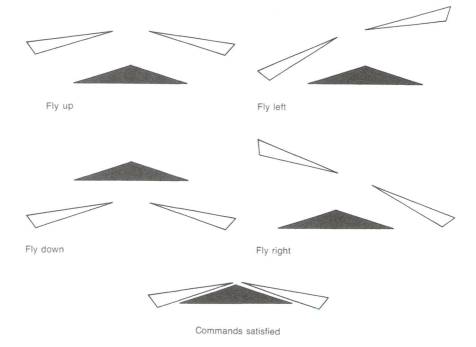

Figure 9.5 Command bars deflection.

Fly up

Fly left

Fly down

Fly right

Commands satisfied

Operation

As far as aircraft attitudes are concerned, the ADI displays the information in two ways: 1. as a primary attitude reference, and 2. as command attitude changes.

Primary attitude reference

The interconnection between the VGU and the ADI are shown in simplified form in Fig. 9.6.

When the gyroscope is operating and is stabilized with its axis vertical, both CX synchros are at their 'null' positions and so there is no output from their stator windings. The ADI will therefore indicate a level flight attitude. If a displacement of the aircraft occurs about, say, its pitch axis, the outer gimbal ring is also displaced and carries the pitch synchro stator around its stabilized rotor. This produces a displacement signal voltage which is then supplied to the stator of the CT synchro in the ADI. Because at that moment an unbalanced condition exists between rotors and stators of both synchros, then the output from the CT synchro rotor is an induced error signal voltage of a phase and magnitude appropriate to the direction and degree of displacement.

As may be seen from the diagram, this error signal voltage is supplied to the pitch channel of the instrument amplifier, and after amplification it drives the corresponding servomotor in the ADI to position the attitude tape upwards or downwards appropriate to the aircraft's displacement about the pitch axis. At the same time, the

213

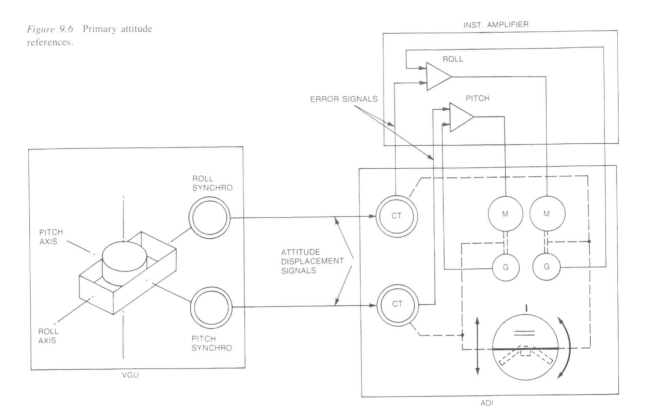

Figure 9.6 Primary attitude references.

INST. AMPLIFIER

ROLL

ERROR SIGNALS

PITCH

ROLL SYNCHRO.

PITCH AXIS

ATTITUDE DISPLACEMENT SIGNALS

ROLL AXIS

PITCH SYNCHRO.

VGU

CT

M   M

G   G

CT

ADI

servomotor repositions the CT synchro rotor until displacement of the aircraft ceases, at which point no further error signal is induced. The servomotor also drives a tachogenerator which produces an output proportional to the motor speed and out of phase with the error signal voltage. This output is fed back to the pitch channel of the instrument amplifier to provide damping of motor rotation and attitude tape movement.

A similar operating sequence takes place when an aircraft is displaced about its roll axis, except, of course, that the ADI's attitude tape is rotated left or right in response to signals from the roll CX synchro.

If an aircraft is displaced simultaneously about the pitch and roll axes, as for example in a climbing turn, both channels will operate and, by means of the differential gearing in the drive system, the ADI's attitude tape will be positioned to indicate such a turn.

Command attitude changes
As mentioned earlier, command attitude changes are indicated by the movements of the command bars in the required directions. The method of driving the bars is shown in Fig. 9.7, and as will be noted it is a little more complex than that adopted for primary attitude references. The principal reason for this is that command signals can

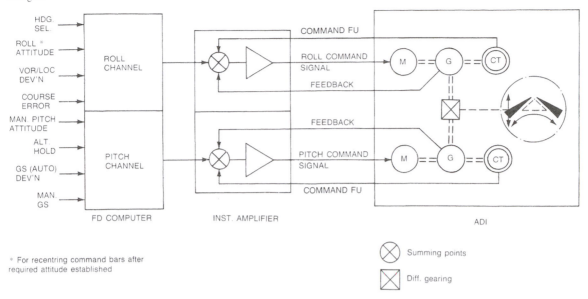

Figure 9.7 Command attitude changes.

HDG. SEL.
ROLL ATTITUDE *
VOR/LOC DEV'N
COURSE ERROR
MAN. PITCH ATTITUDE
ALT. HOLD
GS (AUTO) DEV'N
MAN. GS

ROLL CHANNEL

PITCH CHANNEL

FD COMPUTER

COMMAND FU
ROLL COMMAND SIGNAL
FEEDBACK

FEEDBACK
PITCH COMMAND SIGNAL
COMMAND FU

INST. AMPLIFIER

M G CT

M G CT

ADI

* For recentring command bars after required attitude established

⊗ Summing points

⊠ Diff. gearing

be supplied from several external sources as governed by selected modes of FDS operation. Details of these modes and their command signals will be covered at a later stage, but for the moment we may consider one of them, namely the 'manual pitch attitude mode', in order to see how the signals are generated, and also how the command bars are operated.

The manual pitch attitude mode is one which enables a pilot to select a pitch attitude reference at any time there is no other mode selected for controlling about the pitch axis. An example of this would be the selection of a desired climb attitude that is to be maintained after take-off. The selection is made before take-off by rotating a pitch command knob on the FDS mode controller. The control knob also rotates the rotor of a synchro which then supplies an error signal to the pitch channels of the computer and instrument amplifier as shown in Fig. 9.7. The amplified signal then drives a motor in the ADI to position the command bars above the fixed aircraft symbol. At the same time, the motor drives a tachogenerator that provides a rate feedback signal for motor speed control purposes, and it also positions the rotor of a CT synchro. The output signal from this synchro serves as a follow-up to the command signal, and to null it out so that the motor and the command bars are stopped at the selected pitch up command position. When the aircraft takes off and rotates to the desired climb attitude, the pitch attitude tape of the ADI will be positioned to indicate the climb in the manner explained earlier, and the command bars will be recentred over the fixed aircraft symbol.

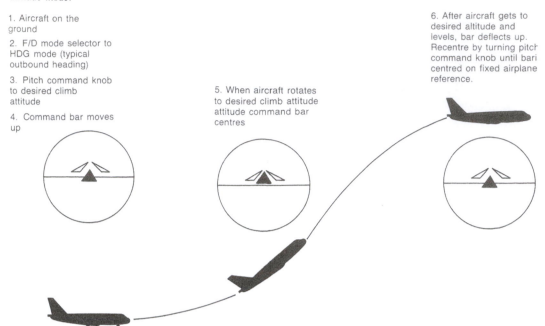

Figure 9.8 Manual pitch attitude mode.

1. Aircraft on the ground

2. F/D mode selector to HDG mode (typical outbound heading)

3. Pitch command knob to desired climb attitude

4. Command bar moves up

5. When aircraft rotates to desired climb attitude attitude command bar centres

6. After aircraft gets to desired altitude and levels, bar deflects up. Recentre by turning pitch command knob until bari centred on fixed airplane reference.

After the aircraft reaches its required altitude and levels off, the command bars deflect upwards, and are then recentred by turning the pitch command knob back to its zero position. The foregoing sequence is shown pictorially in Fig. 9.8.

Horizontal situation indicator (HSI)

This indicator derives its name from the fact that its display, as can be seen from Fig. 9.9, presents a pictorial plan view of an aircraft's situation in the horizontal plane in the form of its heading, VOR/LOC deviation, and data relating to flight to and from a VOR station. In addition, it displays deviations from the GS beam and distance from a distance measuring equipment (DME) station.

The aircraft symbol is fixed at the centre of the display and it indicates the position and heading of an aircraft in relation to the compass card and the VOR/LOC deviation bar. This bar is also sometimes called a lateral deviation bar. Selector knobs at the bottom corners of the indicator permit the setting of a desired magnetic heading and a VOR/LOC course.

Heading display

The primary display element of the indicator is that related to an aircraft's magnetic heading, and so it is integrated with a magnetic

Figure 9.9 Horizontal situation indicator.

heading reference system (MHRS). In aircraft equipped with an inertial navigation system (INS) the indicator is also integrated with the computer of that system so that it can be selected to display either true heading or magnetic heading.

Figure 9.10 shows the arrangement of the heading display section in simplified form. It consists of an azimuth or compass card which is mounted on a ring gear driven through a gear train by a servomotor. Headings are indicated by the position of the card with respect to a fixed lubber line. The servomotor also drives a tachogenerator that provides rate feedback signals for motor speed control. In order to select a magnetic heading, a heading marker is provided, and can be positioned relative to the compass card by rotating the heading selector knob. The differential gear shown is to permit relative movement between the marker and card, and also to allow the marker to rotate with the card when a change in heading takes place.

Heading signals are supplied from the MHRS via its RMI (or

217

Figure 9.10 Heading selection
and display.

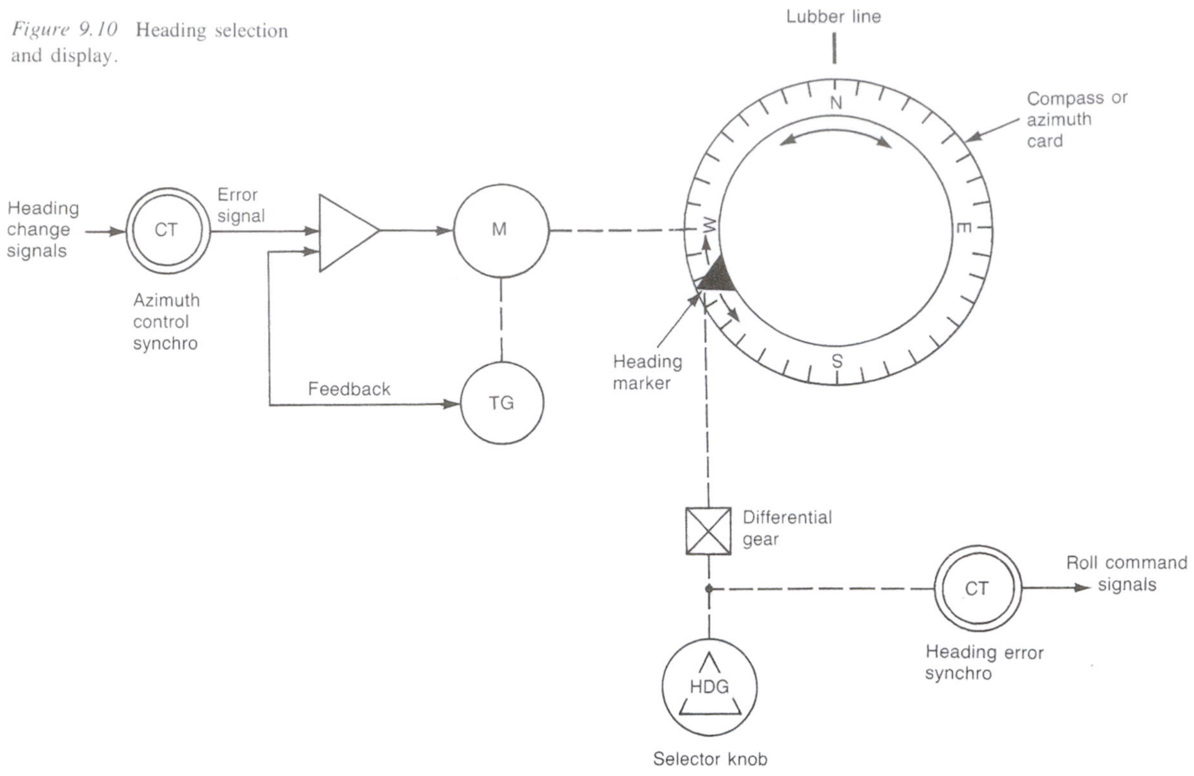

compass coupler unit in the case of integration with an INS) and as
will be noted they are fed to a CT azimuth synchro in the HSI. On a
constant heading, the synchro is at 'null' with that of the MHRS, and
so the compass cards of both the RMI and HSI indicate the same
heading.

When it is required to change an aircraft's heading, the FDS is
operated in the heading (HDG) mode to provide roll commands. To
select a heading change the heading select knob is rotated to position
the heading marker against the corresponding graduation mark on the
compass card. At the same time, the rotor of a heading error CT
synchro is rotated to produce an error signal in its stator. This signal
is supplied to the roll channel of the FDS computer and then to the
ADI as a roll command. The command bars are therefore deflected in
the manner already described (see Fig. 9.5), to indicate the direction
in which the aircraft is to be turned onto the new heading.

As the aircraft turns, a heading change signal is produced by the
MHRS to rotate the RMI compass card. The signal, as shown in Fig.
9.10, is also supplied to the azimuth CT synchro in the HSI, but as
the rotor of this synchro will, at that moment, be desynchronized in
relation to the MHRS synchro, it produces an error signal. This
signal is then amplified and supplied to the servomotor that is
coupled through a gear train to the compass card. Thus, the card and
also the heading marker are rotated in the appropriate direction. It

will also be noted from Fig. 9.10 that the motor drives the azimuth CT synchro rotor; this is done to reduce the error signal progressively as the aircraft turns onto the selected heading. In other words, the motor brings the synchro into synchronism with the MHRS so that the HSI compass card 'repeats' the heading indication of the RMI. When the aircraft has levelled off on the new heading, synchronism is attained and the heading is indicated by alignment of the heading marker with the lubber line.

Aircraft position with respect to VOR/LOC beams

The secondary section of an HSI display relates to an aircraft's lateral position with respect to a VOR station and to the localizer beam of an ILS; the basic arrangement of this section is shown in Fig. 9.11. It consists of a deviation bar that can be deflected left or right over a scale plate, and in relation to a course pointer or marker. The deflections perform a directive command function, i.e. a deflection to the left is a 'fly left' command to capture a beam, and a deflection to the right is a 'fly right' command.

Each dot shown on the scale corresponds to a 1° deflection. The bar is deflected by a dc meter movement supplied with signals from the radio navigation receiver. When there are no deviation signals

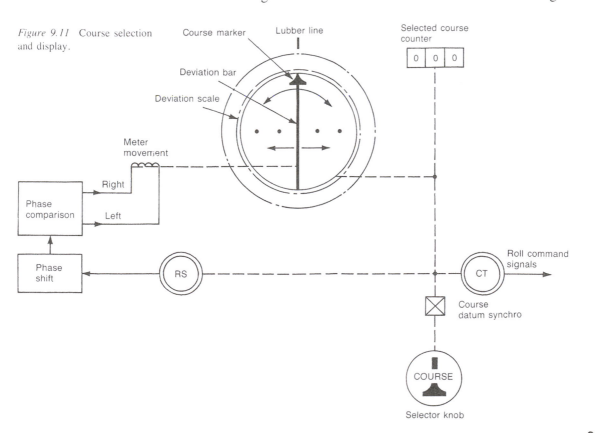

Figure 9.11 Course selection and display.

present, the bar is aligned with the course marker as shown. In addition to deflection, the bar, its scale and course marker can be rotated: (i) relative to the compass card whenever the course to a VOR station or ILS localizer is selected, and (ii) rotated with the compass card when the aircraft turns onto the selected course.

Selection of a desired VOR radial or localizer course is carried out by rotating the course selector knob until the course marker coincides with the desired value on the compass card. The deviation bar, its meter movement and scale also rotate with the marker. At the same time, the control knob drives a digital counter to the corresponding course indication. The gear train comprising the drive from the selector knob is coupled to the rotor of a course resolver (RS) synchro associated with the VOR/LOC navigation receiver, and to the rotor of a course datum CX synchro. Both rotors are, therefore, set to some angular position with respect to their stators when the course selector knob is rotated.

In the case of selecting a course to fly onto a desired VOR radial, changing the position of the course RS synchro rotor causes it to shift the phase of the low-frequency (30 Hz) reference modulating signal received from the station by the aircraft's radio navigation receiver. The signal is then compared with the station's variable modulating signal (also 30 Hz) in a phase comparator circuit, the output of which is supplied to the meter movement to deflect the deviation bar left or right as appropriate.

The course datum CX synchro performs the same function as that of the heading error synchro, i.e. it produces a roll command signal which is supplied to the ADI command bars to deflect them in the direction in which the aircraft is to be turned to intercept the VOR beam radial. As the aircraft turns in response to the roll command, the compass card rotates in response to the signals sensed by the MHRS, and through the differential gearing the card also rotates the deviation scale, the bar and the course marker. When the beam is being approached, the signals from the RS synchro, and the phase comparator, to the deviation bar meter movement are being reduced and so the bar deflection is towards the fixed aircraft symbol. The output from the course datum CT synchro also changes to deflect the ADI command bars in the opposite direction, thereby commanding that the aircraft be rolled out to a wings-level attitude and on course to the VOR station.

Flight along the beam is indicated by alignment of the deviation bar with the course arrow. The effects of any cross-winds during the flight along the beam are automatically corrected by a compensating 'wash-out' circuit in the FDS computer, to establish the 'crab angle' necessary for the aircraft to stay on course. This angle is also indicated on the HSI by the position of the deviation bar relative to the fixed lubber line.

To−from indicators

Indication of whether an aircraft is flying to or from a station is provided by an arrow-shaped marker that is positioned by a dc meter movement. The meter is supplied with signals from a phase comparator circuit when the VOR station frequency is tuned in. Referring to Fig. 9.10 once again, the marker is positioned in the direction of the course marker, thus indicating that the selected course is to the station selected. When an aircraft flies from the station, the meter movement deflects the arrow through 180°. In some types of FDS, the HSI has two separate meter movements and arrow-shaped markers.

LOC mode

When the FDS is operating in the LOC mode, the HSI functions in the same manner as that just described for the VOR mode, but with two exceptions. Firstly, the output to the meter movement controlling the deviation bar results from amplitude comparison of signals either side of localizer beam centre, and secondly, the to−from arrow remains out of view since no to−from signals are transmitted in the LOC mode.

Warning flags

As in the case of the ADI, red warning flags are provided in an HSI to indicate that GS, LOC and VOR signals are unreliable or have completely failed. In addition a flag is provided to give similar indications in the case of the MHRS.

DME indicator

This indicator receives signals from the interrogator of the distance measuring equipment (DME) carried in an aircraft, and displays the distance in nautical miles to be flown to a selected DME ground station. If the system is not valid the indicator display is obscured by an electrically-operated shutter.

Radio altitude

In some HSIs an indicator light is provided and is connected to a radio altimeter system such that it illuminates when an aircraft reaches a specified minimum altitude, referred to as a 'decision height', during the final stages of an automatically-controlled approach.

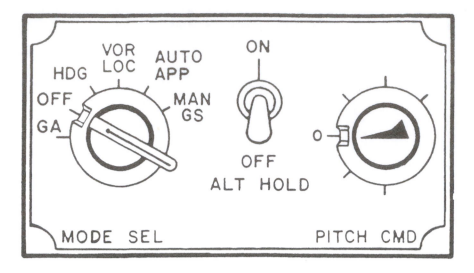

Figure 9.12 Mode controller.

Mode controller

A typical controller is shown in Fig. 9.12; it provides three principal control functions: (i) mode selection, (ii) altitude hold, and (iii) manual pitch command.

Mode selection is done by means of a rotary selector switch which can be manually placed in any of the indicated positions. In all positions except GA (go-around) a pin drops into a detent to maintain the switch in position. If the switch is in either the 'AUTO APPR' or the 'MAN' position, and the GA manoeuvre has to be carried out (see page 227), the pins are automatically retracted from their detents, and a spring returns the switch to the GA position.

The six positions of the switch and the system's response in each case are as follows:

GA	Causes the ADI command bars to display a pitch-up command and a wings-level attitude.
OFF	Disables the pitch and roll outputs from the computer causing the ADI command bars to be deflected out of view.
HDG	Allows selection of a magnetic heading which, as indicated earlier, is done by means of the selector knob on the HSI. While in this mode, the computer pitch channel can be operated in the 'MANUAL PITCH' mode by means of the pitch command selector, or the 'ALT HOLD' mode by attitude command signals generated and supplied by an ADC.
VOR/LOC	Allows selection of a VOR radial or a localizer beam for lateral guidance, so that input command signals can be produced in the manner already described. The pitch channel of the computer can be operated in either the 'MANUAL PITCH' mode or the 'ALT HOLD' mode.

AUTO/APPR Allows the use of the ILS beams (LOC and GS) for lateral and vertical guidance during an automatic approach. In the roll channel, the input commands are LOC deviation, course error and roll attitude. The pitch channel is also armed for capture of the GS beam. Before capture of this beam, however, the pitch channel can be operated in the 'MANUAL PITCH' or 'ALT HOLD' modes. At GS capture, both these modes are automatically cancelled, and pitch command signals are generated within the computer by a combination of GS deviation and pitch attitude signals.

MAN/GS This mode is selected for use under three different conditions: (i) to establish a fixed intercept angle of either a VOR or LOC beam; (ii) to force a GS beam capture condition when making an approach which is above the beam; and (iii) to force a LOC or GS beam capture condition if it is known that the beam-sensing circuits of the computer are inoperative.

The function of the 'ALT HOLD' switch is to establish a command signal that positions the ADI command bars in pitch so that they are referenced to the altitude sensed by an ADC. The switch is held in the 'ON' position by a solenoid provided valid signal conditions exist. If the switch is on at GS capture, it will automatically be returned to the 'OFF' position by de-energizing of the solenoid. Further details of the altitude hold function will be given later.

As described earlier, the 'PITCH COMMAND' knob permits selection of a desired pitch attitude of an aircraft when the FDS computer is operating in the 'MAN PITCH' mode. In a typical system, a minimum of 15° of pitch-up command or 10° pitch-down command can be generated for the input to the pitch channel.

FDS/AFCS mode controllers

The basic attitude and navigational data displayed by the indicators of an FDS, together with the selection of operating modes, are features that are also common to the operating requirements of an AFCS. They can, therefore, be developed as a natural complement to each other, and by matching their characteristics with those relating to the aerodynamics and flight control of any one type of aircraft, they can be fully integrated to serve as an overall flight guidance system. As far as the operation of such a system is concerned, one of the items of FDS 'hardware' having a common purpose is the mode controller. In some systems this may be provided as a separate unit (as, for example, the one shown in Fig. 9.12) for use in combination with the control panel of an AFCS. It is, however, a more logical approach to

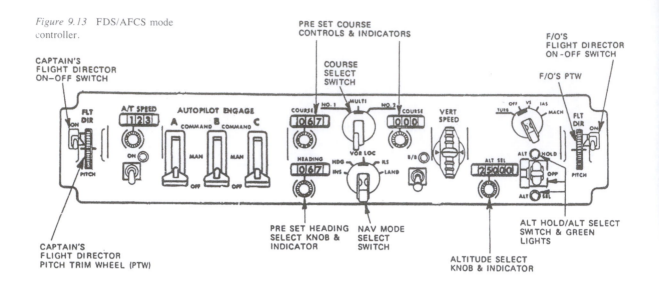

Figure 9.13 FDS/AFCS mode controller.

eliminate the panel by combining its selective functions and switching circuits within the AFCS control panel itself, and this is a method now adopted in the majority of present-day systems. Figure 9.13 illustrates an example of an FDS/AFCS mode controller, or mode select panel, based on that used in some series of Boeing 747. The controls annotated perform functions associated with the FDS as follows:

Flight director switches	Switch on an output to the command bars of their respective ADIs, and also control power to the annunciator panel. They do not control power to the computers.
Pitch control wheels	Control pitch command bars of their respective ADIs when no other pitch mode has been selected. They are disabled electrically when a pitch mode is captured.
Heading selector	This replaces the selector that is normally provided in an HSI, and performs the same function, i.e. it establishes error signals for the purpose of roll control, and also drives the heading marker in the HSI.
Nav. mode selector switch	Apart from the 'INS' and 'LAND' positions, which are specific to the aircraft concerned, the functions of the other positions are the same as those described earlier, namely, 'HDG' selects the signals from the MHRS, 'VOR/LOC'

provides appropriate deviation signals from the VHF navigation receivers, and 'ILS' provides localizer and GS deviation signals from the receivers.

Course selectors	These are connected, one to each of two MHR systems provided in the aircraft concerned, and perform the same function as the selector normally provided in an HSI, i.e. they establish course error signals for roll control, and also drive the HSI course pointer.
Course transfer switch	This is a three-position switch for determining which MHRS and radio navigation receiver are to be used for supplying signals to the roll channels of the FDS/AFCS computers. In either the No. 1 or No. 2 positions, the switch is held in the selected position by a solenoid whenever the nav. mode selector switch is at either the 'INS', 'HDG' or 'VOR/LOC' position. If the 'ILS' or 'LAND' positions are selected, the switch is spring-returned to the 'MULTI' position, in which the distribution of signals between computers is varied, and outputs from a third VHF navigation receiver are introduced.

Flight mode annunciators

The purpose of these units is to annunciate, by coloured lights, the conditions appropriate to the flight modes selected on a mode controller. The lights are grouped so that those at the left of the panel relate to an FDS, while those at the right relate to an AFCS. Two examples are shown in Fig. 9.14.

The annunciator illustrated at (a) is referred to as an approach progress display and, as will be noted, the FDS group consists of VOR/LOC, glide slope and go-around (GA) lights. If a VOR or LOC mode has been selected, the VOR/LOC light illuminates amber, and it remains on until the appropriate beam has been captured, when it then changes to green. The glide slope light illuminates amber when the approach mode is selected, then changes to green when the GS beam has been captured. The GA annunciator light illuminates green whenever the GA mode is selected as a result of an automatic approach having to be aborted. In addition to VOR/LOC and glide slope, the AFCS group of the display has annunciator lights for altitude select and altitude hold modes.

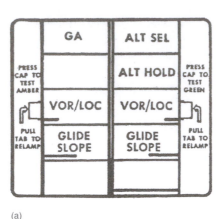

Figure 9.14 Flight mode annunciators.

(a)

ALT SELECT
ANNUNCIATOR

NAV MODE
ANNUNCIATOR

PHOTO-CELL
DIMMING

GLIDE SLOPE
ANNUNCIATOR

GO-AROUND
ANNUNCIATOR

FLARE
ANNUNCIATOR

(b)

The lights may be checked for functioning by individual press-to-test facilities, and provision is also made for bulb replacements.

The annunciator shown in (b) of Fig. 9.14 is of the type used in conjunction with the FDS/AFCS combination adopted in some series of Boeing 747 (see also page 224). The lights are so arranged that captions appropriate to each mode are illuminated white whenever a mode is selected, while the lower half of each light is illuminated green when the desired function has been satisfied. The arrow-shaped symbol of the go-around annunciator light is illuminated green when the associated manoeuvre has been initiated by activation of go-around switches on the engine thrust levers. Testing of the white and green lights is done by depressing the cap of the FDS 'ALT/S' annunciator; this also checks amber lights in the AFCS and autothrottle annunciators in the top row of the panel. This group of annunciators also contains red lights which, together with a blue light in the 'TEST' annunciator, may be checked by depressing the cap of the AFCS 'ALT/S' annunciator. Dimming of the lights is controlled by two photocells which operate in conjunction with a separate master switch when selected to a 'dim' position.

Go-around switches

When an automatic approach has to be aborted, engine thrust has to be increased in order to climb the aircraft into what is termed a 'go-around' (GA) manoeuvre. The FDS also has to go into the GA mode of operation, and this is done by switches (one for each pilot's FDS) located on the engine power levers as shown in Fig. 9.15.

The switches are of the 'momentary press' type so that they can be conveniently operated by the palm of the hand as the power levers are pushed forward. Closing of the switches causes the mode selector switches on the mode control panels to move to the 'GA' position from either the 'AUTO APPR' position or 'MAN GS' position, whichever mode was in operation at the time. This action places the computer in the GA mode, and it then provides pre-set bias signals to each pilot's ADI, causing the command bars to deflect to a 'pitch-up' and 'wings-level' command position. The amount of command bias required varies between different types of aircraft and their flight characteristics; in one particular type the command is 14° pitch up.

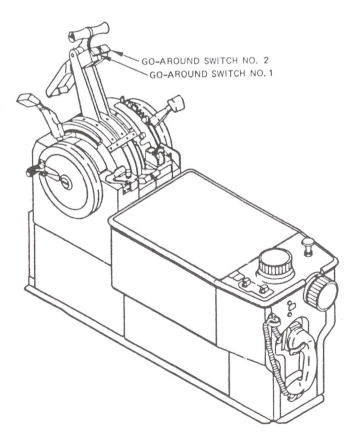

Figure 9.15 Go-around switches.

GO-AROUND SWITCH NO. 2
GO-AROUND SWITCH NO. 1

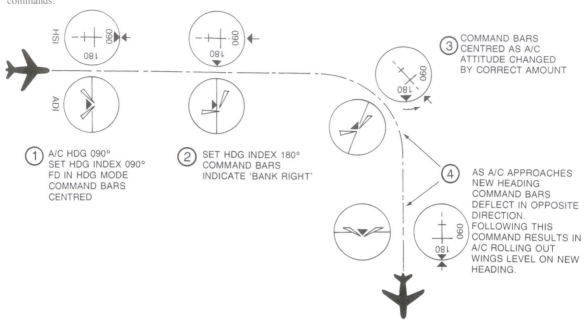

Figure 9.16 Heading select commands.

① A/C HDG 090°
SET HDG INDEX 090°
FD IN HDG MODE
COMMAND BARS
CENTRED

② SET HDG INDEX 180°
COMMAND BARS
INDICATE 'BANK RIGHT'

③ COMMAND BARS
CENTRED AS A/C
ATTITUDE CHANGED
BY CORRECT AMOUNT

④ AS A/C APPROACHES
NEW HEADING
COMMAND BARS
DEFLECT IN OPPOSITE
DIRECTION.
FOLLOWING THIS
COMMAND RESULTS IN
A/C ROLLING OUT
WINGS LEVEL ON NEW
HEADING.

Operating sequences

Heading changes

Figure 9.16 illustrates in pictorial form the sequence of ADI and HSI display presentations (as would be seen in the direction of flight) when, with the FDS operating in the 'HDG' mode, it is required to change an aircraft's magnetic heading. For purposes of explanation it is assumed that the change is to be from 090° to 180°.

Position 1 The HSI heading marker is aligned with the lubber line to show the present heading of 090°. There are no command signals to the ADI and so it displays a level flight attitude.

Position 2 The heading select knob is now rotated to position the heading marker at 180° on the compass card. Since the aircraft must be turned to the right to attain this heading, the heading error CT synchro in the HSI sends a command signal to the ADI, via the computer roll channel, and in response the command bars are deflected to the right.

Position 3 The aircraft is headed into the turn and, at the correct bank attitude, the command bars are centred and the ADI then indicates this attitude. The MHRS detects the heading change, and its signals to the HSI produce rotation of the compass card and marker towards the lubber line in the manner described on page 218. The compass card rotates in the opposite direction to the turn because the MHRS is continuously being slaved to magnetic North. As the aircraft approaches the new heading, and since it must be rolled out to a wings-level attitude, the ADI command bars now receive a signal that deflects them to a 'fly left' command position.

Position 4 The aircraft has rolled out to wings level as shown by the ADI display, and is now flying on the selected heading as indicated by alignment of the heading marker and lubber line of the HSI.

Flying to a VOR station

Assuming that an aircraft is flying on a magnetic heading of 180°, and that it is then required to fly to a VOR station on a transmitted beam radial of 090°, the display sequence would be as shown in Fig. 9.17.

Position 1 The aircraft is being flown with the FDS operating in the 'HDG' and 'ALT HOLD' modes, and so the heading marker of the HSI is aligned with the lubber line and compass card to indicate the magnetic heading of 180°. The ADI indicates the level flight attitude.

Figure 9.17 Flying to a VOR station.

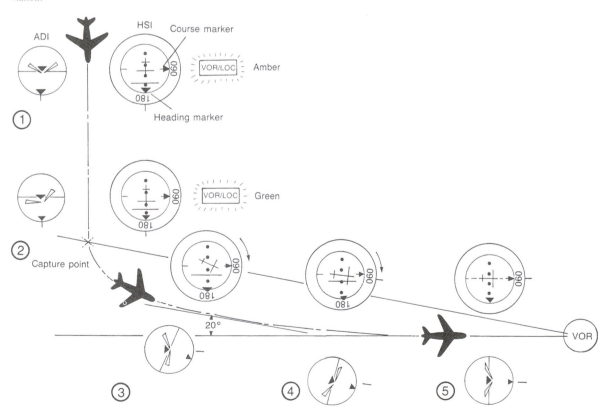

In order to fly onto the required VOR beam radial, the course selector knob is rotated to set the course marker and deviation bar at 090°. The VOR/LOC mode is then selected on the mode controller, and the corresponding light on the appropriate mode annunciator unit will be illuminated; e.g., on an approach progress display type of unit, the VOR/LOC light will illuminate amber. At the same time, the deviation bar will respond to the signals transmitted by the VOR station, and it will be deflected to the extreme right of the course marker to give advance indication of the fact that after turning towards the station, the aircraft must subsequently 'fly right' to attain the selected course.

Position 2 The beam is captured, the annunciator light illuminates green, and magnetic heading control is automatically switched out since control is taken over by the VOR. The course datum synchro in the HSI supplies a 'fly left' command to the ADI command bars. At beam capture, the course error and beam deviation signals are summed in the computer roll channel circuit so that as the aircraft turns, the selected course, i.e. the beam centre, is intercepted at a specific angle which, typically, is 20°. The to−from arrow in the HSI is activated to indicate flight to the station.

Position 3 The HSI indicates the turn to the left by relative movement between the lubber line and compass card, and the deviation bar deflection is progressively reduced to indicate that the aircraft is flying towards the beam centre. The ADI command bars are centred to indicate that the correct bank angle for the turn is established.

Position 4 The aircraft is approaching beam centre, and the ADI command bars are deflected to the right, commanding that the aircraft be rolled out in this direction to a wings-level attitude.

Position 5 The HSI and ADI indicate the final course situation.

If the aircraft encounters a cross-wind when 'on course', the effects are compensated by a 'wash-out' circuit in the computer, and the crab angle necessary to stay on course is established and indicated on the HSI.

LOC mode

Operation in this mode relates to the approach of an aircraft to an airport runway along the beam transmitted by the ILS localizer transmitter. The mode is selected on the appropriate mode controller;

Figure 9.18 Localizer approach commands.

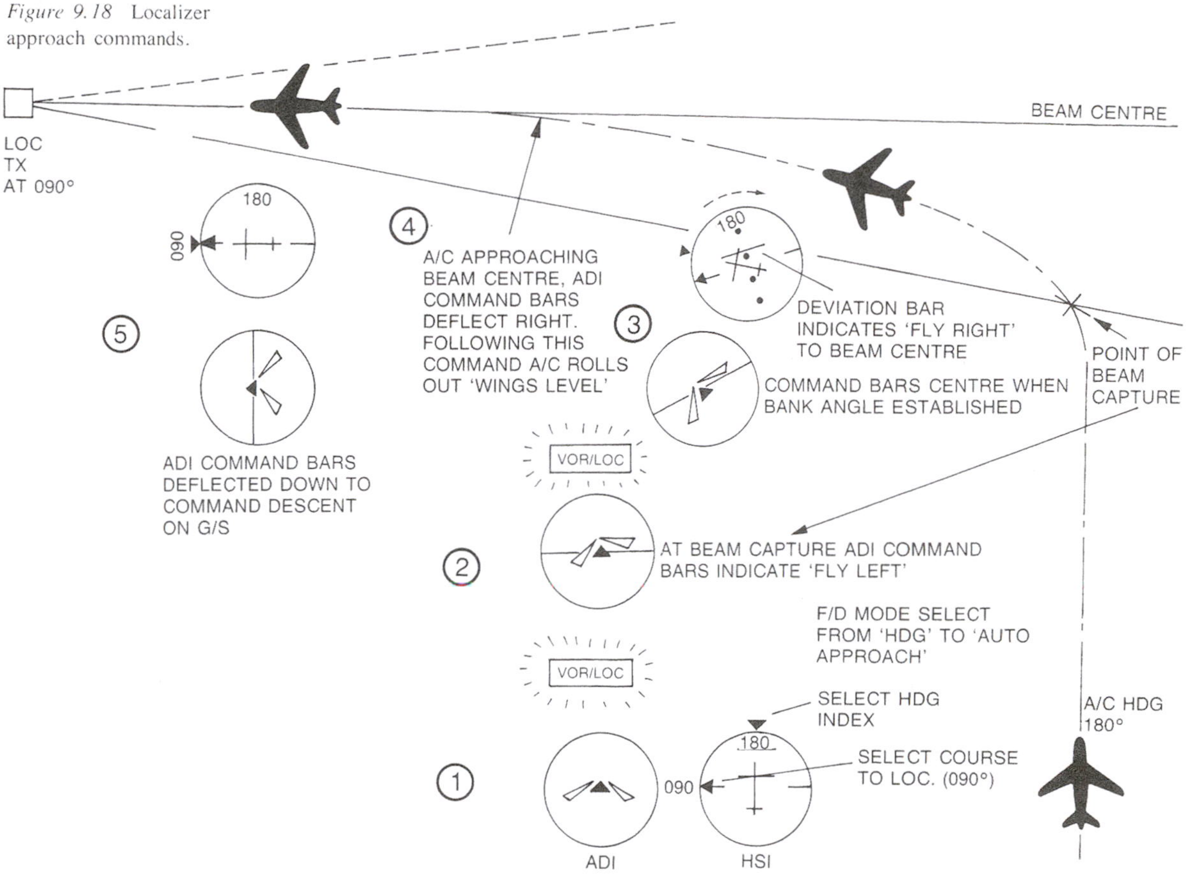

in the case of the unit shown in Fig. 9.12, the selector knob is set at the 'AUTO APPR' position. Although it is called a LOC mode, it also provides the control commands for capture of, and flight along, the GS beam since on selection it automatically 'tunes in' to the GS transmitter.

Figure 9.18 illustrates, also in pictorial form, typical localizer approach display sequences, and their similarity to those of the VOR mode is very noticeable. The principal differences are: (i) the changeover at position 1 from 'HDG' mode to 'AUTO APPR' mode; (ii) the localizer capture point (position 2) varies with rate-of-change of beam deviation, whereas capture of a VOR beam radial is at some fixed value, typically 5° deviation; and (iii) the 'ALT HOLD' mode is automatically switched out when the GS beam is captured, for the obvious reason that the aircraft must subsequently commence its descent for landing. At position 5, the ADI command bars are deflected to a 'fly down' command position in order to descend on the glide slope.

The profile of the GS stage of approach is shown in Fig. 9.19. Prior to beam capture, 'ALT HOLD' is on, and the GS annunciator

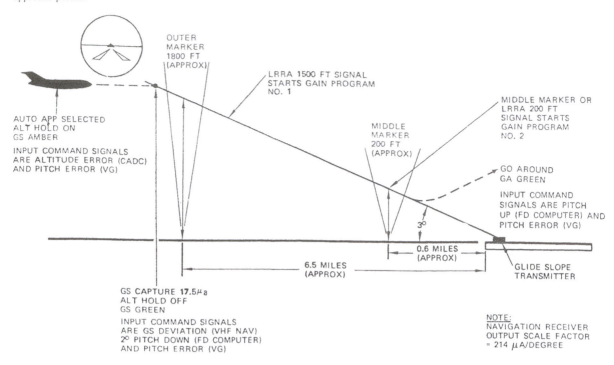

Figure 9.19 Glide slope approach profile.

OUTER
MARKER
1800 FT
(APPROX)

LRRA 1500 FT SIGNAL
STARTS GAIN PROGRAM
NO. 1

MIDDLE MARKER OR
LRRA 200 FT
SIGNAL STARTS
GAIN PROGRAM
NO. 2

AUTO APP SELECTED
ALT HOLD ON
GS AMBER

INPUT COMMAND SIGNALS
ARE ALTITUDE ERROR (CADC)
AND PITCH ERROR (VG)

MIDDLE
MARKER
200 FT
(APPROX)

GO AROUND
GA GREEN

INPUT COMMAND
SIGNALS ARE PITCH
UP (FD COMPUTER) AND
PITCH ERROR (VG)

3°

0.6 MILES
(APPROX)

6.5 MILES
(APPROX)

GLIDE SLOPE
TRANSMITTER

GS CAPTURE 17.5μa
ALT HOLD OFF
GS GREEN

INPUT COMMAND SIGNALS
ARE GS DEVIATION (VHF NAV)
2° PITCH DOWN (FD COMPUTER)
AND PITCH ERROR (VG)

NOTE:
NAVIGATION RECEIVER
OUTPUT SCALE FACTOR
= 214 μA/DEGREE

light illuminates amber. At beam capture, the light illuminates green, 'ALT HOLD' is automatically disconnected, and a pitch-down bias signal from the computer pitch channel is supplied to the ADI to deflect the command bars 2° down, thereby commanding that the aircraft's attitude be changed from that established at the time.

During the descent along the beam, any deviations from it are provided by the GS pointers in the HSI and ADI, and they are corrected by flying the aircraft in the directions commanded by the pointers.

As can be seen from Fig. 9.19, the GS beam converges towards the runway and this means that as an aircraft gets closer to landing, less pitch control is required to counteract deviations from the beam. To allow for convergence, therefore, the gain or response to GS deviation signals as a function of time is automatically reduced by gain-programming circuits in the pitch channel of the FDS computer. Programming is carried out in two stages, and is activated by signals from an aircraft's radio altimeter. The first stage is activated at about 1500 ft radio altitude, and activation of the second stage is at about 200 ft. Gain reductions against time (based on a typical system) are shown pictorially in Fig. 9.20.

The go-around manoeuvre indicated in the diagram is established for the reasons, and in the manner, already described (see page 227).

Figure 9.20 Gain programming.

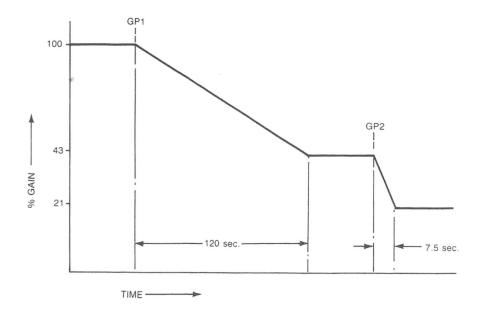

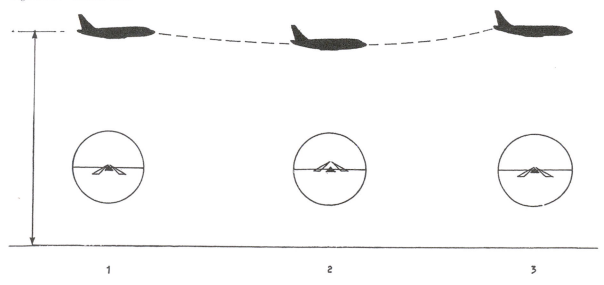

Figure 9.21 Altitude hold.

Altitude hold

In a typical FDS, this mode may be utilized when the system is selected for operation in either the 'HDG', 'VOR/LOC' or 'AUTO APPR' modes.

When an aircraft is flying at the altitude it is required to maintain, as at position 1 in Fig. 9.21, the altitude switch is placed in the 'ON' position and is held there by a solenoid. At the same time, a 'hold' signal is supplied to the ADC. In the case of an analog type of ADC, the signal energizes a clutch between the altitude module sensing

233

element and its synchro; the synchro, therefore, becomes locked at the reference altitude and there is no output from its stator. Since the stator is connected to the ADI via the pitch channel of the FDS computer, the command bars indicate a level flight attitude.

If the aircraft deviates from the reference altitude (as at position 2) the altitude sensing element repositions the synchro rotor, causing it to produce an error signal that deflects the command bars. In our example, the bars command a climb to return to the reference altitude, and they are centred again when this is attained as at position 3.

Dual flight director systems

In most aircraft, dual systems are installed, one for the captain and the other for the first officer. Each system consists of the principal units described in this chapter, and they are integrated in such a way that display indications will be available in the event of failure of the sources supplying the relevant command signals. This applies particularly to the outputs from attitude-sensing units and radio navigation receivers. In addition, electrical power is supplied from separate busbars. As a typical example we may consider the dual system arrangement shown in Fig. 9.22, which is based on that adopted in some series of Boeing 737.

As far as pitch and roll attitude references are concerned, the captain's ADI is, under normal operating conditions, supplied with signals from VGU No. 1, while the first officer's ADI is supplied from VGU No. 2. The signals are transmitted through transfer relays, which are normally de-energized. In the event of failure of, say, VGU No. 1, the No. 1 transfer relay is energized by operating a transfer selector switch so that through a second set of contacts in the relay, the captain's ADI is supplied with attitude signals from VGU No. 2. Similarly, if VGU No. 2 should fail, switching of its appropriate transfer relay would permit the first officer's ADI to derive attitude signals from VGU No. 1.

In some dual system installations an auxiliary VGU is also provided and may be selected to provide attitude reference signals to either or both ADIs in the event of failure of the main VGUs. The interconnection of the relevant components is shown in Fig. 9.23. The 'INOP' annunciator on the transfer switch panel monitors the auxiliary VGU at all times, and provides the appropriate indication only in the event of failure of this VGU.

Each FDS is supplied with the required radio navigation receivers from separate VHF navigation receivers. The signals also pass through transfer relays to ensure that FD systems are supplied in the event of failure of one or other of the receivers.

Magnetic heading signals are supplied to the computer of each FDS

Figure 9.22 Dual FD system.

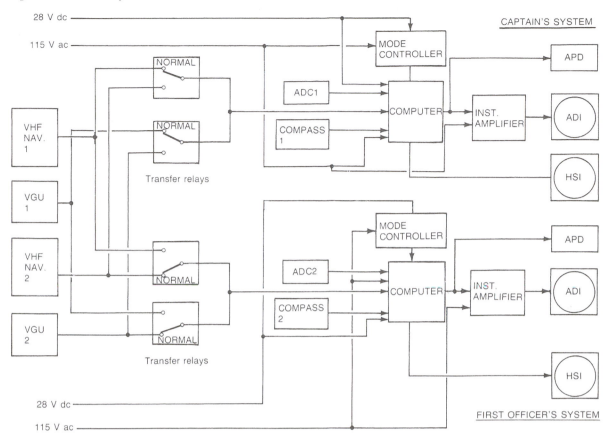

from separate MHR systems (usually designated 'COMPASS 1' and 'COMPASS 2'), and so in the event of failure of either one of them, a transfer switching and relay system is also provided to ensure that both HSIs receive steering commands from the MHRS remaining in operation.

The selection switches for the transfer of VHF navigation receivers and MHR systems are shown in Fig. 9.24.

Instrument comparator and warning system

This system is used in some dual FDS arrangements, e.g. the one just described, its purpose being to compare pitch and roll attitude signals, and also certain signal inputs from interfacing radio navigation systems, and to illuminate corresponding warning lights whenever a significant difference between signals exist. The system consists of a comparator unit and annunciator panels interconnected as shown in Fig. 9.25.

The front panel of the comparator contains six amber lights which are connected in parallel with amber lights on the annunciator panels;

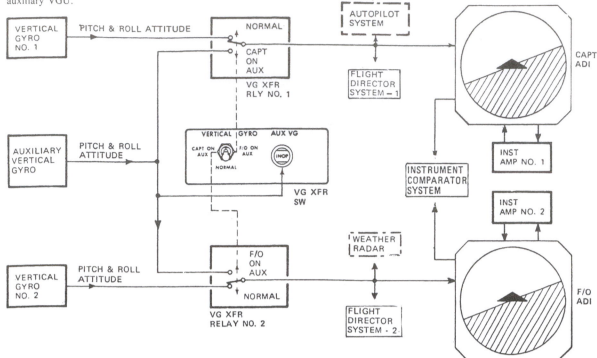

Figure 9.23 FDS with
auxiliary VGU.

these lights are placarded 'HDG', 'PITCH', 'ROLL', 'GS', 'LOC'
and 'ALT'. Failure of a comparator power supply is indicated by a
red 'PWR/MON' light which is also connected in parallel with a
corresponding light on the annunciator panels. A momentary 'PUSH
TO TEST' switch is provided for checking the serviceability of all
the lights (except the 'PWR/MON' light). The switch is paralleled
with a test switch on a flight deck panel to permit similar checks to
be carried out by the crew.

The annunciator panels are connected in parallel and they each
contain seven amber lights placarded as indicated in Fig. 9.25. If a
light illuminates, it can be dimmed by pressing its cover.

Operation

The operation of the annunciator panel lights is as follows:

1. **HDG**   The comparator supplies 26 V ac excitation to differential
resolver synchros in the captain's HSI and ADI. The
captain's HSI transmits heading signals to the first
officer's HSI. If the positions of the compass cards in both
indicators do not agree, the first officer's HSI transmits an
error signal to the comparator. This, in turn, produces a
signal to illuminate the 'HDG' lights on both annunciator
panels.

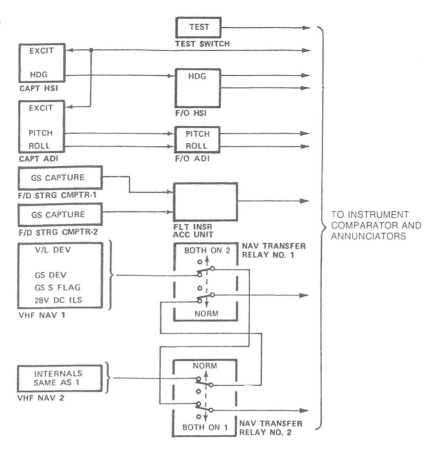

Figure 9.24 VHF nav. transfer switching.

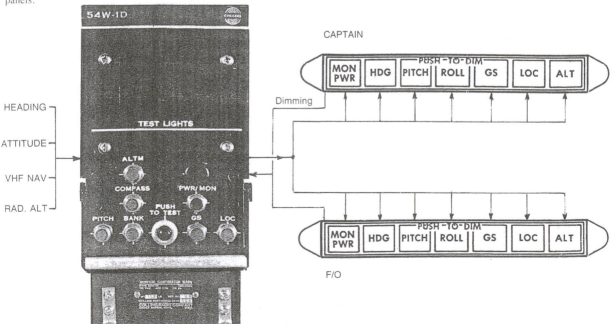

Figure 9.25 Warning comparator and annunciator panels.

2. **ROLL** The captain's ADI transmits roll attitude signals to the first officer's ADI. If the attitude indications do not agree, the latter ADI transmits an error signal to the comparator and so produces a signal to illuminate the 'ROLL' lights on both annunciator panels. Since a roll attitude change initiates a change in heading, a separate roll attitude signal input is supplied to the comparator to modify the threshold at which the comparison between heading indications operates (the greater the roll angle the larger the heading threshold). This helps to reduce 'nuisance' warnings due, for example, to precession of an MHRS directional gyroscope unit, and system tracking during manoeuvring of an aircraft.

3. **PITCH** This light is illuminated in a simlar manner to that for 'ROLL'.

4. **LOC** Illumination of this light occurs whenever ILS localizer or VOR deviation signals from navigation receivers do not agree. A 28 V dc (ILS) signal is also supplied to the comparator by the receivers, in order to activate the comparison between deviation signals.

5. **GS** This light is illuminated whenever ILS glide slope deviation signals from the navigation receivers are not in agreement. In addition to the 28 V (ILS) signal noted above, a 'GS flag' signal is supplied from both receivers to activate the comparison between GS deviation signals.

6. **ALT** This light is illuminated whenever the altitude signals from dual radio altimeter transmitter/receivers do not agree.

Typical threshold levels required to illuminate the annunciator lights are given in Table 9.1.

If the excitation voltage from the comparator unit is lost, the 'HDG', 'ROLL' and 'PITCH' annunciator lights may also illuminate simultaneously. In such a case the comparison between the signals from the reference sources is, of course, unreliable.

FDS interfacing with inertial navigation systems (INS)

As in the case of an MHRS (see page 198), an FDS can also be interfaced with an INS so that its ADI can then utilize the attitude references established by the gyroscopically-stabilized inertial platform, instead of it being dependent on a VGU. The interconnection arrangement is, as far as attitude error signal transmission is concerned, basically similar to that described earlier, and as may be seen from Fig. 9.26.

An HSI and its interconnections, however, differ somewhat from

Table 9.1

Light	Conditions	Threshold level
HDG	Before GS capture at 0° roll	6°
	Before GS capture at 20° roll	9°
	After GS capture at 0° roll	4°
PITCH & ROLL	Before GS capture	4°
	After GS capture	3°
LOC	At 0 dots deviation	30m V
	At 2 2/3 dots deviation	54m V
GS	At 0 dots deviation	40m V
	At 2 2/3 dots deviation	64m V
ALT	The altitude threshold levels increase from 5 ft at touchdown in accordance with the graph below.	

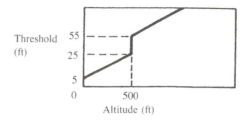

Figure 9.26 INS attitude references.

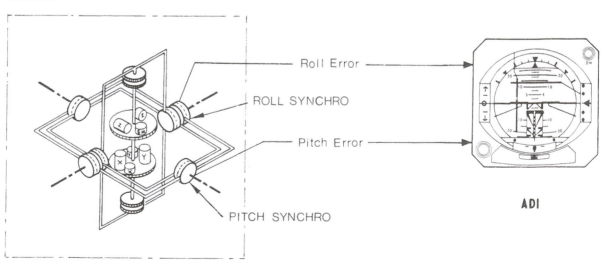

INS PLATFORM

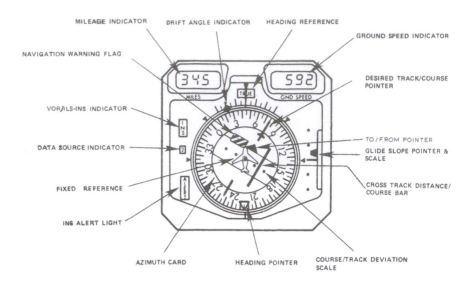

Figure 9.27 HSI interfaced with INS.

MILEAGE INDICATOR DRIFT ANGLE INDICATOR HEADING REFERENCE

GROUND SPEED INDICATOR

NAVIGATION WARNING FLAG

DESIRED TRACK/COURSE POINTER

VOR/ILS-INS INDICATOR

DATA SOURCE INDICATOR

TO / FROM POINTER

GLIDE SLOPE POINTER & SCALE

FIXED   REFERENCE

CROSS TRACK DISTANCE/ COURSE BAR

INS ALERT LIGHT

AZIMUTH CARD   HEADING POINTER   COURSE/TRACK DEVIATION SCALE

the normal in that in addition to the navigational data supplied by an MHRS and VHF navigation receivers, they can also display the data computed by the navigation unit of an INS, as shown in Fig. 9.27. The display elements, therefore, serve a dual role which, in the case of the system utilizing this indicator, can be selected under mode designations 'RADIO' and 'INS'. Details of the data displayed appropriate to each mode are given in Table 9.2.

The HSI incorporates three warning flags: (i) navigation warning flag which comes into view when there is a loss of input from an operating radio navigation receiver or INS; (ii) a 'HEADING' flag which obscures the heading reference whenever there is an invalid heading reference (magnetic or true); and (iii) a 'GS' flag which obscures the glide slope scale whenever the relevant signal is invalid.

Data transfer switching

The ADI and HSI are part of two independent FD systems, i.e. one for the captain and the other for the first officer, and so they are also interconnected to ensure that in the event of failure of a data input source, a display of data will still be available. The data input transfer switching arrangements for the ADI and HSI are shown in Figs 9.28 to 9.31; these are based on the application to some series of Boeing 747 which has triple AFC and IN systems. The switches related to each transfer function are mounted on panels located at each side of the main instrument panel.

Attitude data (Fig. 9.28)
Under *normal* operating conditions, the captain's transfer switch is set at position 1 so that his ADI is supplied with attitude data from INS

Table 9.2

Display element	Data displayed	
	'RADIO' mode	*'INS' mode*
Azimuth card	Heading referenced to magnetic North	Heading referenced to true North
Heading pointer	Heading selected on mode controller	Biased to six o'clock position and does not function
INS 'Alert' light	—	Illuminates when aircraft is within 2 minutes of a waypoint while navigating along an INS track. Does not function at ground speed less than 250 knots
Data source indicator	Shows source of data, e.g. System 1 or 2 Radio Nav. or INS	
VOR/ILS indicator	'VOR/ILS'	'INS'
Mileage indicator	Distance to DME station	Distance to next waypoint
Drift angle indicator	Drift Angle	Drift Angle
Heading reference	'MAG'	'TRUE'
Ground speed indicator	—	Ground speed in knots as computed by INS
Desired track/course	Course selected on mode controller	Desired track
Cross track distance/ course bar	Displacement from VOR or ILS course	Displacement from track
Course/track deviation scale	Deviation in degrees from VOR or localizer beam	Cross track distance in miles (2 dots = $7\frac{1}{2}$ miles)

No. 1. The first officer's transfer switch is set at its position 2 to supply his ADI with data from INS No. 2.

In the event of *failure* of data from INS No. 1, the data established by INS No. 3 can be transferred to the captain's ADI by setting the switch to position 3 and energizing its associated relay. Similarly, the first officer can also transfer data from INS No. 3 to his ADI by resetting his transfer switch from the No. 2 to the No. 3 position.

Radio navigation data (Fig. 9.29)
For the transmission of this data, the main selector switches placarded 'RADIO/INS' must each be selected to the 'RADIO' position. In *normal* operation of both FD systems, the captain's radio transfer switch is set at position 1 so that his ADI and HSI are supplied with the relevant deviation signals from the VHF navigation receiver No. 1. The first officer's transfer switch is set at its position

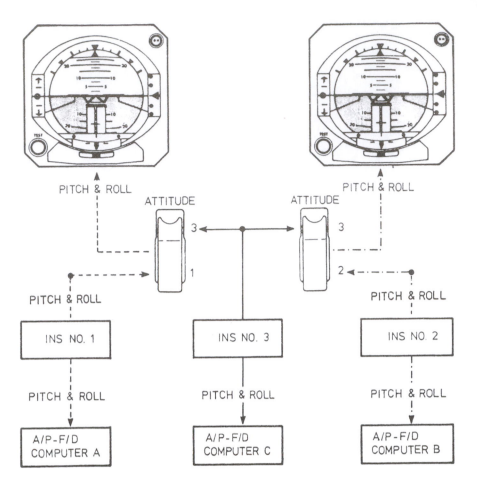

Figure 9.28 Attitude data switching.

PITCH & ROLL

ATTITUDE

3

PITCH & ROLL

ATTITUDE

3

1

2

PITCH & ROLL

PITCH & ROLL

INS NO. 1

INS NO. 3

INS NO. 2

PITCH & ROLL

PITCH & ROLL

PITCH & ROLL

A/P-F/D COMPUTER A

A/P-F/D COMPUTER C

A/P-F/D COMPUTER B

2 and so his ADI and HSI are supplied with the signals from the No. 2 navigation receiver.

In the event of *failure* of receiver No. 1, the captain's transfer switch is set to the No. 2 position so that the No. 2 receiver can then supply his ADI and HSI with deviation signals in addition to the first officer's indicators. Similarly, the first officer's indicators can be supplied from receiver No. 1 by the selection of his transfer switch to the No. 1 position.

The transfer switches are electrically interlocked such that once a transfer of data has been selected from one side, a transfer from the other side cannot be effected.

Heading data (Fig. 9.30)

Heading data may be displayed on each pilot's HSI as either magnetic or true depending on the setting of the 'RADIO/INS' selector switches. Both RMIs always indicate magnetic heading since they are part of each MHRS (see Chapter 8).

The transfer of data between MHR and IN systems is accomplished by means of two switches (placarded 'COMPASS' and 'INS') in the

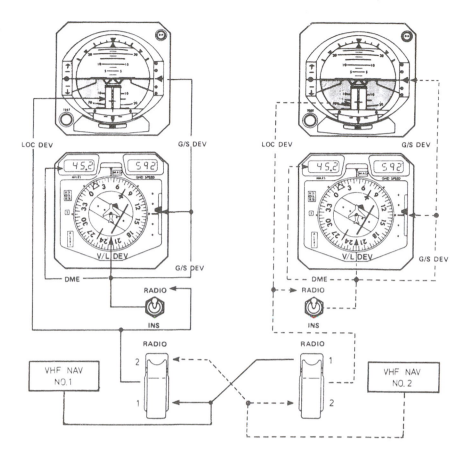

Figure 9.29 Radio navigation
data switching.

input circuits to each HSI. If, in *normal* operation, it is required that
magnetic heading be displayed on both HSIs, then both 'RADIO/INS'
selector switches are set at the 'RADIO' position. The captain's
'COMPASS' switch is set at its No. 1 position so that his HSI and
RMI are supplied with magnetic heading data from MHRS No. 1,
and the first officer's 'COMPASS' switch is set at its No. 2 position,
enabling his HSI and RMI to be similarly supplied from MHRS
No. 2.

In the case of *failure* of either MHRS No. 1 or No. 2, then the
captain's or first officer's transfer switches respectively would be set
at their No. 2 and No. 1 positions.

If, under *normal* operating conditions, it is required that one HSI
should display true heading and the other magnetic heading, then the
respective 'RADIO/INS' switches are selected at 'INS' and 'RADIO'
The diagram illustrates the case in which true heading is to be
displayed on the captain's HSI, and magnetic heading on that of the
first officer. True heading data is supplied from INS No. 1 via the
captain's 'INS' transfer switch when set at its No. 1 position. The
heading pointer of the HSI is biased to the 6 o'clock position, the

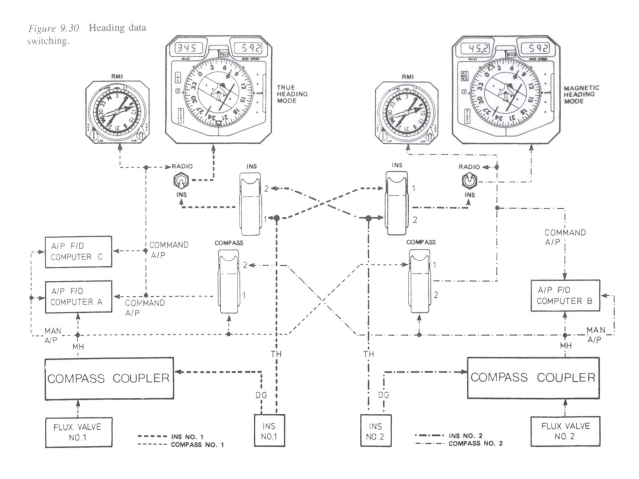

Figure 9.30 Heading data switching.

'VOR/ILS—INS' indicator displays 'INS' and the heading reference indicates 'TRUE'.

Magnetic heading data is supplied to the first officer's HSI from MHRS No. 2 via the No. 2 position of the compass transfer switch. The HSI heading reference in this case indicates 'MAG'. If it is required for this indicator to display true heading, in addition to that of the captain, the associated 'RADIO/INS' selector switch would also be placed in the 'INS' position.

In the case of *failure* of data from either INS No. 1 or No. 2, the captain's or first officer's transfer switches respectively would be set at their No. 2 and No. 1 positions.

As in the case of radio navigation data transfer, the 'COMPASS' and 'INS' transfer switches are also electrically interlocked.

INS data (Fig. 9.31)
In addition to true heading, the other data shown in the diagram is also supplied to both HSIs when the 'RADIO/INS' selector switches are set to 'INS' (see also Table 9.2). The transfer of this data is effected in the same manner as that described above.

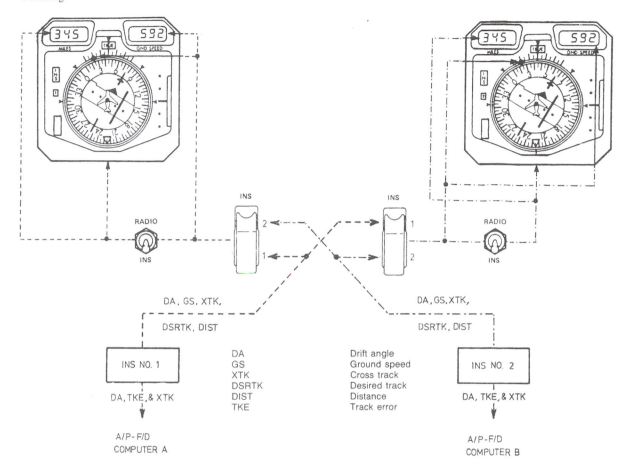

Figure 9.31 INS data switching.

DA	Drift angle
GS	Ground speed
XTK	Cross track
DSRTK	Desired track
DIST	Distance
TKE	Track error

DA, GS, XTK,

DSRTK, DIST

INS NO. 1

DA, TKE, & XTK

A/P-F/D
COMPUTER A

DA, GS, XTK,

DSRTK, DIST

INS NO. 2

DA, TKE, & XTK

A/P-F/D
COMPUTER B

10 Inertial navigation/ reference systems (INS/IRS)

The navigation requirement of an aircraft is quite simply that of determining its position in relation to its point of departure and points en route in order to reach a known destination. In practice, however, the fulfilment of this requirement is made somewhat complicated by having to provide for a lot of basic data, principally the following: time, speed, distance between points, longitude, latitude, magnetic heading, wind speed and direction, and bearings relative to known points on the earth's surface.

The provision of such data can be, and is in fact, made by a variety of navigational aids, a number of which are dependent on an external reference source of one form or another. Although these aids provide reasonably accurate answers to navigational task problems, there are certain limitations: for example, radio navigation aids that provide direction-finding and position-fixing capabilities require extensive networks of ground stations and are subject to both natural and man-made interference. An INS or IRS overcomes such limitations by utilizing operating principles that make them entirely independent of external references.

Systems and units

As the title of this chapter infers, there are two classifications of inertial systems which, although performing the same basic navigational functions, vary extensively in the manner in which they process data, and in their capability of satisfying the needs for such data by other interfaced systems.

The *INS*, which is the forerunner of systems and is still currently used in some types of aircraft, utilizes analog and digital signal-processing techniques, mechanical arrangements such as gimballed platforms, and synchronous servo transmission loops. A system consists of the four principal units shown in Fig. 10.1, together with their interconnection, data outputs and the other aircraft systems with which it is generally interfaced.

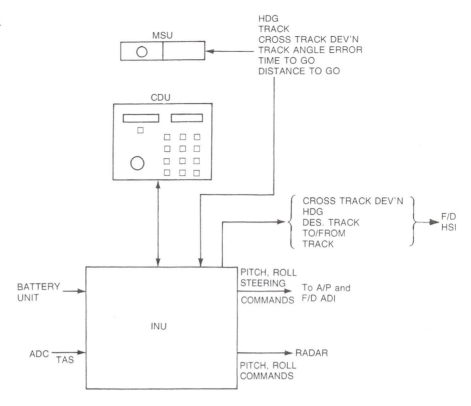

Figure 10.1 Units of an INS.

1. *Inertial navigation unit* (INU): This unit contains an inertial section consisting of accelerometers, gyroscopes and gimballed platforms, a digital computer and all associated circuit module cards, and a battery charger unit.

2. *Control and display unit* (CDU): This allows all associated data to be inserted into the computer, and to be read out from it by means of segmented LED displays.

3. *Mode selector unit* (MSU): This unit controls all the modes in which the system can be operated.

4. *Battery unit*: This unit provides dc power for turning the system on, and is also used as back-up in the event that power from an aircraft's sytem is interrupted.

Although this type of system is highly accurate, the levels of accuracy demanded for the navigation of those aircraft that are designed for operation under what may be termed the 'computer chips with everything' philosophy preclude its application to such aircraft in favour of its more sophisticated descendant, namely, the *inertial reference system* (IRS). It performs the same basic navigational functions as an INS, but, as its fully digital computer can also be pre-programmed with other relevant reference data, there was some justification in changing its name.

The system consists of only two principal units. The outputs are supplied to a greater number of interfacing systems, and since the majority of them are also individually controlled by digital computers, signal transmissions are via an ARINC 429 data bus (see Chapter 6) as opposed to conventional 'hard wiring'.

The inertial reference unit (IRU) also contains accelerometers, gyroscopes and the computer, but here, its similarity with the INU referred to earlier ends. The major differences are: (i) the gyroscopes are of the ring laser type (see page 270) instead of the spinning rotor type; (ii) the complex mechanical arrangement of a gimbal system and synchronous transmission loops is replaced by a mathematical equation program so that acceleration and attitude signals required for navigation are directly computed; (iii) the unit is directly mounted to an airframe, i.e. it is of the 'strapdown' type so that the aircraft itself becomes the inertial platform (see page 279); (iv) magnetic and true headings are derived from a program of known data related to the position data loaded into the computer, so that headings can be computed without the aid of MHRS flux detector units (in fact, these units are no longer required in aircraft equipped with an IRS); and (v) no battery unit and charger is used.

The inertial mode reference panel (IMRP) combines the functions of mode selection and control and display of data.

Multi-installations

In order to ensure on-going navigation capability, and operation of interfacing systems, adequate 'back-up' must be provided to safeguard against system failures. Dual or triple IN/IR systems, depending on size and operating category of an aircraft, are therefore installed with the appropriate system transfer switching arrangements.

Power supplies

Both ac and dc power is required for system operation which must be maintained in the event of failures occurring. The sources from which power is derived can vary depending on the type of system, but a common feature is that after starting up, the system can be maintained in operation *from either of the sources*. This is effected by the integration of power supply monitor and conversion circuits in the navigation unit of a system.

In a typical gimballed-platform INS, the ac power is supplied from an essential busbar, and the dc power from a nickel-cadmium battery unit which is part of the system installation. The unit provides auxiliary power for the initial start-up, and also the power to maintain system operation in the event of ac power failure, or a reduction in voltage level. Under these conditions, the battery unit will sustain system operation in any operating mode for periods up to 15 minutes' duration. Indication that battery power is in use is provided by illumination of an amber 'BATT' light on the control and display unit.

The battery unit has a direct connection to the system's mode selector switch so that when this is set to the positions for initial starting of the system, battery power is used momentarily for energizing a relay, the contacts of which are connected in the circuit from the aircraft's ac busbar. Thus, ac power is supplied to the navigation unit via the relay which is then held in the energized state by the dc produced by the power conversion unit. The battery supply remains on for a short period (typically 10 seconds), enabling it to be checked during alignment of the system (see page 279). On completion of this check, the battery is isolated from the system and is on standby until there is an interruption of the ac power supply. In the event that an external power source is disconnected from an aircraft while the INS in on, battery power will automatically be transferred to the system, and some warning of this is required in order to protect the battery against an inadvertent discharge. In one example of warning system, a horn is located in the nose wheel bay of an aircraft, and is activated 30 ± 10 seconds after power transfer, thereby alerting the ground crew.

The inertial navigation unit is provided with a battery charger circuit which automatically comes into operation when the battery is not in use, and whenever its voltage drops below 26.5 V. The charger is disconnected when the on-charge voltage increases to 29.5 V.

In multi-system installations, and after interruption of an aircraft's power supply to the systems, switching arrangements are provided which enable battery units to be paralleled in order to sustain the operation of one of the navigation units. For example, in a triple installation the battery units of Nos 1 and 3 systems can be paralleled to supply the navigation unit of the No. 1 system.

In aircraft equipped with IR systems, the use of battery units is eliminated since dc power from the busbar of the aircraft's battery system is utilized for the starting up of a system. This supply is also automatically switched on in the event of a loss of ac power.

Navigation fundamentals

Before going into details of the principles involved in the operation of these systems, it is useful at this juncture to consider some aspects relating to the form of the earth, and also to define some of the terms associated with navigation over its surface.

Form of the earth

The earth is not a true sphere; its equatorial diameter of 6884 nautical miles exceeds its polar diameter by about 23 nautical miles.

The 'flattening' at the polar regions gives rise to a more precise definition of the earth's form, which is known as an *oblate spheroid*. For practical navigation purposes, however, the earth can be considered as a sphere.

Direction on the earth

This is measured in degrees clockwise from north, and when the datum is the direction of the north end of the earth's axis, it is referred to as *true* direction. North is one of four points known as *cardinal points*; the other three are south, east and west.

North and south define the axis about which the earth rotates from west to east. To avoid ambiguity, a three-figure group is always used to indicate direction, e.g. north — 000°; south — 180°; east — 090° and west — 270°.

Great circle

This is a circle on the surface of a sphere whose centre and radius are those of the sphere itself. Relating this to the earth, the equator and all the lines joining the north and south cardinal points (the earth's poles) are examples of great circles.

On a plane surface, the shortest distance between two points is, of course, a straight line which joins them. On a sphere, the shortest distance between two points is the smaller arc of the great circle which passes through both points.

Small circle

This is a circle on the surface of a sphere whose centre and radius are not those of the sphere. With the exception of the equator, all lines of latitude are small circles; they do not represent the shortest distance between two points.

Longitude and latitude

These form a reference system for the position of points on the earth's surface, and in determining the in-flight position of an aircraft with respect to the earth.

Firstly, the datum is established by a great circle through the north and south poles which passes through Greenwich. That half of the circle which passes through Greenwich is known as the *prime or Greenwich meridian* and is 000°. The other half is called the *anti-meridian* and is 180°. Other great circles in the form of meridians, or *lines of longitude* as they are called, are established to the east and to the west of the prime meridian.

The next step is to have a datum point for positions in the direction north and south. This is obtained by dividing the earth by a great

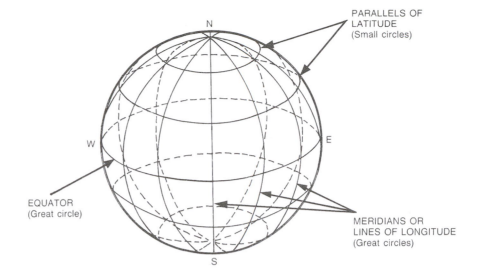

N

PARALLELS OF
LATITUDE
(Small circles)

W

E

EQUATOR
(Great circle)

MERIDIANS OR
LINES OF LONGITUDE
(Great circles)

S

circle midway between the poles. This circle is the *equator* and is *0°
latitude*.

From the foregoing, we can derive more precise definitions of
longitude and latitude as follows:

Longitude The longitude of any point is the shortest distance in
the arc along the equator between the prime meridian and the
meridian through the point. It is expressed in degrees and minutes
and is annotated east or west according to whether the point lies east
or west of the prime meridian.

Latitude The latitude of any point is the arc of the meridian
between the equator and the point. It is also expressed in degrees and
minutes, and is annotated north or south according to whether the
point lies north or south of the equator.

The whole network of meridians (longitude and parallels of
latitude), imagined to cover the earth, is called a *graticule*. Thus, as
shown in Fig. 10.2, meridians or lines of longitude start from the
prime meridian or 0° and go right round up to 180° E and 180° W.
Similarly, the parallels of latitude start from the equator as 0° and go
up to 90° N and 90° S.

When giving a position, it is always quoted in the sequence *latitude
and longitude*; e.g. the latitude of London Heathrow is the arc of the
meridian between the equator and Heathrow, and is *51 degrees and
28 minutes N*. Its longitude is the shorter arc of the equator between
the prime meridian and Heathrow, and is *00 degrees and 27 minutes
W*. It is expressed as: 51° 28′ N 00° 27′ W.

Convergency

Because the meridians converge towards each other to the poles, then
any line or track cutting successive meridians will do so at different

Figure 10.3 Convergency.

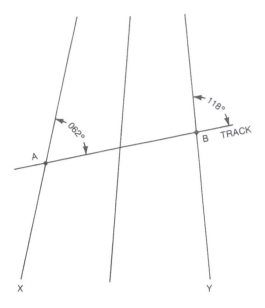

angles; *this inclination is called convergency* and it equals the angular difference between the measurements of the line or track at each meridian. If, as shown in the example of Fig. 10.3, a track passes through a point 'A' on meridian 'X' at an angle of 062°, then in passing through point 'B' on meridian 'Y' it will cut this at an angle of 118°. The convergency is, therefore, equal to 118 − 062 or 056°.

If two places are on the same latitude, convergency may be obtained from the formula:

Convergency = Change in longitude × sine of latitude.

Convergency is 0 at the equator (the meridians cutting it at 90°) and increases to maximum at the poles.

Change of longitude

This is the smaller arc of the equator intercepted between the meridians of the reference points, and is named east or west according to the direction of the change. It is abbreviated as 'ch long' (E or W).

Change of latitude

This is the arc of the meridians intercepted between the parallels of the two places and is named north or south according to the direction of the change. It is abbreviated as 'ch lat' (N or S).

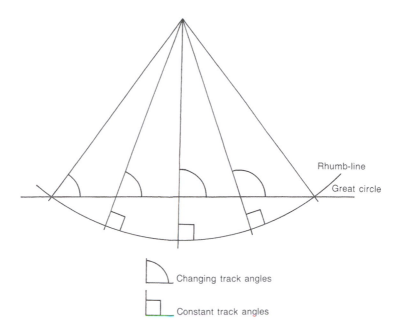

Figure 10.4 Rhumb-line.

Rhumb-line
Great circle

⌐ Changing track angles

⌐ Constant track angles

Rhumb-line

The ideal line to fly would be a great circle, since the shortest distance between any two places is along the circle. There are, however, two disadvantages: (i) the great circle from one point to another will cross the converging meridians at different angles, and (ii) because meridians form the basis of track angle measurements, continuous alterations to these angles would be necessary as a flight progressed.

In order to overcome these disadvantages, a curved line is therefore followed which joins points along it and crosses each meridian at a constant angle; such a line is called a *rhumb-line* (see Fig. 10.4). Distances between points along this line are greater than those along a great circle.

The meridians and the equator are the only examples of great circles which are also rhumb-lines. Parallels of latitude are rhumb-lines because they cut all meridians at 90°.

Distances on the earth

Nautical mile The distance on the earth's surface which subtends an angle of one minute of arc at the centre of the earth.

One nautical mile (nm) equals one minute of latitude and is an average distance of 6080 ft. 1° latitude = 60 nm. A change of latitude from the equator to a pole is therefore equal to 90 × 60 = 5400 nm.

Statute mile Equal to 5280 ft.

Kilometre 1/10 000th of the average distance from the equator to either pole and is accepted as being equal to 3280 ft.

Navigation terms

The following definitions are of navigation terms associated with INS/IRS operation; they are also shown pictorially in Fig. 10.5.

Heading (HDG) The direction in which the nose of an aircraft is pointing; it is measured in degrees (000−360) clockwise from true, magnetic, or compass north, designated as Hdg (T), Hdg (M), and Hdg (C). Hdg (T) is the only one of the three which is plotted.

Track (TK) The direction in which an aircraft is moving over the earth; it is also

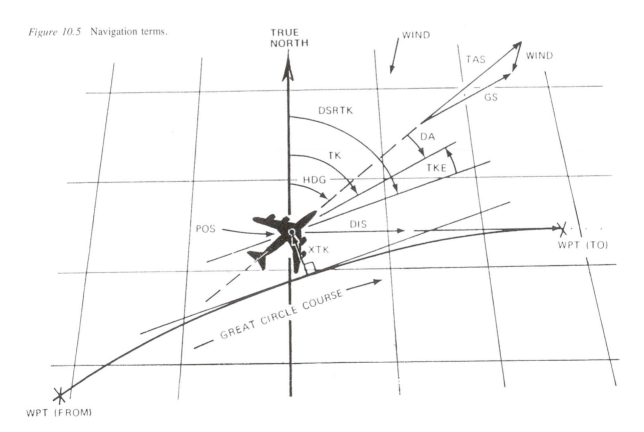

Figure 10.5 Navigation terms.

measured in degrees from true or magnetic north. Only true TK is plotted. If there were no wind, there would be no drift and TK would be the same as HDG; also the case with a direct head- or tail-wind.

Desired track (DSR TK)   The planned direction over the earth in which it is intended the aircraft shall move.

Drift (DA)   The angle between HDG and TK due to the effect of wind. The direction of drift is always from HDG to TK. Each may be true or magnetic but never mixed. If TK is less than HDG, drift is to the left, and if TK is greater than HDG, it is to the right as shown in Fig. 10.6.

Ground speed (G/S)   The actual speed (in knots) of an aircraft over the ground, i.e. speed relative to the earth. If there were no wind, GS would be equal to true air speed (TAS).

Wind direction (W/D)   The angle, measured in degrees clockwise from true north, with respect to the direction from which the wind is blowing.

Wind speed (W/S)   The speed, in knots, at which the air is moving relative to the ground.

Position (POS)   Air: The position of an aircraft

Figure 10.6 Drift angle.
(*a*) TK less than HDG;
(*b*) TK greater than HDG.

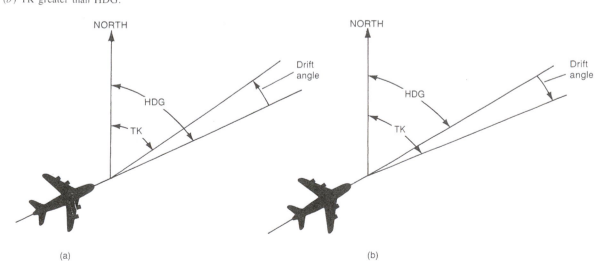

(a)

(b)

255

	relative to the air at a particular time.
	Ground: The position of an aircraft relative to the ground directly beneath it at a particular time.
Track angle error (TKE)	The angle (left or right) between the DSR TK and the actual TK of an aircraft. It is always measured from DSR TK to TK.
Cross track distance (XTK)	Is the distance in nm (left or right) measured from the nearest point on the DSR TK line to the aircraft.
Waypoint (WPT)	This is a point of navigational significance on an air route. Typically, routes are divided up into convenient lengths or legs, defined at each end by a WPT. The end WPT of one leg is the beginning of the next leg as illustrated in Fig. 10.5. The 'FROM' WPT is the one defining the beginning of the flight plan leg currently being flown. The 'TO' WPT is the one defining the end of the current leg of the flight plan. The 'NEXT' WPT in this convention is the one defining the end of the next leg to be flown, i.e. after passing the 'TO' WPT.

Fundamental principle of a system

The operating principle is derived directly from the Newtonian laws of mechanics relating to velocity, acceleration and inertia. The relationships may be summarized as follows:

1. *Velocity* is the rate of change of displacement with respect to time for a moving object, and so is composed of both speed and direction. If the speed of an object is constant, but its direction is changing, then its velocity changes.

2. A change in velocity, either in magnitude or direction of motion, is an *acceleration* (or deceleration). A body accelerates (or decelerates), i.e. changes its state of motion, only if it is acted upon by an external force.

3. All matter tends to return to its existing state of motion and consequently resists any changes to that state; this property is known as *inertia*. The rate of acceleration of a body is proportional to the magnitude of its inertia. The inertial force displayed by a body under a given rate of acceleration gives a measure of the mass of that body.

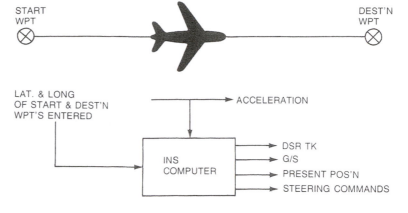

Figure 10.7 Computer input and output data.

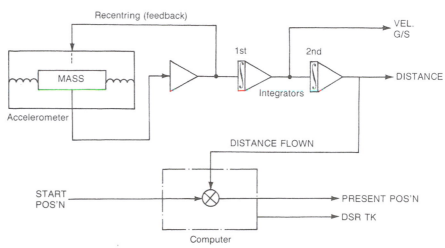

Figure 10.8 Accelerometer operation.

In order for an IN/IR system to navigate an aircraft, its computer must first have knowledge of the latitude and longitude of the starting point of its flight, its final destination and, where necesary, a number of intermediate waypoints. This data is inserted into the computer at the time of initially starting and aligning a system prior to commencement of a flight. The outputs computed are shown in Fig. 10.7.

Since acceleration forms the whole basis of computing the in-flight navigation parameters from the pre-set data, then appropriate sensor units are required. Let us consider the operation of one such unit as shown schematically in Fig. 10.8. In this example, the mass of the unit is normally centred by two springs between two pick-off transformers.

When the aircraft accelerates, the mass is displaced, and a phase-related signal is induced in the pick-off transformers. This signal is amplified and is applied as feedback to the accelerometer via what is termed a force generator, which consists of a coil which moves with the mass between the field of two permanent magnets. The effect of this device is to recentre the mass, and the current needed to achieve

257

this is, therefore, a measure of the acceleration. The use of this principle gives rise to the term 'force rebalancing accelerometer'.

The 'acceleration signal' is supplied to an integrator circuit which relates acceleration to time, and therefore produces a signal corresponding to ground speed (G/S). The G/S signal, in turn, is supplied to a second integrator circuit which then produces a signal corresponding to distance flown. This double integration process solves the basic equation relating distance s travelled in a time t by a body moving at a velocity v as given by:

$$s = \int_0^t v \, dt \qquad\qquad [1]$$

Since an input voltage (V_i) is made proportional to the velocity of motion, and an output voltage (V_o) is proportional to the distance travelled, then in terms of electrical integration, equation [1] becomes:

$$V_o = -\frac{1}{CR_1} \int_0^t V_i \, dt$$

The product CR_1 is referred to as the time constant of the circuit.

If the 'distance flown' signal is then applied to a point within the computer, and summed with one corresponding to the aircraft's starting point, the result is a signal which gives the aircraft's present position. If the latter together with the desired track (DSR TK) are known, the computer can develop steering commands to keep the aircraft on the DSR TK so that it can reach any desired destination.

Dual-axis accelerometers

Because an aircraft can fly in any direction, two acceleration sensors ('X' and 'Y') are required, and are mounted on a platform in horizontal planes to sense accelerations 90° apart, as shown in Fig. 10.9. The designations 'X' and 'Y' relate respectively to accelerations in the horizontal plane and to the east, and to similar accelerations in the direction of the local or north meridian. The outputs are vectorially added to determine a total acceleration (A_t) which after integration gives the actual direction and the distance flown.

The accelerometers are aligned relative to the pitch and roll axes of an aircraft, and *not* oriented geographically N−S and E−W. Their output signals, however, can be related to a NE coordinate system, and while the computer has 'knowledge' of their orientation, it will always determine an aircraft's present position in terms of the latitude and longitude of that position.

Three coordinate systems are involved in the computations performed by an INS and these are shown in Fig. 10.10. The *XY*

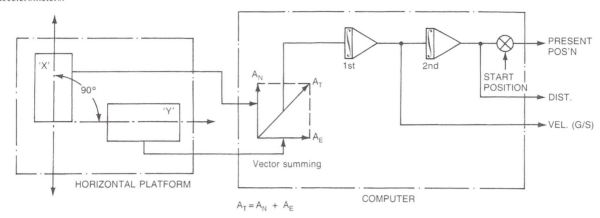

Figure 10.9 Dual-axis accelerometers.

$A_T = A_N + A_E$

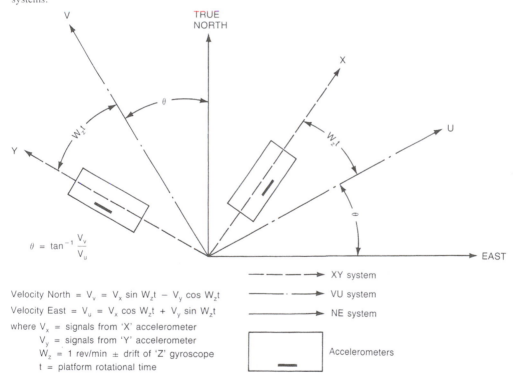

Figure 10.10 Coordinate systems.

$\theta = \tan^{-1} \dfrac{V_v}{V_u}$

Velocity North $= V_v = V_x \sin W_z t - V_y \cos W_z t$

Velocity East $= V_u = V_x \cos W_z t + V_y \sin W_z t$

where V_x = signals from 'X' accelerometer

V_y = signals from 'Y' accelerometer

W_z = 1 rev/min ± drift of 'Z' gyroscope

t = platform rotational time

- - - - - - → XY system

- · - · - → VU system

——————→ NE system

▭ Accelerometers

coordinate system is established by the X—Y platform which, as will be described later, is continuously rotated at 1 rev/min; all acceleration sensing is done in this system.

The second system is the non-rotational *vu coordinate system*, and since the computer always assumes that 'v' and 'u' are related to velocities in the north and east directions respectively, it utilizes this system to perform all its calculations. The system is established by supplying the accelerometer signals to the computer in terms of their sine and cosine components. These components are derived from a resolver synchro controlled by the X–Y platform, and thus correspond to the initial position of the platform (relative to an aircraft's longitudinal axis) before it begins to rotate. Since the X–Y platform can start rotating from any position, then the XY and vu coordinate systems will be in error with respect to the earth's non-rotating coordinate system (NE) by an angle θ which is determined during the alignment mode of the INS.

The *NE coordinate system* is the third involved in computation and is the one in which position data are finally displayed in terms of latitude and longitude. In order, therefore, to attain final alignment of the platform with this system, and as may be seen from Fig. 10.10, the accelerometer signals are resolved from X and Y through an angle $W_z t$, and a correction factor equal to the angle θ is applied.

The formulae relating to the coordinate systems are also given in Fig. 10.10.

Gyro-stabilized platform

For precision operation of an INS, it is essential for the 'X' and 'Y' acceleration-sensing axes to be maintained normal to a local vertical with the earth's centre at all times. If this were not done, false gravitational forces would be sensed, giving rise to errors in the computed distance flown and in the present position of an aircraft. These forces and errors are overcome by mounting the accelerometers in such a way that any displacements are detected by gyroscopic-type sensors.

There are two mounting arrangements: the gyro-stabilized platform, and the 'strapdown', which will be described later in this chapter. In the first arrangement, the accelerometers are mounted on a platform which is supported in gimbal rings and stabilized by gyroscopes and torque motor systems; the platform is referred to as the *X–Y platform*.

The principle of this arrangement, and also that of 'strapdown', is based on the *Schuler* theory of a pendulum whose bob is at the centre of the earth, and supsended from a point above its surface. If, then, the suspension point were accelerated around the earth, the bob, being at the centre of the earth's gravity, would always remain vertically below its suspension point. A platform mounted on the suspension point tangential to the earth's surface would, therefore, also remain horizontal irrespective of acceleration. If, for any reason, the pendulum bob became displaced from the earth's centre, it would start to oscillate with a period of 84.4 minutes; this is the value obtained by substituting the earth's radius (in feet) for the pendulum length l in the basic formula for calculating the time period of a

Figure 10.11 Alignment of
gyroscope axes.

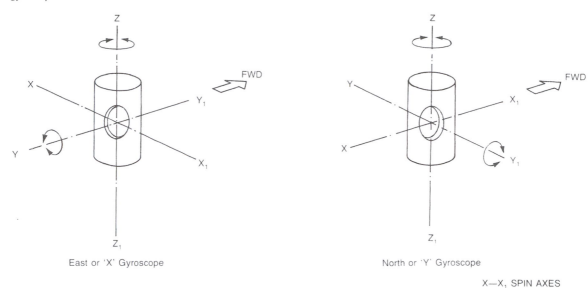

East or 'X' Gyroscope   North or 'Y' Gyroscope

X—X$_1$ SPIN AXES
Y—Y$_1$ INPUT AXES
Z—Z$_1$ OUTPUT AXES

pendulum. Thus, by mechanizing an INS platform to remain horizontal, an analogue of the Schuler earth pendulum with a period of 84.4 minutes is produced, and the platform is then said to be *Schuler tuned*.

The gyroscopes are of the integrating rate type, meaning that they sense movement about only one axis, and that the rate changes are integrated to give distance changes. The input and output axes of the gyroscopes are positioned so that they relate directly to the north and east coordinate system of 'X' and 'Y' accelerometer positioning and, as shown in Fig. 10.11, they are designated as north ('Y') and east ('X') gyroscopes. The 'Y' gyroscope has its input axis aligned with an aircraft's roll axis, while that of the 'X' gyroscope is aligned with the pitch axis; the manner in which they sense attitude changes depends on aircraft heading. Thus, if an aircraft and its INS platform are heading north, the 'X' gyroscope senses pitch attitude changes, and the 'Y' gyroscope senses roll attitude changes. The converse of this is true, however, when an aircraft and platform are heading east.

The gyroscope shown in Fig. 10.12 is, for explanatory purposes, drawn to represent sensing of roll attitude changes. In such an attitude, therefore, the inertial platform and the spin axis of the appropriate gyroscope will be deflected causing precession about its output axis. The angular movement of the gyroscope operates an electrical signal pick-off element which then transmits a signal, via an amplifier, to the corresponding torque motor which then drives the platform back to its level position.

Figure 10.12 Attitude sensing.

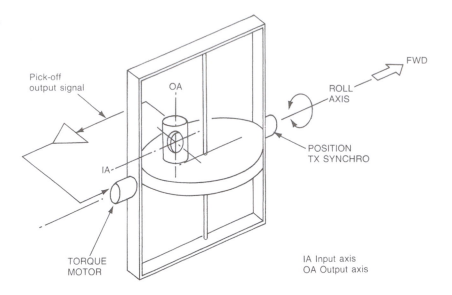

A TX synchro is mounted on the gimbal system, its purpose being to provide position signals proportional to aircraft attitude change for use by other systems, e.g. automatic flight control and flight director systems.

For a pitch attitude change, sensing and platform levelling is accomplished in a similar manner and through a second gimbal ring.

A third gyroscope is also provided and is mounted on a second platform. Its purpose is to sense changes about the local vertical (designated 'Z') and to keep the X–Y platform in the same position relative to space and the N–E coordinates, i.e. to maintain its north datum. This gyroscope and its platform are also designated 'Z'. Any change of platform or azimuth relative to the inner gimbal ring corresponds to an equivalent heading change, and so by connecting a signal pick-off element to the 'Z' gyroscope signals corresponding to such change can be produced. These signals are supplied to an azimuth torque motor which rotates the 'Z' platform in the opposite direction to the heading change, and at the same time positions the rotor of a TX synchro. The output signals from this synchro are transmitted to the HSI of a flight director system which will then indicate true heading (see also Chapter 9).

It will be apparent from the foregoing that between the headings north and east, both the 'X' and 'Y' gyroscopes will exercise control, the magnitude of which must be determined by heading. This is accomplished by connecting a resolver synchro between the 'X' and 'Y' gyroscopes as shown in Fig. 10.13, and then positioning the synchro rotor by means of the azimuth torque motor so that the attitude signals transmitted to the pitch and roll torque motors are modified by heading-related error signals.

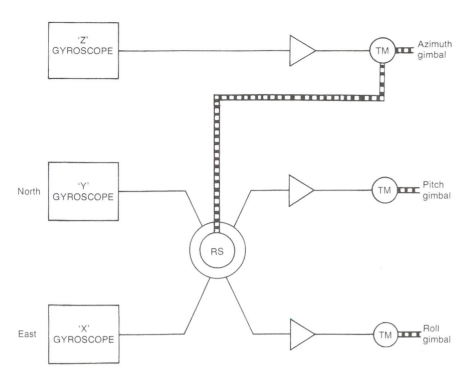

Figure 10.13 Heading control of pitch and roll torque motors.

'Z' GYROSCOPE — Azimuth gimbal

North 'Y' GYROSCOPE — Pitch gimbal

East 'X' GYROSCOPE — Roll gimbal

Transport rate and earth rate compensation

As we learned from Chapter 4, gyroscopes must be compensated for the effects of the rate at which they are transported over the earth's surface, and for those of earth's rotation (earth rate). This compensation, which is necessary to maintain the required earth reference orientation, is also relevant to the gyroscopes of gimballed-type INS platforms; in other words, they must be Schuler tuned (see also page 260). The effects of both these rates is to cause a platform to tilt from the required horizontal position with respect to the earth's surface, thereby causing the accelerometers to produce false output signals. The gyroscopes must, therefore, be subjected to equal and opposite forces so that corresponding signals can be produced for maintaining the platform level by means of its torque motor system.

The principle of compensation is illustrated in Fig. 10.14 which, although drawn to represent the 'Y' accelerometer loop, applies equally to that of the 'X' accelerometer. The signals from the accelerometers are supplied to an electronic resolver circuit which converts the signals to the 'vu coordinate' system of the computer. The signals are then integrated to produce the N−S and E−W velocity signals, and these, in turn, are divided by a value $R + h$, where R = the mean radius of the earth, and h = aircraft altitude. In other words, $R + h$ is the distance from the centre of the earth to the X−Y platform, and since it corresponds to the radial path flown by

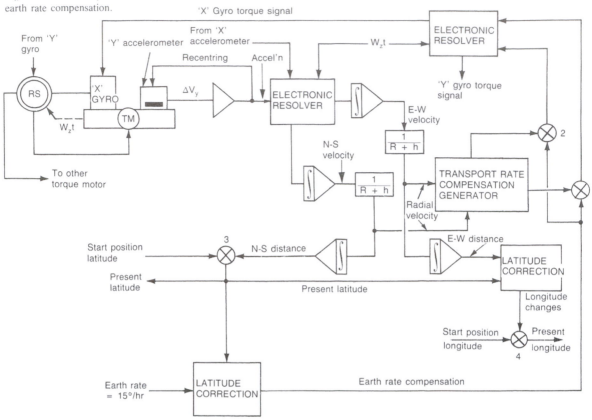

Figure 10.14 Transport and earth rate compensation.

an aircraft around the earth, then after dividing the coordinate velocity signals by $R + h$, signals corresponding to radial velocity are obtained. The value of R is a constant that is pre-programmed into the navigation computer, while h is, of course, a variable that is supplied to this computer from an ADC.

The coordinate velocity signals are supplied to a transport rate compensation generator circuit for conversion to transport rate signals, which are then supplied via summing points 1 and 2 to a second electronic resolver circuit. The purpose of this circuit is to convert these signals so that they relate to velocities in the XY coordinate system. They are then transmitted to the torquer coils of the 'X' and 'Y' gyroscopes, thereby precessing them in order to establish angular distance output signals in their respective pick-off coils, which are connected to a coordinate resolver. This synchro, therefore, separates the signals into pitch and roll components, and supplies them to the corresponding levelling torque motors, which then tilt the inertial platform back to a level position, through an angle determined by the transport rate.

The effects of earth rate on a gyroscope are dependent on the latitude in which it is operating at any one moment, i.e. its present position. Compensation signals related to this latitude must, therefore, be generated and, as will be noted from Fig. 10.14, this is accomplished by supplying the corresponding signal from summing point 3 to a latitude correction circuit. The compensating signal output from this circuit corresponds to a rate equal to $15°$ cos of latitude and is summed at points 1 and 2. The outputs from these points to the electronic resolver, and the resulting signals supplied to the torquer coils of the gyroscopes, and to the platform torque motors, therefore provide for combined transport rate and earth rate compensation.

Gimballed platform arrangement

The arrangement adopted in a typical IN system is shown in Fig. 10.15.

The two platforms are supported by an inner roll (IR) gimbal within a pitch gimbal which is, in turn, supported by an outer roll (OR) gimbal; this gimbal supports the whole system within the structure of the system's navigation unit. The reason for having two roll gimbals oriented as shown is to prevent the condition known as 'gimbal lock' from occurring (see also page 106). The angular movement of the IR gimbal is limited to $\pm 10°$.

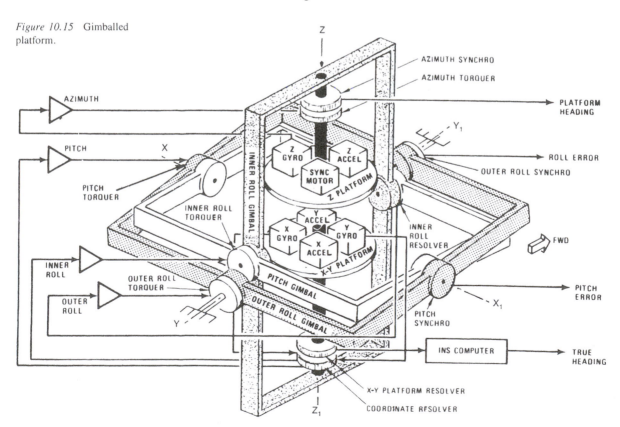

Figure 10.15 Gimballed platform.

Table 10.1

Component	Location		Function
Torque motors:			
Outer roll	Fixed to INU casing		Drives OR gimbal.
Inner roll	Fixed to pitch gimbal		Drives IR gimbal.
Azimuth	Fixed to IR gimbal		Drives 'Z' platform in a direction opposite to aircraft heading change.
Synchros:	Stator	Rotor	
Pitch	Fixed to OR gimbal	Fixed to pitch gimbal	Provides pitch and roll attitude signals to AFC and FD systems.
Outer roll	Fixed to INU casing	Fixed to OR gimbal	
Azimuth	Fixed to IR gimbal	Driven by azimuth torquer	Provides true heading signals to HSI of FDS.
Resolvers:			
Inner roll	Fixed to pitch gimbal	Fixed to IR gimbal	Supplies roll error signals to OR torquer.
X–Y platform	Fixed to IR gimbal	Fixed to platform	Supplies signals to computer which are related to platform heading, and in terms of sine and cosine components.
Coordinate	Fixed to IR gimbal	Fixed to platform	Receives signals from 'X' and 'Y' gyroscope pick-offs, separates them into pitch and roll components, and supplies them to the pitch and IR torquers respectively.

A synchronous-type servomotor is mounted on the 'Z' platform and it rotates the 'X–Y' platform at the rate of 1 rev/min. The reason for rotating the platform is to modulate errors generated by misalignment of the X–Y accelerometers and gyroscopes, scale errors of the accelerometers, and drift of the gyroscopes so that the computer 'sees' a minimal error condition. Since the platform is rotating, the 'X' and 'Y' gyroscopes sense pitch and roll attitude changes alternately.

The locations and functions of the various torque motors, synchros, and resolvers are detailed in Table 10.1.

Accelerometer construction

The constructional arrangement of a typical accelerometer is shown in Fig. 10.16.

The mass and its force coil are suspended by two flat springs so that they can move linearly along the acceleration-sensing axis. A ferrite armature is fixed at each end of the mass and in the proximity of two pick-off transformers supplied with 5 V ac at 12.8 kHz. When the mass is centred, the space between the armatures and transformers will be equal and so the coupling between primary and secondary in each transformer will also be equal. The secondaries are connected in series, so that the signal induced in the secondary of one transformer is in-phase with primary voltage, while that in the other

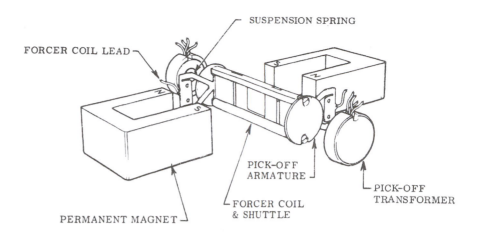

Figure 10.16 Typical accelerometer arrangement.

SUSPENSION SPRING

FORCER COIL LEAD

PICK-OFF
ARMATURE

PICK-OFF
TRANSFORMER

FORCER COIL
& SHUTTLE

PERMANENT MAGNET

transformer's secondary is out-of-phase with primary voltage. Thus, with equal coupling, the net output from both secondaries will be zero.

When the mass is deflected as a result of an acceleration, the coupling between the armatures and transformers will no longer be equal, and so the secondaries produce an output related to phase and position of the mass. The 'mass position' output signal is supplied to an amplifier from which it is fed back to the force coil which is also deflected within the field of two permanent magnets. The current flowing in the coil interacts with the field of the magnets and establishes a force that recentres the mass and coil to the 'null' position. The current required to do this is therefore directly proportional to the force that originally caused deflection; in other words, it is a measure of acceleration.

The accelerometer assembly is enclosed within an hermetically-sealed case filled with a low-density fluid to provide damping of the pendulous mass. The temperature of the fluid is regulated by a thermistor-controlled heater element.

In a number of systems currently in use, the accelerometers adopt capacitance-type pick-offs. The design of their sensing element simplifies the construction and, in addition, eliminates the need for fluid damping. The arrangement of such a unit is shown in Fig. 10.17.

The sensing element is a ceramic disc which is suspended by four flexible metal 'hinges', and by means of a mass subassembly the disc is made pendulous so that it can move through an arc under the influence of an acceleration force. The disc is coated on each side with a metallized pattern to form a capacitance plate. The plates are also used to form the suspension 'hinges' which, in turn, are used for the connection of electrical leads. The other capacitor plates are formed by the surfaces of upper and lower permanent magnets which, together with a coil that is also supported on the disc assembly,

267

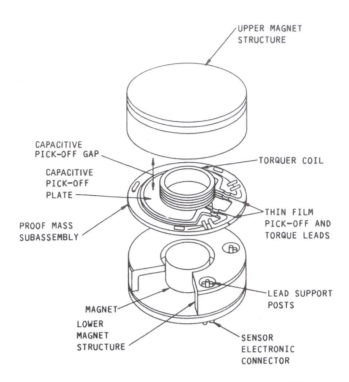

Figure 10.17 Operation of capacitance-type accelerometer.

UPPER MAGNET
STRUCTURE

CAPACITIVE
PICK-OFF GAP

CAPACITIVE
PICK-OFF
PLATE

TORQUER COIL

THIN FILM
PICK-OFF AND
TORQUE LEADS

PROOF MASS
SUBASSEMBLY

LEAD SUPPORT
POSTS

MAGNET

LOWER
MAGNET
STRUCTURE

SENSOR
ELECTRONIC
CONNECTOR

forms an electrical force or torquer device. The two capacitors of the pick-off are incorporated in an ac bridge circuit.

When subjected to an acceleration, the rotation of the sensing element about its suspension axis causes the capacitance of one half of the disc to be increased, while the capacitance of the other half is decreased. The bridge circuit is, therefore, unbalanced, and a signal is developed and supplied to a differentiating operational amplifier which produces a phase-sensitive signal proportional to sensing element displacement. This signal is then demodulated and filtered, and after amplification it is supplied to the torquer coil which then develops a current proportional to the acceleration force, and restores the sensing element to a 'null' position. The coil current is measured as a voltage drop across a precision scaling resistor.

The compensation circuit module is provided for the purpose of making the accelerometer insensitive to temperature variations with regard to acceleration force scale factors, thereby eliminating thermal lag.

Gyroscopes and construction

In systems utilizing electrically-operated gyroscopes, two-phase synchronous/hysteresis-type motors are adopted, which operate at about 24 000 rev/min from an input of 115 V ac at 1.2 kHz supplied via rotary transformers. An example of a typical unit is shown in Fig. 10.18.

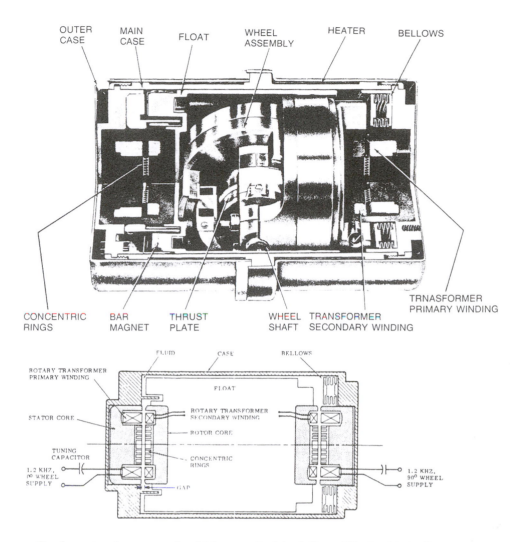

Figure 10.18 Floated gyroscope.

Labels in top diagram: OUTER CASE, MAIN CASE, FLOAT, WHEEL ASSEMBLY, HEATER, BELLOWS, CONCENTRIC RINGS, BAR MAGNET, THRUST PLATE, WHEEL SHAFT, TRANSFORMER SECONDARY WINDING, TRNASFORMER PRIMARY WINDING

Labels in bottom diagram: FLUID, CASE, BELLOWS, ROTARY TRANSFORMER PRIMARY WINDING, FLOAT, STATOR CORE, ROTARY TRANSFORMER SECONDARY WINDING, ROTOR CORE, TUNING CAPACITOR, CONCENTRIC RINGS, 1.2 KHZ, 0° WHEEL SUPPLY, GAP, 1.2 KHZ, 90° WHEEL SUPPLY

Each motor is mounted within a cylindrical float filled with helium so that in operation the motor 'rides' on a thin layer of the gas. The float assembly is, in turn, mounted within a cylindrical main case filled with a dense viscous fluid which supports the weight of the float and so provides damping, and protection against vibration or shock loading. The float is suspended radially and axially (on its output axis) by a magnetic field generated by permanent magnets and by tuning the inductance of the rotary transformers located at each end of the assembly. This 'floated gyro' concept therefore eliminates the mechanical forms of suspension adopted for conventional gyroscopes.

Any angular displacement of the float assembly with respect to the case is detected by a position pick-off which then supplies an analog output signal to torquer coils which return the float assembly to a 'null' position. This is accomplished as a straightforward application of a force, and not of precession, and since the null position provides

269

the datum from which any tilt of the inertial platform is sensed, then in functional terms the torquer coils may be considered analogous to a restraining spring used for the rate gyroscope of a turn-and-bank indicator.

In respect of gyroscopes that are used for the stabilizing of an X—Y platform, the torquer coils are also supplied with constant current signals, the purpose of which is to compensate for the effects of transport rate and earth rate (see page 263).

Signals resulting from precession of the gyroscopes during platform attitude changes are also derived from the position pick-offs, and after amplification they are supplied to the respective gimbal ring torque motors.

Each gyroscope is maintained at an operating temperature of 76°C by a 30 W blanket-type heating element wrapped around the main casing. The heater circuits and gyroscope temperatures are controlled by thermistor-type sensors.

Ring laser gyroscope (RLG)

This type of unit is essentially a rate sensor, and it is only in this context that it can justify the name 'gyroscope'. It has no rotating mass or gimbal system, and therefore does not possess the conventional characteristics of rigidity and precession. Since it has no gimbal system it is, and can only be, used in a 'strapdown' configuration (see page 279) so that its attitude output signals can be supplied direct to the navigation computer, and it eliminates the use of complex platform levelling systems. It has many other advantages such as wide dynamic range, allowing of very short alignment times, and high reliability factors, and is adopted as standard in the inertial reference systems (IRS) currently in use in many types of civil aircraft.

The basis of a typical sensor is a triangular block of specially fabricated glass ('Cervit' glass) that is extremely hard and does not expand or contract under varying temperature conditions. By means of computerized ultrasonic diamond-drilling techniques, a cavity is formed within the whole block. A precision-made mirror is fitted at each corner of the block, and a cathode and two anodes are located as shown in Fig. 10.9. The mirrors serve as both reflectors and optical filters, reflecting the light frequency for which they are designed, and absorbing all others.

The cavity is filled with a lasing medium (typically helium-neon) and when excited by an electrical potential across the cathode and anodes, the medium is ionized and is transformed into light in the orange—pink part of the visible spectrum. By design, two light beams resonating at a single frequency are emitted and are made to travel in opposite directions around the cavity. Since the beams travel at the same constant speed and are 'bounced off' the mirrors, then in a

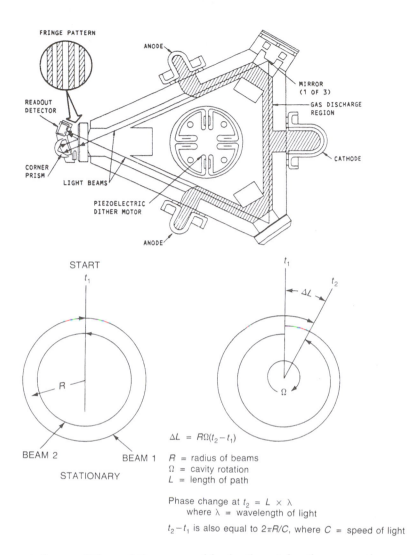

Figure 10.19 Ring laser gyroscope.

$\Delta L = R\Omega(t_2 - t_1)$

R = radius of beams
Ω = cavity rotation
L = length of path

Phase change at $t_2 = L \times \lambda$
where λ = wavelength of light

$t_2 - t_1$ is also equal to $2\pi R/C$, where C = speed of light

static condition of the sensor block, they take the same time to complete a closed path in inertial space around the cavity.

Although the frequency is determined by the gas that is 'lasing', it can be varied somewhat by changing the path length over which the light waves have to travel; for a given length there are an integral number of waves occurring over the complete path. If the length is altered, the waves will either be compressed or expanded, and this results, respectively, in an increase or decrease of their frequency. Both beams combine in an optical sensor, or readout detector, located at one corner of the block. The sensor operates on the interferometer principle, i.e. it contains a prism that deflects the beams so that they 'interfere' with each other in order to form what is termed a fringe pattern.

As in the case of a gimballed platform-type of INS, three sensors are required to be mounted in an aircraft so as to detect attitude changes about the pitch, roll and yaw axes. When the aircraft is in straight and level flight, all three RLG sensors are in a static condition, and so the resonant frequencies of the beams are equal.

If now the aircraft's attitude is changed about, say, the pitch axis, the corresponding sensor will also be rotated about its axis perpendicular to the plane of the beams. Since the beams are travelling at a constant speed on paths in inertial space, the bodily rotation of the sensor will then be with respect to inertial space. This means, therefore, that the beam travelling from one mirror to the next in the direction of sensor rotation will move through a greater distance than the second beam that is travelling in the opposite direction. Thus, the times taken for the beams to travel around the cavity of the sensor will now differ. As already pointed out, a change in path distance produces a change in frequency of wave propagation, and so by measuring the frequency difference resulting from rotation of an RLG sensor, the angular rate at which it does so can be determined.

The spacing of the light and dark portions of the fringe pattern referred to above depend on the angle between the interfering of the beams and their frequency. When both beams are of the same frequency, the pattern is stationary, and constant signals are produced by two photo-diode type detectors which are spaced an odd number of quarter wavelengths apart in the pattern. Thus, when rotation of the sensor causes a difference between frequencies, the fringe pattern moves across the detectors, and due to the spacing of the beams, one detector will receive maximum light when the other is at half intensity. Each detector then converts this fringe pattern movement into signal pulses, the phasing of which give the direction of sensor rotation, while the frequency is proportional to the angular rate of rotation. The signals, which are in digital format, are transmitted to the appropriate attitude computing software within the IR computer.

The relationship between the input rate of rotation and the output frequency is a linear one, and ideally it should remain so throughout a full rotation of an RLG sensor. At low input rotation rates tending towards zero, however, the output frequency can become non-linear, and at a certain threshold value can drop abruptly to zero. This phenomenon is known as 'lock-in', and is due to small amounts of energy from the beams being back-scattered into each other, the energy causing the beam frequencies to be pulled together until eventually the beams synchronize. Since extremely low rotation rates (typically 0.001°/hr) are required to be measured in IN/IR systems, 'lock-in' can result in undesirable errors. In order therefore to circumvent these effects, a technique known as 'dither' is introduced, and is effected by means of a piezo-electric motor. This motor is mounted on the sensor in such a way that it vibrates the laser ring about its input axis through the 'lock-in' region, thereby unlocking the beams and enabling the optical sensor to detect the smaller movement of the fringe pattern. The motions caused by the dither motor are decoupled from the output of an RLG.

Mode selection

In order to control the modes of IN/IR system operation, mode selector units are provided. In systems utilizing gimballed inertial platforms, the unit is located on a flight deck panel, while in others it is integrated with the control and display unit (CDU) to form what is termed an inertial reference mode panel.

Figure 10.20 illustrates the controls of a separately located unit; the modes that can be selected are as follows:

STBY This mode, which is for ground use only, selects power onto the system and allows it to 'warm up' and to run the gyroscopes up to speed. The aircraft's present latitude and longitude are inserted in the CDU when in this mode, and an auto-alignment sequence commences. *The INS is not affected by movements of the aircraft while in this mode.*

ALIGN This mode allows the INS to align automatically to its true north point. When alignment is completed the 'READY NAV' light illuminates green to indicate that the system is ready to go into the 'NAV' mode. The aircraft *must remain stationary* in the 'ALIGN' mode.

NAV This is the normal in-flight operating mode in which the data required for navigating the aircraft and stabilizing the

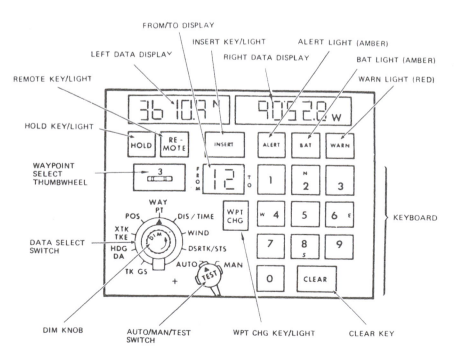

Figure 10.20 Control and display unit.

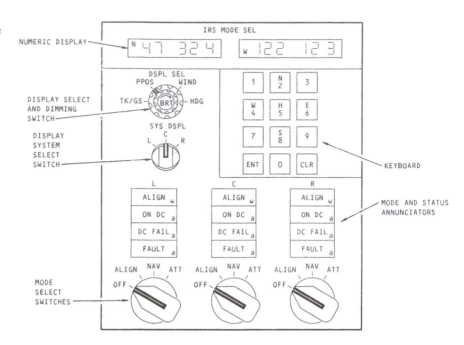

Figure 10.21 Inertial reference mode panel.

NUMERIC DISPLAY

DISPLAY SELECT AND DIMMING SWITCH

DISPLAY SYSTEM SELECT SWITCH

KEYBOARD

MODE AND STATUS ANNUNCIATORS

MODE SELECT SWITCHES

INS are computed. *It must be selected before the aircraft moves from its parked position.* The 'READY NAV' light is extinguished on selection. Initial track selected may be made and started when in this mode.

ATT REF Selects pitch, roll and platform heading stabilization outputs only; no display is presented on the CDU. This mode disables the navigational capability of the computer for the remainder of a flight, because once turned off, the computer cannot be switched on again. Normally, this selection is only made when a computer failure has occurred.

The selector switch is provided with two mechanical stops: one between the 'STBY' and 'ALIGN' mode positions, and the other between the 'NAV' and 'ATT' mode positions. In order for the selector knob to move over the stops, it must be pulled out. The reason for having the stops is to prevent the 'NAV' mode from being inadvertently switched out.

The red 'BATT' light is illuminated when back-up dc voltage is being used and is less than the minimum required (typically 18 V) to operate the system; the system is automatically switched off.

An inertial reference mode panel as used in a typical IR system is shown in Fig. 10.21, from which it will be noted that the mode selector switches appropriate to the normal triple systems installation are grouped together instead of having separate space-consuming units of the type just described.

The three modes that can be selected and their functions are the same as those already outlined. A 'STBY' mode is not required for the reason that the application of comprehensive digital signal-processing techniques, and of ring laser gyroscopes, eliminates the need to allow for 'warm-up' and gyroscope 'run-up'.

Four mode and status annunciator lights are provided for each system as follows:

ALIGN	Illuminates white when a system is in the alignment mode. In the event of alignment procedure failure, it flashes on and off.
ON DC	This illuminates amber to indicate that power to the system has automatically changed over from the normal 115 V ac to 28 V dc power from the battery system.
DC FAIL	Illuminates amber when the battery power source drops below 18 V.
FAULT	Illuminates amber when failures in the system are detected.

Longitude and latitude data from any of the three systems (Left — Centre — Right) are selected as appropriate, for readout on the single display at the top of the panel.

Control and display unit (CDU)

This unit serves as the primary interface between the flight crew and the inertial system computer, in that it contains the controls necessary for the selection and display of all essential navigational data. The panel layout of a unit can vary between systems, but the one illustrated in Fig. 10.20, and used in conjunction with the computer of a gimballed platform-type of INS, serves to illustrate control functions and selection methods that are generally applicable.

Control switches

DATA SELECTOR	This switch selects the navigational data listed in Table 10.2 for presentation in the upper left and right numerical displays of the unit.
WPT SELECTOR	This switch is of the thumbwheel type, and when the 'DATA SELECTOR' switch is set to 'WPT', it enables WPTs 1 to 9 to be selected for latitude and longitude insertion, or selection of WPTs 0 to 9 for presentation of their coordinates on the upper displays. It is also used for inserting and displaying latitude, longitude, altitude and frequency of up to nine DME stations.

DATA KEYBOARD	This contains ten push-button key switches (0−9) for entering present position and WPT coordinates, 'FROM/TO' WPTs, desired XTK effect, and TK hold. Each key illuminates white when pressed. For the display of DME station latitude, longitude, altitude and frequency, the following key selections are made:

3 and 9 enable altitude and frequency to be displayed.

2 or 8 enable altitude to be loaded.

4 or 6 enable frequency to be loaded.

7 and 9 enable latitude and longitude to be loaded.

The 'FROM' display remains blank, and the 'TO' display flashes the station number when DME data is being used.

INSERT	Operation of this push-button switch transfers entered data into the computer.
CLEAR	This push-button switch is used to erase data loaded into displays but not yet loaded into the computer; it is illuminated white.
AUTO/MAN/TEST	In 'AUTO' the system makes automatic sequential TK leg changes, and permits manual TK leg changes. In 'MAN', TK leg changes can only be initiated manually. The 'TEST' position enables checks to be made on the displays and annunciators.
WPT (or TK) CHANGE	Allows initiation of manual TK leg change. Illuminates white when pressed, and is extinguished when the 'INSERT' or 'CLEAR' keys are pressed.
HOLD	This permits a position check and up-date to be made, and also a display of malfunction codes. It also illuminates white when pressed.
REMOTE	Enables semi-automatic (loading and insertion of WPT and DME data into more than one INS) autofill operation, remote ranging, and display of XTK offset. It illuminates amber when pressed.

Displays

There are three electronic displays of the segmented type: left and right for displaying the parameters listed in Table 10.2, and a

Table 10.2

Data selector switch position	Left data display	Right data display
TK/GS	TK	GS
HDG DA	True HDG (See Note 2)	DA (See Note 2)
XTK TKE	XTK distance (See Note 1)	TKE (See Note 1)
POS	Lat. of present position	Long. of present position
WPT	Lat. of selected WPT Lat. or alt. of a selected DME station	Long. of selected WPT Long. or freq. of a selected DME station
DIS/TIME	Distance to WPT Distance to selected DME station	Time to WPT — Blank —
WIND	Wind direction	Wind speed
DSR TK/STS	DSR RK — Blank —	— Blank — Action codes and alignment status during 'ALIGN' and 'NAV' modes

Notes: 1 An 'R' or 'L' also displayed to indicate that present position is to right or left of DSR TK, and present TK angle to right or left of DSR TK.
2 An 'R' or 'L' also displayed to indicate that present TK is to right or left of aircraft's HDG.

'FROM/TO' display to indicate the number of the 'FROM' and 'TO' WPTs on the TK leg being navigated. The characters of this display will flash, if the 'REMOTE' switch has been pressed, to indicate the remote ranging leg, remote direct ranging leg, or remote ranging along the flight path to the desired WPT.

Annunciators
There are three annunciators as follows:

ALERT This illuminates amber two minutes before reaching a 'TO' WPT. The operation of this annunciator also depends on the settings of the 'AUTO/MAN/TEST' switch. Thus, if it is at 'AUTO', the annunciator is extinguished when the 'FROM/TO' display changes to the next two WPT numbers. If the switch is at 'MAN', the light flashes as the aircraft flies over the WPT. The annunciator is operable at a G/S greater than 250 knots.

BATT Illuminated amber when the INS is operating on battery power.

WARN Illuminated red when a system malfunction occurs, or during 'ALIGN' mode it flashes to indicate system degradation, or that an alignment failure has occurred. It will not extinguish unless the fault is corrected or the INS is switched off.

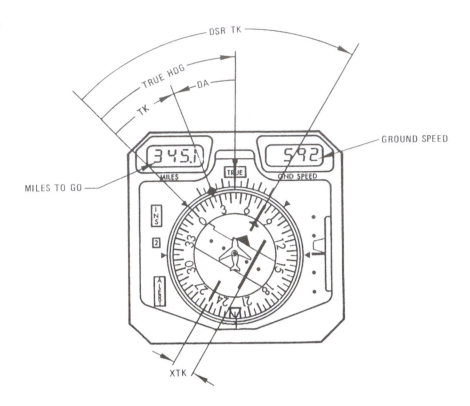

Figure 10.22 Navigation data display on an HSI.

A certain amount of the data listed in Table 10.2 can also be derived from the indications that are normally displayed by the HSI of a conventional type of flight director system as illustrated in Fig. 10.22. They are interpreted as follows:

TK is the compass card reading when referenced against the diamond-shaped 'bug' or cursor, 025° in this case.

DA is the reading of the scale above the compass card when referenced against the diamond-shaped 'bug'. In this case DA is 020° to the left.

XTK is indicated by the deflection of the deviation bar with respect to its scale. Each dot corresponds to 3.75 nm.

'Miles to go' is always displayed on the top left-hand indicator even if the CDU is displaying the miles between WPTs, or DME miles.

DSR TK is the compass card reading referenced against the pre-set course arrow, 075° in this case.

TRUE HDG is the compass card reading referenced against the lubber line: 045° as shown. An annunciator flag above the DA scale also displays 'TRUE'.

GS is always displayed on the indicator at the top right-hand corner of the HSI.

'Strapdown' configuration	This applies to the installation of the reference unit of an IRS which, in dispensing with the complex mechanical gimballed platform arrangement for its accelerometers and ring laser gyroscopes, enables them to become part of the unit's fixture to an aircraft's structure. Thus, the aircraft itself becomes a platform for the sensors, and in moving with it, the signals they produce are used by the computer to extract aircraft attitude, and to resolve body axis accelerations into navigation axis displacements. For this purpose, mathematical equations are programmed into the computer which, in essence, are a functional replacement of a gimballed platform.

Alignment sequencing	The accuracy of an IN/IR system is dependent on precise alignment of its inertial platform with respect to the latitude and longitude of the ground position an aircraft is in at the time of 'starting up' the system. The computer must, therefore, be programmed to carry out a self-alignment calibration procedure over a specific time period before the system is ready to navigate an aircraft. This procedure can only be carried out on the ground, and *during the actual alignment stage the aircraft must not be moved*.

The technique and the time period involved can vary between types of system installed in an aircraft, but the following, which is based on a gimballed-platform system, is generally representative:

1. The selector switch on the mode select unit is set to the 'STBY' mode position, and the data select switch on the CDU is set to 'POS'. The latitude and longitude of the present position is then set in the appropriate displays from the keyboard.

The 'INSERT' pushbutton on the CDU will illuminate, and remains illuminated, until the present position is 'loaded' into the computer. During this mode, heater power is applied to the navigation unit until it attains its operating temperature, and the gyroscopes are run up to speed; the platform and gimbal system are 'caged' with respect to the aircraft's axes.

2. The mode selector switch is then set to the 'ALIGN' mode position. The present position of the aircraft is loaded into the computer by pressing the 'INSERT' switch button; the switch light is then extinguished. The computer also carries out a check which compares the present position with that last displayed.

For purposes of sequencing, the 'ALIGN' mode consists of submodes designated by numbers that decrease in sequence as follows:

9. Standby.
8. Course levelling.
7. Course azimuth. ⎫
6. Fine alignment. ⎬ Gyrocompassing
5. Gyrocompassing complete.

The numbers are displayed by the fourth digit of the CDU's right-hand display, when the CDU select switch is at the 'DSR TK/STS' position. At the same time, the first and fifth digits of the display are a '0' and a '5' respectively, to indicate that the system is not in the 'NAV' mode. This last digit is referred to as a *performance index (PI) number*.

If the system has not reached operating temperature the *sub-mode number 9* is displayed when the mode select switch is moved from 'STBY' to 'ALIGN'. During this sub-mode, the platform and gimbal system remains caged.

When operating temperature is reached, and the gyroscopes are up to speed, the alignment sequence goes into:

sub-mode 8 and the CDU display changes accordingly. During this mode, the 'BATT' annunciator in the CDU will illuminate to indicate battery power availability, the gimbal system is uncaged, and the platform is aligned to the local horizontal utilizing the accelerometers to detect any 'out-of-level' orientation. At the end of the mode time period, the CDU display changes to indicate:

sub-mode 7. This mode provides an initial estimate of the azimuth orientation of the 'X' and 'Y' accelerometers' input axes with respect to true north. Local latitude is also computed and compared to the present position loaded into the computer. At the end of the time period, the CDU indicates that the sequence goes into:

sub-mode 6, during which the initial determination of true north is refined, and the computer provides a bias signal for the gyroscopes and corrects for earth rate at the present position latitude. This, together with sub-mode 7, is normally referred to as '*gyro-compassing*'. When completed, the CDU display again changes to indicate:

sub-mode 5. The 'READY NAV' light on the CDU will illuminate and the 'NAV' mode can then be selected on the mode select unit. When selected, confirmation of entering the 'NAV' mode is indicated by the first digit of the CDU display changing from a 0 to a 1, and by extinguishing of the 'READY NAV' light.

When a pre-flight calibration of the 'Z' gyroscope is required, the system can be left in 'ALIGN' after gyrocompassing is complete, i.e. sub-mode 5 is displayed. As the calibration takes place, the PI number referred to earlier decreases in sequence from 5 to 0, the latter indicating the best calibration.

As noted earlier, alignment time periods can vary; typically they would be ten minutes for an IRS and nineteen minutes for an INS.

Slewing

This is a procedure that provides the capability of moving the gimballed platform of an INS about the pitch and roll axes, thereby simulating in-flight attitude changes, in order to observe the resulting changes in the displays of the units and instruments interfaced with the system. The procedure, which can be carried out without the use of a tilt table or removal of the navigation unit from an aircraft, allows attitude changes of $\pm 71°$ pitch and roll, $\pm 180°$ azimuth to be made. The slew rate of pitch and roll is approximately $2°/\text{min}$.

Since attitude changes are transmitted to interfacing systems, slewing can be used to verify the integrity of wiring, to calibrate a weather radar antenna and flight director HSI and ADI displays, or to verify such maintenance activities as replacement of HSIs, ADIs, weather radar antennae and AFCS pitch and roll computers.

System malfunctions

In the event of any malfunctions occurring in flight, appropriate annunciations are made in a manner which depends on the type of system installed. In the case of an INS, a 'WARN' annunciator light on the CDU is illuminated, and in order to determine the cause of the warning, and what action is to be taken, a corresponding code numbering system is programmed into the computer. The code numbers are made to appear in the right-hand display of the CDU by moving the display selector switch to the 'DSR TK/STS' position, and also by operating the 'TEST' switch. There are two groups of code numbers designated as: (i) *action codes* and (ii) *malfunction codes*.

Action codes

These are always displayed first, after the display selector switch is moved to the 'DSR TK/STS' position. The codes and their interpretation are as follows:

01. Complete system is inoperative.

02. Failure of the computer; the INS is no longer used for navigation of the aircraft, but may be used to supply attitude data.

03. Does not illuminate the 'WARN' annunciator, but is detected by monitoring the INS display on the FD system HSI. The computer feeds data to the CDU and to a digital-to-analog converter in the navigation unit. The code indicates that the INS may be used, but the HSI and the AFCS should not be used in the INS modes.

04. Indicates abnormality that may be eliminated if the system is realigned (see page 279).

05. Data from the automatic data entry unit is not reasonable. The system should be switched off, then back on and automatic data loading again be attempted. If this causes warning again, the data may be loaded manually.

Before complying with an action code, the malfunction causing the warning must be determined by operating the 'TEST' switch on the CDU. Successive operation of the switch causes all existing malfunction code numbers to appear in sequence in the right-hand display.

Malfunction codes

These appear in place of the action codes. After the last malfunction code is displayed, either an action code number is again displayed or the display remains blank. If the latter is the case, the malfunction is momentary and the corresponding logic circuits are reset; the 'WARN' light is also extinguished. If, however, the action code number reappears, the recommended procedure associated with it must be carried out.

Any malfunction that occurs is stored in the computer memory and their code numbers, of which there are 31, can be retrieved for maintenance check-out and rectification purposes.

In an IRS, malfunctions are indicated by an amber 'FAULT' light on the IRMP (see Fig. 10.22), a yellow 'fault ball' on the inertial reference unit, and also appropriate messages displayed on the screen of the interfaced engine indicating and crew alerting system (EICAS). An invalid 'bit' is also transmitted to the data busses connecting other user systems so that they may also generate fault messages. For example, messages relating to IRS malfunctions can be generated and displayed on the screens of an ADI and HSI comprising the electronic flight instrument system (EFIS) and also on a maintenance control and display panel (MRCP) unit which is integrated with all interfacing computer-controlled systems in an aircraft. The IR computer stores the appropriate message status words in a non-volatile memory, and by means of the controls on the MRCP, these can then be extracted and identified for the purpose of ground testing and fault isolation.

11 Electronic (CRT) displays

Displays of this type, which are based on the electron beam scanning technique, have been in use in aircraft for very many years. For example, during World War II military aircraft used equipment developed from the then existing ground-based radar systems. With the aid of such equipment, and depending on an aircraft's specific operational role, crews were able to navigate by 'radar mapping' of terrain, to identify ground target areas, and also to detect the positions of hostile intercepting aircraft.

As far as civil aircraft are concerned, this display technology first came into prominence in 1946, with the introduction of weather radar systems to satisfy the operational requirements for transport category aircraft, and it has continued to be an essential part of the 'avionics fit' of this and other categories of aircraft.

The situation, however, of a weather radar display indicator remaining as an isolated item of video equipment was to undergo considerable change, largely as a result of systems analysis, exploration of the versatility of the CRT, and also investigation into methods whereby not only weather data, but also that associated with the many other utilities systems of an aircraft, could be programmed into computers. These had reached such high levels of sophistication and capacity for data processing that it became possible for a single CRT display unit, under microprocessor control, to project the same quantity of system status data which would otherwise have to be displayed by a very large number of conventional-type instruments. Furthermore, the introduction of CRTs and circuits capable of producing a wide range of colours made it possible to differentiate between significant parts of a display, and in particular, to lay emphasis on information of an advisory, cautionary, or warning nature.

The development of such multi-data display technology for both civil and military aircraft was also influenced by the fact that by integrating all computers via a data 'highway' bus, the scene was set for the management of all aspects of in-flight operation to be fully automated while still enhancing flight safety. This also led to improvements in levels of systems' redundancy, changes in the layouts of transport aircraft flight decks, and a reduction in crew complement with the attendant changes in their role and workloads.

The first of the 'new technology' transport aircraft (generally dubbed as 'glass cockpit' aircraft) were the Boeing 757, 767 and

Figure 11.1 Flight deck layout
of the Boeing 757.

Airbus A310. All three were launched as design projects in 1978,
and both the B757 and B767 first entered commercial service in the
US in December 1982. The first A310 services were operated by two
European airlines in April 1983. These aircraft, and several of their
descendant types, are now in service world-wide, together with many
types of smaller aircraft, including helicopters, in which the foregoing
technology has also satisfied an operational need.

Figures 11.1 and 11.2 show the flight deck layouts and CRT
display locations of the B757 and A310 respectively.

Principle of the CRT

A CRT is a thermionic device, i.e. one in which electrons are
liberated as a result of heat energy. As may be seen from Fig. 11.3,
it consists of an evacuated glass envelope, inside which are positioned
an electron 'gun' and beam-focusing and beam-deflection systems.
The inside surface of the screen is coated with a crystalline solid
material known as a phosphor. The electron 'gun' consists of an

Figure 11.2 Flight deck layout of the Airbus A310.

indirectly-heated cathode biased negatively with respect to the screen, a cylindrical grid surrounding the cathode, and two (sometimes three) anodes. When the cathode is heated, electrons are liberated and in passing through the anodes they are made to form a beam.

The grid is maintained at a negative potential, its purpose being to control the current and so modulate the beam of electrons passing through the hole in the grid. The anodes are at a positive potential with respect to the cathode, and they accelerate the electrons to a high velocity until they strike the screen coating. The anodes also provide a means of focusing, which, as will be noted from Fig. 11.3, happens in two stages.

285

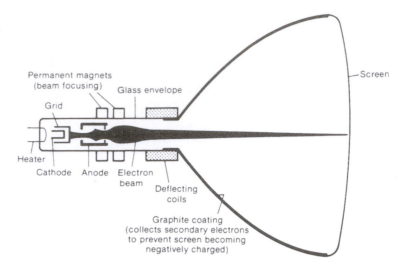

Figure 11.3 Cathode ray tube.

The forces exerted by the field set up between the grid and the first anode bring the electrons into focus at a point just in front of the anode, at which point they diverge, and are then brought to a second focal point by the fields in the region between the three anodes. A focus control is provided which by adjustment of the potential at the third anode makes the focal point coincide with the position of the screen. When the electrons impact on the screen coating, the phosphor material luminesces at the beam focal point, causing emission of a spot of light on the face of the screen.

In order to 'trace out' a luminescent display, it is necessary for the spot of light to be deflected about horizontal and vertical axes, and for this purpose a beam-deflection system is also provided. Deflection systems can be either electrostatic or electromagnetic, the latter being used in the tubes applied to the display units of aircraft systems.

The manner in which an electromagnetic field is able to deflect an electron beam is illustrated in Fig. 11.4. A moving electron constitutes an electric current, and so a magnetic field will exist around it in the same way as a field around a current-carrying conductor. In the same way that a conductor will experience a deflecting force when placed in a permanent magnetic field, so an electron beam can be forced to move when subjected to electromagnetic fields acting across the space within the tube. Coils are therefore provided around the neck of the tube, and are configured so that fields are produced horizontally (X-axis fields) and vertically (Y-axis fields). The coils are connected to the signal sources whose variables are to be displayed, and the electron beam can be deflected to the left or right, up or down, or along some resultant direction depending on the polarities produced by the coils, and on whether one alone is energized, or both are energized simultaneously.

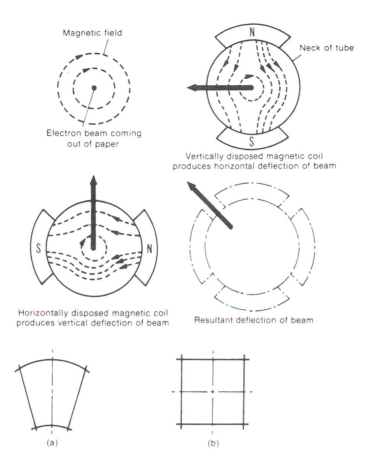

Figure 11.4 Electron beam deflection.

Magnetic field

Electron beam coming out of paper

Neck of tube

Vertically disposed magnetic coil produces horizontal deflection of beam

Horizontally disposed magnetic coil produces vertical deflection of beam

Resultant deflection of beam

Figure 11.5 Data cells. (*a*) Rho-theta; (*b*) X−Y coordinate.

(a)

(b)

Colour CRT displays

These are used in weather radar display units, and are the norm for those units designed for the display of data associated with the systems installed in the types of aircraft referred to earlier. In these display units weather data is also integrated with the other data displays, and since there is a fundamental similarity between the methods through which they are implemented, the operation of a weather radar display unit serves as a useful basis for study of the display principles involved.

The video data received from a radar antenna is conventionally in what is termed rho-theta form, corresponding to the 'sweeping' movement of the antenna as it is driven by its motor (see Fig. 11.5(a)). In a colour display indicator, the scanning of data is somewhat similar to that adopted in the tube of a television receiver, i.e. *raster* scanning in horizontal lines. The received data is still in rho-theta form, but in order for it to be displayed it must be converted into an X−Y coordinate format as shown in Fig. 11.5(b)). This format also permits the display of other data in areas of the screen where weather data is not displayed. In addition it permits a doubling-up of the number of data cells, as indicated by the dotted lines in the diagram.

Each time the radar transmitter transmits a pulse, the receiver begins receiving return echoes from 'targets' at varying distances (rho) from the transmitter. This data is digitized to provide output levels in binary-coded form, and is supplied to the indicator on two data lines. The binary-coded data can represent four conditions corresponding to the level of the return echoes which, in turn, are related to the weather conditions prevailing at the range in nm preselected on the indicator. The data are stored in memories which, on being addressed as the CRT is scanned, will at the proper time permit the weather condition to be displayed. The four conditions are displayed as follows:

Blank screen: Zero or low-level returns.
Green: Low returns (lowest rainfall rate).
Yellow: Moderate returns (moderate rainfall rate).
Red: Strong returns (high-density rainfall rate).

Scan conversion

The principle of conversion from rho-theta form to an X−Y coordinate scan is shown in Fig. 11.6. With a 'target' at point P, at a range R and an angle θ, it will have coordinates: $X = R \sin \theta$ and $Y = R \cos \theta$. Thus, for an echo received at an azimuth angle of, say, 30° and a range of 235 nm, the coordinates will be: $X = 235 \sin 30° = 117.5$ nm, and $Y = 235 \cos 30° = 203.5$ nm. The conversion is performed by a microprocessor on the indicator's display circuit board.

Screen format

The coordinate system format of the screen is shown in Fig. 11.7, and from this example it will be noted that the screen is divided into two halves representing two quadrants in the coordinate system. The origin is at the bottom centre, so that values of X are negative to the left and positive to the right; all values of Y are positive. The screen

Figure 11.6 Scan conversion.

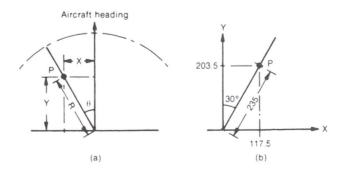

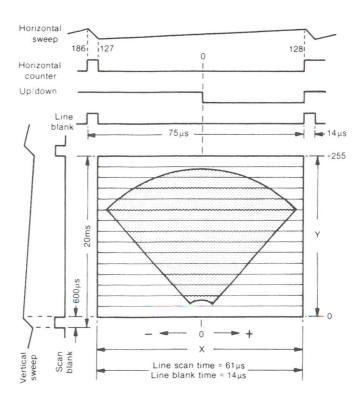

Figure 11.7 Screen format.

is scanned in 256 horizontal lines, and there are 256 'bits' of information displayed on each line.

Each line is located by a value of Y and each bit by a value of X; the screen therefore has a 256 × 256 matrix. The X and Y values are used to address the memory and display the information stored there as the appropriate time in the scan occurs. The memory for the weather data is in two parts which store the bits of the data words that represent the colours red, yellow or green and the corresponding weather conditions. Each part of the memory contains one address for every bit on every line in the display; each memory, therefore, is also a 256 × 256 matrix, and allows the entire weather display to be stored continuously.

As the screen is scanned, the memory is addressed at each point on each line by two counters: a horizontal or X counter for addressing the rows in the memory, and a vertical or Y counter for addressing the columns. The X counter generates an output for each of the 256 bits on a line, and counting is started by a 'high' state output signal from an up/down divider circuit. The counter is caused to count down, i.e. left to right, from the number 186 to 0 at the centre of the screen. When it reaches 0, the divider circuit changes to a 'low' state output, thereby causing the counter to count up to the number 128 at the end of the line, at which point a 'line blank' pulse of 14 μs duration is generated. The line scan time is about 61 μs, and so the total time for each line is 75 μs. The divider circuit again changes to a 'high' state to cause the counter to start down for the next line, and is a process that is repeated for all remaining lines.

An output from the X counter is also applied to the Y counter, which counts to 256 (one for each line) plus eight counts for a scan blank time to allow for the CRT beam 'spot' to return to the upper left corner of the screen. This process is repeated, and since there are 256 lines in the display it takes 20 ms to scan the entire screen (19.4 ms for the 256 lines and 600 μs for the scan blank time). The vertical and horizontal sweep circuits are synchronized by the triggering of the line and scan blank pulses.

In addition to the foregoing raster scanning technique, which produces sections of a CRT screen in 'solid' colour, a *stroke* scanning technique is also used for producing displays of symbols and of data in alphanumeric format. Details of this will be given later in this chapter.

Colour generation

A colour CRT has three electron guns, each of which can direct an electron beam at the screen which is coated with three different kinds of phosphor material. On being bombarded by electron beams, the phosphors luminesce in each of the three primary colours red, green and blue.

The screen is divided into a large number of small areas or dots, each of which contains a phosphor of each kind as shown in Fig. 11.8. The beam from a particular gun must only be able to strike screen elements of one colour, and to achieve this a perforated steel sheet called a *shadow mask* is accurately positioned adjacent to the coating of the screen. The perforations are arranged in a regular pattern, and their number depends on the size of screen; 330 000 is typical.

Beams emitted from each gun pass through the perforations in the mask and they cause the phosphor dots in the coating to luminesce in the appropriate colour. For example, if a beam is being emitted by the 'red' electron gun only, then only the red dots will luminesce, and if the beam completes a full raster scan of the screen, then as a result of persistence of vision by the human eye, a completely red screen will be observed. In the display units of electronic instrument systems, a number of other colours are also required and these are derived by independent circuit control of the three guns and their beam currents, so that as the beams strike the corresponding phosphor dots, the basic process of mixing of primary colours takes place (see Fig. 11.9). In other words, an electronic form of 'paint mixing' is carried out.

Referring once again to the weather radar indicator application, the data readout from the memory, apart from being presented at the

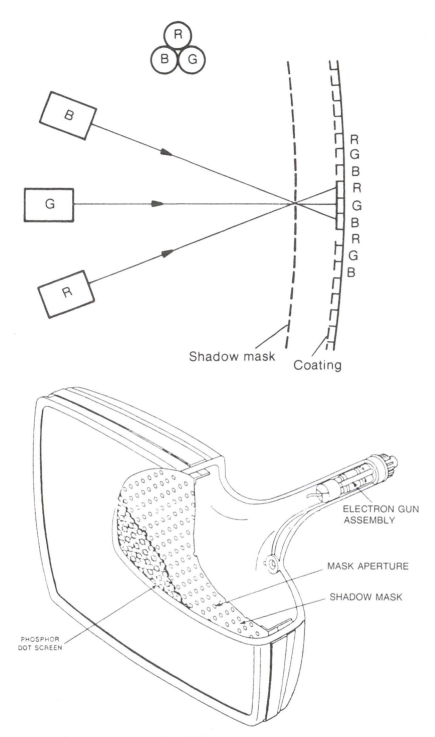

Figure 11.8 Colour CRT.

appropriate location of the CRT screen, must also be displayed in the colours corresponding to the weather conditions prevailing. In order to achieve this, the data is decoded to produce outputs which, after amplification, will turn on the requisite colour guns; the data flow is shown in Fig. 11.10. The memory output is applied to a data

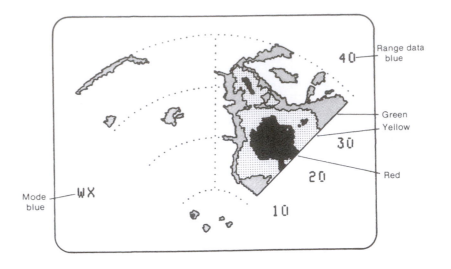

Figure 11.9 Weather data display.

Range data blue

Green
Yellow

Red

Mode blue

WX

40

30

20

10

Table 11.1

Inputs		Outputs			Colour
M	L	A_1	A_2	A_3	
0	0	1	1	1	Black (off)
0	1	0	1	1	Green
1	0	1	0	1	Yellow
1	1	1	1	0	Red

Table 11.2

Outputs to guns			Resulting colours
B_1 Green	B_0 Blue	B_2 Red	
1	1	1	Black (off)
0	0	0	White
0	0	1	Yellow
0	1	1	Red
1	0	0	Light blue
1	0	1	Green

demultiplexer whose output corresponds to the most significant and least significant bits (M and L) of the two-bit binary words and is supplied to a data decoder. The inputs are decoded to provide three-bit output words corresponding to the colours to be displayed, as shown in Table 11.1. The outputs are then applied to the colour decoder and primary encoder circuit, and this in turn provides three outputs, each of which corresponds to one of the colour guns as shown in Table 11.2.

The 'low' state outputs turn on the guns, and from Table 11.2 it

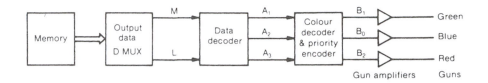

Figure 11.10 Data flow for gun operation.

can also be seen how simultaneous gun operation produces other colours from a mix of the primary colours. Figure 11.9 illustrates a typical weather data display together with associated alphanumeric data, namely ranges in nm, and an operating mode which in this case is WX signifiying 'weather' mode.

Alphanumeric displays

The display of data in alphanumeric and in symbolic form is extremely wide-ranging. For example, in a weather radar indicator it is usually only required for range information and indications of selected operating modes to be displayed, while in systems designed to perform functions within the realm of flight management, a very much higher proportion of information must be 'written' on the screens of the relevant display units. This is accomplished in a manner similar to that adopted for the display of weather data, but additional memory circuits, decoders, and character and symbol generator circuits are required.

Raster scanning is also used, but where datum marks, arcs or other cursive symbols are to be displayed, a *stroke pulse* method of scanning is adopted. The position of each character on the screen is predetermined and stored in a memory matrix, typically 5×7, and when the matrix is addressed, the character is formed within a corresponding matrix of dots on the screen by video signal pulses produced as the lines are scanned.

Figure 11.11 illustrates how, for example, the letters 'WX' and the number '40' are formed. One line of dots is written at a time for the area in which the characters are to be displayed, and so for a 5×7 matrix, seven image lines are needed to write complete characters and/or row of characters. As will be noted from Fig. 11.9, the characters are displayed in blue, so only the 'blue' electron gun is active in producing them. Spacing is necessary between individual characters and also between rows of characters, and so extra line 'blanking bits', e.g. three, are allocated to character display areas.

In the example of the weather radar indicator, the characters each have an allocation of eight bits (five for the characters and three for the space following) on each of 21 lines (14 for the character and seven for the space below). The increase in character depth to 14 lines is derived from an alphanumeric address generator output that writes each line in a character twice during line scanning. The

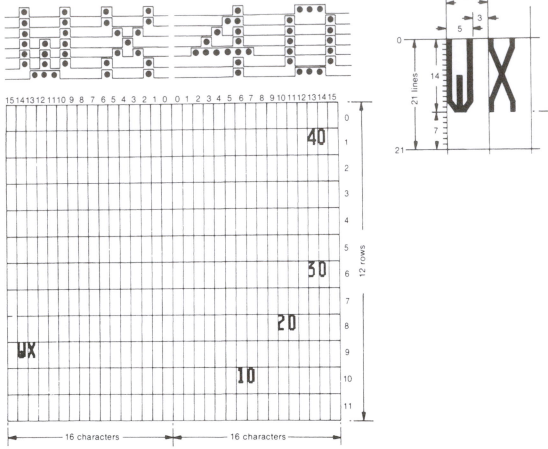

Figure 11.11 Alphanumeric display.

character format in this case permits the display of 12 rows each of 32 characters.

The CRT display units of the more comprehensive electronic instrument systems (see Chapters 12 and 16) operate on the same fundamental principles as those described, but in applying them, more extensive microprocessing circuit arrangements are required in order to display far greater amounts of changing data in quantitative and qualitative form.

The microprocessor processes information from the data 'highway' bus and, from the memory circuits, it is instructed to call up sub-programs, each of which correspond to the individual sets of data that are required to be displayed. Signals are then generated in the relevant binary format, and are supplied to a symbol generator unit. This unit, in turn, generates and supplies signals to the beam deflection and colour gun circuits of the CRT, such that its beams are raster and stroke scanned, to present the data at the relevant parts of the screen, and in the required colour.

The displayed data is in two basic forms: fixed and moving. Fixed data relate in particular to such presentations as symbols, scale markings, names of systems, datum marks, names of parameters being measured, etc. Moving data are in the majority, of course, since they present changes occurring in the measurement of all parameters essential for in-flight management. The changes are indicated by the movement of symbolic pointers, index marks, digital counter presentations, and system status messages, to name but a few.

12 Electronic flight instrument systems

As far as the pure basic functions and number of display units are concerned, this system, which is generally referred to as 'EFIS' (pronounced ee-fiss), may be considered as being similar to the types of flight director system described in Chapter 9. However, since it is fully integrated with digital computer-based navigation systems, and utilizes colour CRT types of ADI and HSI, then it is far more sophisticated not only in terms of physical construction, but also in the extent to which it can present attitude and navigational data to the flight crew of an aircraft.

Units of a system

As in the case of conventional flight director systems, a complete EFIS installation is made up of left (Captain) and right (First Officer) systems. Each system in turn is comprised of two display units: an electronic attitude director indicator (EADI) and an electronic horizontal situation indicator (EHSI), a control panel, a symbol generator (SG), and a remote light sensor unit. A third (centre) SG is also incorporated so that its drive signals may be switched to either the left or right display units in the event of failure of the corresponding SGs. The signal switching is accomplished within the left and right SGs, using electromechanical relays powered from an aircraft's dc power supply via pilot-controlled switches. The interface between EFIS units, data busses, and other systems is shown in Fig. 12.1.

Display units

Each display unit consists of the sub-units shown in Fig. 12.2. The power supply units provide the requisite levels of ac and dc power necessary for overall operation; the supplies are automatically regulated and monitored for undervoltage and overvoltage conditions.

The video/monitor card contains a video control microprocessor, video amplifiers and monitoring logic for the display unit. The main tasks of the processor and associated ROM and RAM memories are to calculate gain factors for the three video amplifiers (red, blue and green), and perform input and sensor and display unit monitor

Figure 12.1 EFIS units and
signal interfacing.

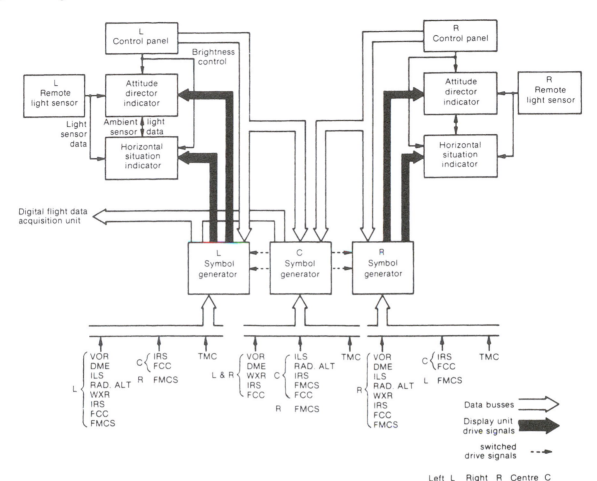

functions. the input/output interface functions for the processor are
provided by analog multiplexers, an A/D converter and a D/A
converter.

The function of the convergence card is to take X and Y deflection
signals and to develop drive signals for the three radial convergence
coils (red, blue and green) and the one lateral convergence coil (blue)
of the CRT. Voltage compensators monitor the deflection signals in
order to establish on which part of the CRT screen the beams are
located (right or left for the X comparator, and top or bottom for the
Y comparator).

Signals for the X and Y beam deflections for stroke and raster
scanning are provided by the deflection amplifier card. The amplifiers
for both beams each consist of a two-stage preamplifier and a power
amplifier. Both amplifiers use two supply inputs, 15 V dc and 28 V

Figure 12.2 Display unit.

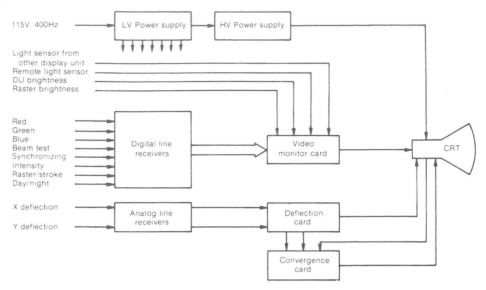

dc; the former is used for effecting most of the stroke scanning or writing, while the latter is used for repositioning and raster scanning.

The interconnect card serves as the interface between the external connector of a display unit and the various cards. Digital line receivers for the signals supplied by the SGs are also located on this card.

In a typical system, six colours are assigned for the display of the many symbols, failure annunciators, messages and other alphanumeric information, and are as follows:

White   Display of present situation information.

Green   Display of present situation information where contrast with white symbols is required, or for data having lower priority than white symbols.

Magenta All 'fly to' information such as flight director commands, deviation pointers, active flight path lines.

Cyan   Sky shading on an EADI and for low-priority information such as non-active flight plan map data.

Yellow   Ground shading on an EADI, caution information display such as failure warning flags, limit and alert annunciators and fault messages.

Red   For display of heaviest precipitation levels as detected by the weather radar.

Symbol generators (SGs)

These provide the analog, discrete and digital signal interfaces between an aircraft's systems, the display units and the control panel,

Figure 12.3 Symbol generator
and card interfacing.

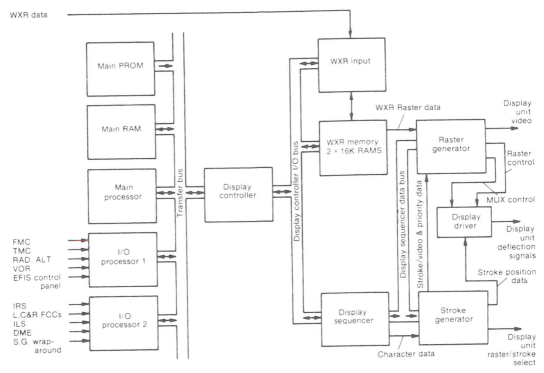

and they perform symbol generation, system monitoring, power
control and the main control functions of the EFIS overall. The
interfacing between the card modules of an SG is shown in Fig. 12.3,
and card functions are given in Table 12.1.

Control panel

A control panel is provided for each system, and, as shown in Fig.
12.4, the switches are grouped for the purpose of controlling the
displays of their respective EADI and EHSI units as listed in Table
12.2.

Remote light sensor

This is a photodiode device which responds to flight deck ambient
light conditions and automatically adjusts the brightness of the CRT
displays to a compatible level.

Display presentations   The EADI displays traditional pitch and roll attitude indications
against a raster-scanned background, and as may be seen from the

299

Table 12.1 Symbol generator card functions

Card	Function
I/O 1 & 2	Supply of input data for use by the main processor
Main processor	Main data-processing and control for the system
Main RAM	Address decoding, read/write memory and I/O functions for the system
Main PROM	Read-only memory for the system
Display controller	Master transfer bus interface
WXR input	Time scheduling and interleaving for raster, refresh, input and standby functions of weather radar input data
WXR memory	RAM selection for single-input data, row and column shifters for rotate/translate algorithm, and shift registers for video output
Display sequencer	Loads data into registers on stroke and raster generator cards
Stroke generator	Generates all single characters, special symbols, straight and curved lines and arcs on display units
Raster generator	Generates master timing signals for raster, stroke, EADI and EHSI functions
Display driver	Converts and multiplexes X and Y digital stroke and raster inputs into analog for driver operation, and also monitors deflection outputs for proper operation

example illustrated in Fig. 12.5, the upper half is in cyan and the lower half in yellow. Attitude data is provided by an IR system. Also displayed are flight director commands, localizer and glide slope deviation, selected airspeed, ground speed, AFCS and autothrottle system modes, radio altitude and decision height.

Figure 12.5 illustrates a display representative of an automatically-controlled approach to land situation together with the colours of the symbols and alphanumeric data produced via the EFIS control panel and SGs. The autoland status, pitch, roll-armed and engaged modes are selected on the AFCS control panel, and the decision height is selected on the EFIS control panels. Radio altitude is digitally displayed during an approach, and when the aircraft is between 2500 and 1000 ft above ground level. Below 1000 ft the display automatically changes to a white circular scale calibrated in increments of 100 ft, and the selected decision height is then displayed as a magenta-coloured marker on the outer scale. The radio altitude also appears within the scale as a digital readout. As the aircraft descends, segments of the altitude scale are simultaneously erased so that the scale continuously diminishes in length in an anti-clockwise direction.

At the selected decision height plus 50 ft, an aural alert chime sounds at an increasing rate until the decision height is reached. At the decision height, the circular scale changes from white to amber

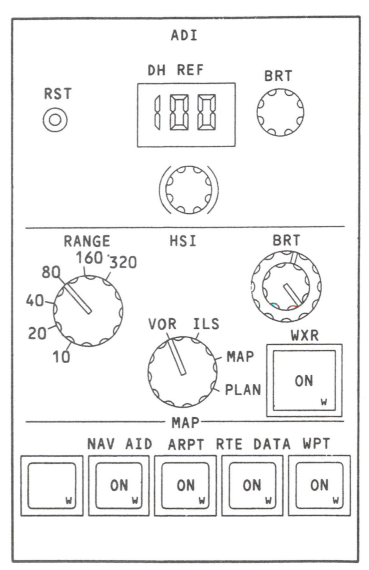

Figure 12.4 Control panel.

ARPT Airport
RTE DATA Route Data
WPT Waypoint

and the marker changes from magenta to amber; both the scale and marker also flash for several seconds. A reset button is provided on the control panel and when pressed it stops the flashing and causes the scale and marker to change from amber back to their normal colour.

If during the approach the aircraft deviates beyond the normal ILS glide slope and/or localizer limits (and when below 500 ft above ground level), the flight crew are alerted by the respective deviation pointers changing colour from white to amber; the pointers also start flashing. This alert condition ceases when the deviations return to within their normal limits.

Table 12.2 Control panel switch functions

Switch	Function
EADI section:	
BRT	Controls levels of display brightness.
DH SET	Setting of decision height.
RST	Manually resets decision height circuits after aircraft has passed through decision height.
EHSI section:	
RANGE	Selects range for displayed WXR and navigation data.
MODE SELECT	Selects display appropriate to mode required.
BRT	Outer knob controls main display brightness; inner knob controls WXR display.
WXR	When pushed in, WXR data displayed during all modes except PLAN.
MAP switches	Used in MAP mode, and when pushed in they cause their placarded data to be displayed. Illuminate white.

Figure 12.5 EADI display.

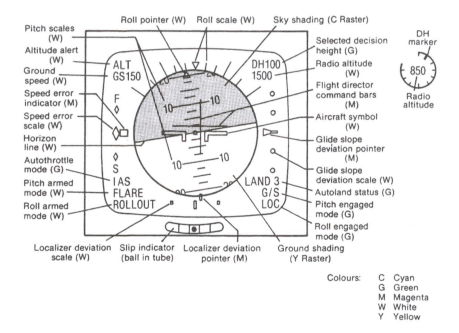

The EHSI presents a selectable, dynamic colour display of flight progress and plan view orientation. Four principal display modes may be selected on the control panel: MAP, PLAN, ILS and VOR. Figure 12.6 illustrates the normally-used MAP mode display which, in conjunction with the flight plan data programmed into a flight management computer, displays information against a moving map background with all elements positioned to a common scale.

The symbol representing the aircraft is at the lower part of the display and an arc of the compass scale, or rose, covering 30° on either side of the instantaneous track is at the upper part of the

Figure 12.6 EHSI in 'MAP' mode.

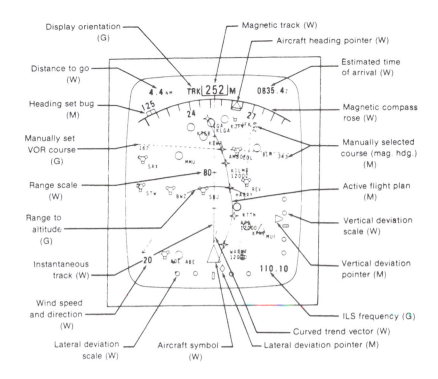

Display orientation (G)

Magnetic track (W)

Aircraft heading pointer (W)

Distance to go (W)

Estimated time of arrival (W)

Heading set bug (M)

Magnetic compass rose (W)

Manually set VOR course (G)

Manually selected course (mag. hdg.) (M)

Range scale (W)

Active flight plan (M)

Range to altitude (G)

Vertical deviation scale (W)

Instantaneous track (W)

Vertical deviation pointer (M)

Wind speed and direction (W)

ILS frequency (G)

Lateral deviation scale (W)

Aircraft symbol (W)

Curved trend vector (W)

Lateral deviation pointer (M)

Symbols:

✦ Waypoints: Active (M) one aircraft currently navigating to.
       Inactive (W) a navigation point making up selected active route.

◯ Airports (C)

⬡ Navaids (C)

╱ Wind direction (W) with respect to map display orientation
       and compass reference.

△ Off-route waypoints (C)

Colours: C Cyan
              G Green
              M Magenta
              W White

display. Heading information is supplied by the appropriate IRS, and the compass rose is automatically referenced to magnetic north (via a crew-operated 'MAG/TRUE' selector switch) when between latitudes 73° N and 65° S, and to true north when above these latitudes. When the selector switch is set at 'TRUE', the compass rose is referenced to true north regardless of latitude.

Tuned VOR/DME stations, airports and their identification letters, and the flight plan entered into the flight management system computer are all correctly oriented with respect to the positions and track of the aircraft, and to the range scale (nm/in) selected on the EFIS control panel. Weather radar 'returns' may also be selected and

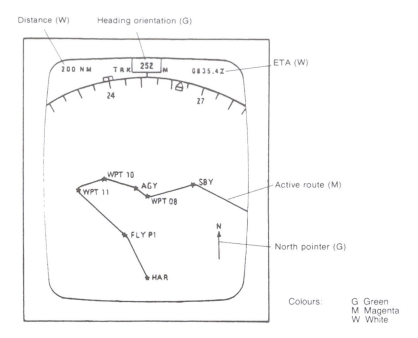

Figure 12.7 EHSI in 'PLAN' mode.

displayed when required, at the same scale and orientation as the map.

Indications of other data such as wind speed and direction, lateral and vertical deviations from the selected flight profile, distance to waypoint, etc., are also displayed.

The map display also provides two types of predictive information. One combines current ground speed and lateral acceleration into a prediction of the path over the ground to be followed over the next 30, 60 and 90 seconds. This is displayed by a curved track vector, and since a time cue is included the flight crew are able to judge distances in terms of time. The second prediction, which is displayed by a range to altitude arc, shows where the aircraft will be when a selected target altitude is reached.

In the PLAN mode, a static map background with active route data oriented to true north is displayed in the lower part of the HSI display, together with the display of track and heading information as shown in Fig. 12.7. Any changes to the route are selected at the keyboard of the flight management system display unit, and appear on the EHSI display so that they can be checked by the flight crew before they are entered into the flight management computer.

The VOR and ILS modes present a compass rose (either expanded or full) with heading orientation display as shown in Fig. 12.8. Selected range, wind information and system source annunciation are also displayed. If selected on the EFIS control panel, weather radar returns may also be displayed, though only when the mode selected presents an expanded compass rose.

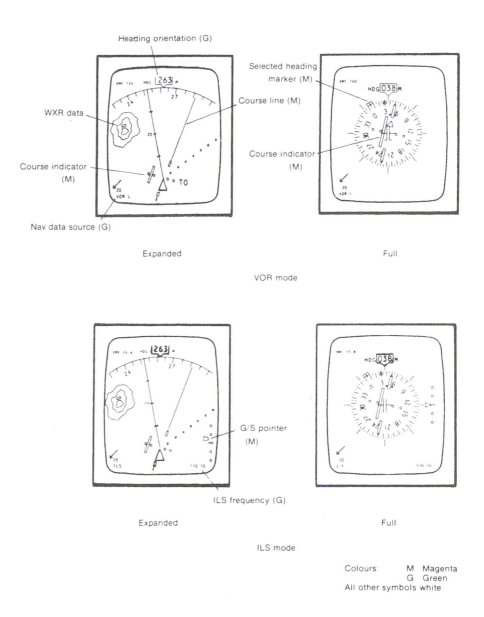

Figure 12.8 VOR and ILS mode displays.

Heading orientation (G)

Selected heading marker (M)

Course line (M)

WXR data

Course indicator (M)

Course indicator (M)

Nav data source (G)

Expanded

Full

VOR mode

G/S pointer (M)

ILS frequency (G)

Expanded

Full

ILS mode

Colours: M Magenta
          G Green
All other symbols white

Failure annunciation

Failure of data signals from such systems as the ILS and radio altimeter are displayed on each EADI and EHSI in the form of yellow flags 'painted' at specific matrix locations on their CRT screens. In addition, fault messages may also be displayed: for example, if the associated flight management computer and weather radar range disagree with the control panel range data, the discrepancy message 'WXR/MAP RANGE DISAGREE' appears on the EHSI.

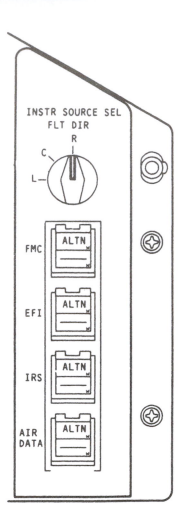

Figure 12.9 Source selector switch panel.

INSTR SOURCE SEL
FLT DIR

R

C

L

FMC   ALTN

EFI   ALTN

IRS   ALTN

AIR
DATA   ALTN

Data source selection

In the type of system described, means are provided whereby the pilots can, and independently of each other, connect their respective display units to alternate sources of input data, e.g. left or right ADCs, flight management computers, flight control computers, and standby IRS.

Each pilot has a panel of selector switches arranged as shown in Fig. 12.9. The upper rotary type of switch connects either the left, centre or right flight control computer to the EADI as the source of attitude data. The other switches are of the illuminated push type and are guarded to prevent accidental switching. In the normal operating configuration of systems they remain blank, and when activated they are illuminated white.

Display of air data

In a number of EFIS applications, the display of such air data as altitude, airspeed and vertical speed is provided in the conventional

Figure 12.10 Flight deck layout of the Boeing 747—400 series aircraft.

manner, i.e. separate indicators servo-operated from ADCs are mounted adjacent to the EFIS display units in the basic 'T' arrangement (see page 21). With the continued development of display technology, however, CRTs with much larger screen areas have been produced and, as may be seen from the Boeing 747—400 aircraft flight deck layout in Fig. 12.10, (*see also* front cover), such displays make it unnecessary to provide conventional primary air data instruments for each pilot.

13 Engine instruments

At the present time there are three principal types of engine in use, namely, piston (unsupercharged and turbocharged), turbopropeller, and pure turbine, and their selection as the means of propulsion for any one type of aircraft depends on its size and operational category. In each case there are certain parameters that are required to be monitored to ensure that they are operated in accordance with their designed performance ratings, and within specific limitations. The parameters involved overall are listed in Fig. 13.1, the actual number required varying, of course, in accordance with the type of engine.

Monitoring is accomplished by means of specifically designed instrument systems, the sensor units of which may, in very basic form, be incorporated within an indicator, or be of the remote type which transmit data in the form of electrical signals to 'clock'-type indicators, or to electronic display units.

Certain of the parameters listed in Fig. 13.1 relate to the operation of engine *systems*, e.g. the pressure and temperature of lubricating oil systems, while others are more of a primary nature in that they relate directly to the performance of engines in terms of their *power* and/or *thrust*.

The instrument systems required for monitoring purposes may, therefore, be broadly grouped into these two main areas, and so the details of operating principles and construction typical of those currently in use are accordingly arranged to form the subject of this chapter and also of Chapter 15. The monitoring of data by means of CRT-type display units will be covered in Chapter 16.

Pressure measurement

Pressure is measured by instrument systems which in the majority of applications are of the remote-indicating type, i.e. their sensor (or

Figure 13.1 Monitored operating parameters.

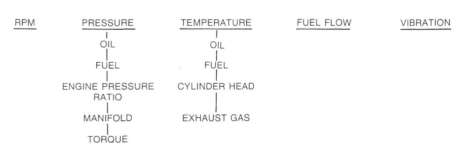

RPM	PRESSURE	TEMPERATURE	FUEL FLOW	VIBRATION
	OIL	OIL		
	FUEL	FUEL		
	ENGINE PRESSURE RATIO	CYLINDER HEAD		
	MANIFOLD	EXHAUST GAS		
	TORQUE			

transmitter) units are connected to a pressure source located at some remote point, and they transmit data through an electrical transmission circuit.

The sensor units contain elements which, depending on the particular design and pressure ranges to be measured, are in the form of either metal capsules, diaphragms or bellows. Another form of element which may be mentioned at this point is the earliest type ever to be adopted, namely, the Bourdon tube. In present-day aircraft, however, its application, if required at all, is limited to certain systems other than those of engines, in which direct-reading indicators may be permitted for monitoring of operation.

The elements are mechanically connected to electrical transmitters which, in some cases, are of the moving core or synchro type of induction device. These, in turn, are connected to indicators which incorporate either a moving coil mechanism, a synchro receiver, or a synchro/servomotor mechanism, as appropriate to the particular design.

Indicating systems

Figure 13.2 illustrates the arrangement of a type of turbine engine oil pressure indicating system that utilizes a TX synchro transmission system (see Chapter 5). The rotor of the TX is in this case connected via a mechanical quadrant and pinion to two bellows; one is sensitive

Figure 13.2 Synchronous transmission-type pressure indicating system.

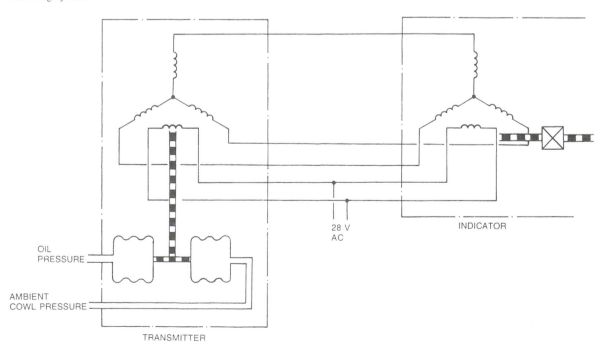

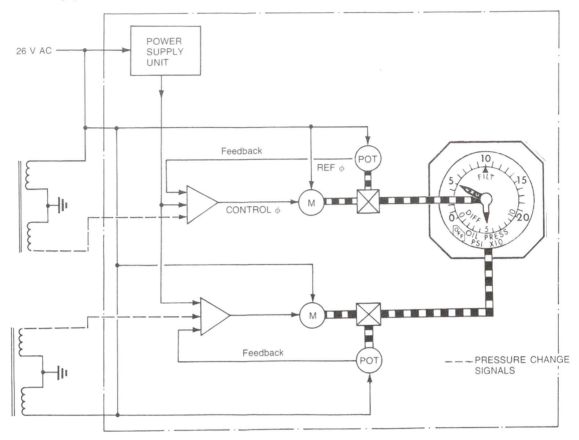

Figure 13.3 Servo-operated pressure indicating system.

to oil pressure, the other to prevailing pressure of the ambient air in the cowled area of the engine. Thus, the rotor is positioned so as to produce output signals proportional to the difference between the two pressures, i.e. gauge pressure. In responding to these signals, the TR rotor positions the pointer over the indicator scale via an appropriate ratio gear train. In the event of failure of electrical power to the system then, due to the gear train, the indicator pointer will remain at the pressure value that was being measured at the time of failure.

An example of a system which employs a capsule and moving core type of inductive sensor and a servo-operated indicator is shown in Fig. 13.3. This system is used in some series of Boeing 747 for the measurement of such parameters as engine oil and oil filter inlet pressures, fuel pump inlet and discharge pressures, and engine breather pressure.

In its application to oil pressure measurement, the indicator contains dual servomechanisms connected to individual transmitters, one of which senses oil system pressure, the other oil filter inlet

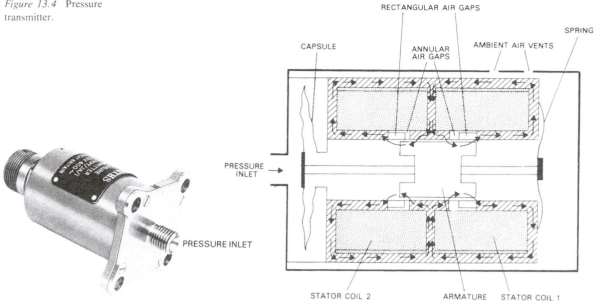

Figure 13.4 Pressure transmitter.

RECTANGULAR AIR GAPS

CAPSULE

ANNULAR AIR GAPS

AMBIENT AIR VENTS

SPRING

PRESSURE INLET →

PRESSURE INLET

STATOR COIL 2   ARMATURE   STATOR COIL 1

pressure. The only difference between the two is their operating pressure ranges, which respectively are from 0 to 100 psi and 0 to 200 psi.

When pressures are applied to the capsules they vary the position of the inductor core, and thereby cause a change in reluctance between the two windings so that output signals proportional to pressure are produced. These signals are supplied to the amplifiers of the indicator servomechanisms. The mechanism connected to the oil pressure transmitter drives a 'double' pointer, one part of which registers against a fixed outer scale. The mechanism connected to the oil filter inlet pressure transmitter drives an inner disc with an index marker that is also registered against the outer scale.

The inner disc also has a scale which registers against the second part of the double pointer so that, in operation, a continuous indication of pressure difference is provided. In the example readings shown in Fig. 13.3, the oil pressure is 50 psi, filter inlet pressure is 100 psi and so the difference indicated is 50 psi.

The servomotor of each mechanism drives a potentiometer which provides feedback signals to balance out those from the transmitters. A solenoid-operated flag is provided in the indicator, and comes into view whenever there is a failure of the 26 V ac power supply to the system.

Another form of ac inductor type of pressure transmitter is shown in Fig. 13.4. It utilizes a capsule which positions an armature core relative to air gaps in the core of a stator. With pressure applied as indicated, the length of the air gaps associated with stator coil 1 is decreased, while that associated with coil 2 is increased. As the

311

reluctance of the magnetic circuit across each coil is proportional to the effective length of the air gap, then the inductance of coil 1 will be increased and that of coil 2 descreased; the current flowing in the coils will, respectively, be decreased and increased. The output signals are supplied to a moving coil type of indicator which operates on the dc ratiometer principle (see page 316).

In several types of 'new technology' aircraft, pressure transmitters are, as far as internal mechanical arrangement is concerned, much simpler in that they use piezoelectric-type sensors as in the case of air data computers (see page 165).

Pressure switches

In the measurement of pressure, it is a requirement in the case of some engine systems that the flight crew be given some positive indication of pressure variations which could constitute hazardous operating conditions. This requirement also applies, of course, to other systems involving liquids and/or gases under pressure which are used in the operation of aircraft. To meet this requirement, therefore, pressure switches are installed in the relevant systems and are connected to indicator or warning lights (in some cases aural warning devices also) located on cockpit or flight deck panels.

The most common type of pressure switch consists of a diaphragm- (or capsule)-type sensor which is exposed on one side to oil or fuel pressure as appropriate, and on the other side to a local ambient pressure. The displacements of the sensor in response to pressure changes cause it to activate a switch, the contacts of which are connected to the relevant annunciator light and remain open under the normal high-pressure operating conditions. If the pressure should fall to a value at which the contacts have been pre-set to 'make', then the circuit is completed for illumination of the annunciator light.

In engine oil systems, valves are provided to bypass oil around filters in the event of their becoming clogged, and as an indication of this bypassing operation pressure switches can also be used. An example of one such application is shown in Fig. 13.5. In this case, one side of the switch sensor is exposed to filter inlet pressure, and the other side to filter outlet pressure, and, as will also be noted, its switch contacts are arranged the reverse way round to that just described, i.e. they keep the annunciator light circuit open under low pressure conditions. Thus, if the pressure difference across the filter increases to the pre-set highest value, the contacts complete the circuit to the annunciator light which illuminates to indicate that the bypass valve is about to open so that the oil can flow around the filter.

The pressure inlets of switch units are normally in the mounting flange, and they may either be in the form of plain entry holes

Figure 13.5 Pressure switch operation.

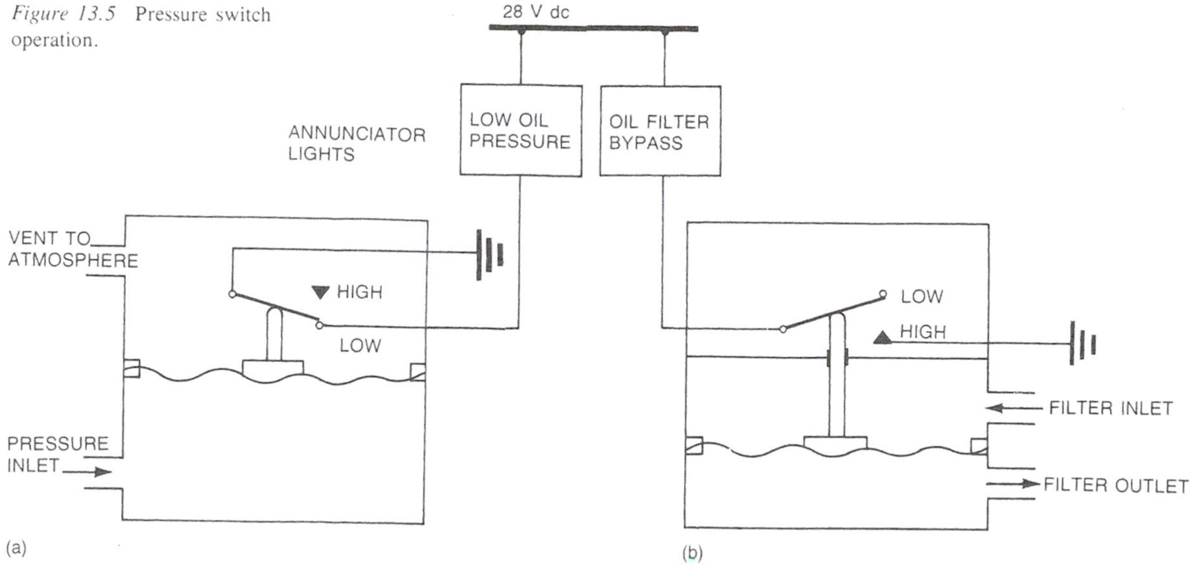

28 V dc

ANNUNCIATOR LIGHTS

LOW OIL PRESSURE

OIL FILTER BYPASS

VENT TO ATMOSPHERE

▼ HIGH

LOW

LOW

▲ HIGH

FILTER INLET

FILTER OUTLET

PRESSURE INLET

(a)

(b)

Figure 13.6 Typical pressure switch.

PRESSURE CONNECTION

directly over the pressure source, spigots with 'O' ring seals as in the example shown in Fig. 13.6, or threaded connectors for flexible or rigid pipe coupling.

Temperature measurement

In most forms of temperature measurement, the variation of some property of a substance with temperature is utilized. These variations may be summarized as follows:

1. Most substances expand as their temperature rises; thus, a measure of temperature is obtainable by taking equal amounts of expansion to indicate equal increments of temperature.

2. When subjected to a temperature rise, many liquids experience such motion of their molecules that there is a change of state from liquid to vapour. Equal increments of temperature may, therefore, be indicated by measuring equal increments of vapour pressure.

3. Substances change their electrical resistance when subjected to varying temperatures, so that a measure of temperature is obtainable by taking equal increments of resistance to indicate corresponding changes of temperature.

4. Dissimilar metals when joined at their ends produce an electromotive force (*thermo-emf*) dependent on the difference in temperature between the two junctions. Since equal increments of temperature are only required at one junction, a measure of the thermo-emf produced will be a measure of the junction temperature.

5. The radiation emitted by any body at any wavelength is a function of the temperature of the body, and is termed its emissivity. If, therefore, the radiation is measured and the emissivity is known, the temperature of the body can be determined; such a measuring technique is known as radiation pyrometry.

In relation to engine instruments, we are particularly interested in variations 3 and 4, since they are utilized extensively for the measurement of temperature of such liquids and gases as oil, fuel, carburettor air, and turbine exhaust gases. In certain types of turbojet engine, the radiation pyrometry technique is adopted for the measurement of actual turbine blade temperature.

Indicating systems

These fall into two main categories: variable resistance and thermoelectric, and so, respectively, they are directly related to the variations 3 and 4 noted earlier.

Variable resistance systems

A system consists of a sensor unit (generally referred to as a 'bulb') and an indicator, connected in a series circuit configuration, and requiring dc power which may be directly supplied from a relevant busbar or, in some cases, by rectification of a single-phase ac supply. Sensor units employ resistance coils of either nickel or platinum wire, and the indicator units are of the moving coil type, having their internal circuits arranged in either the basic Wheatstone bridge configuration, or in the more commonly adopted ratiometer configuration.

Sensor units   The general arrangement of a sensor unit commonly used for the measurement of liquid temperatures is shown schematically in Fig. 13.7. The resistance coil is wound on an insulated former and the ends of the coil are connected to a two-pin

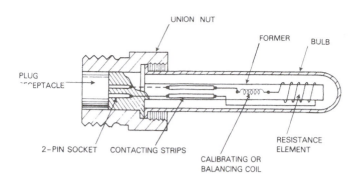

Figure 13.7 Schematic arrangement of a temperature sensor.

socket (or a plug) via contact strips. The 'bulb', which serves as a casing to protect and to seal the coil assembly, is a stainless steel tube closed at one end and secured to a union nut at the other. The union nut is used for securing the complete unit to the connecting point at which the temperature is to be measured.

It will be noted from the diagram that the coil is wound at the bottom end of its former and not along the full length. This ensures that the coil is well immersed in the hottest part of the liquid, thus minimizing errors due to radiation and conduction losses in the 'bulb'.

A calibrating or balancing coil is normally provided so that a standard constant temperature/resistance characteristic can be obtained, thus permitting interchangeability of sensor units. In addition the coil compensates for any slight change in the physical characteristics of the sensor material. The coil, which may be made from Manganin or Eureka wire, is connected in series with the sensor coil and is pre-set in value during initial calibration by the manufacturer.

The resistance of nickel and platinum increases with an increase in temperature, and the corresponding values are given in Table 3 on page 409.

Wheatstone bridge systems   Figure 13.8 shows the arrangement of a system in which measurements are based on the principle of the Wheatstone bridge circuit. Although very old in concept, it is still applied to temperature indicators adopted in some types of aircraft.

The temperature sensor forms the unknown resistance arm R_x of the bridge, while the other three are contained within the moving coil type of indicator; resistance values of each of these three arms are fixed. The moving coil is wound on a former which is pivoted so that it can rotate within the field of a permanent magnet, and, by means of two controlling hairsprings, it is connected across the points B and D of the bridge. The dc supply for the system is connected across points A and C.

When the sensor R_x is subjected to temperature variations, its

Figure 13.8 Wheatstone bridge system.

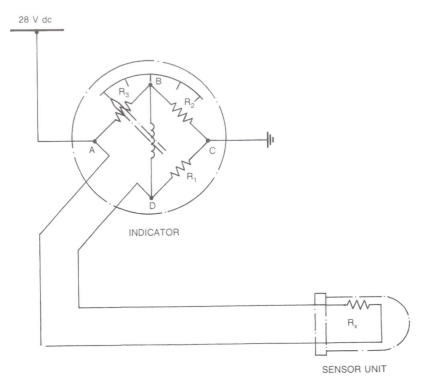

28 V dc

INDICATOR

SENSOR UNIT

resistance will also vary. This upsets the balance of the bridge circuit, and the value of R_x at any particular temperature will govern the amount of current flowing through the moving coil. Thus, for a given value of R_x, the out-of-balance current is a measure of the prevailing temperature. The coil current produces a surrounding magnetic field which, on interacting with that of the permanent magnet, results in rotation of the coil, and movement of its pointer to a scale position that indicates the temperature.

In this type of system, there is only one point at which the circuit is balanced, and at which no current will flow through the moving coil; this is its null point. It is usually denoted on the indicator scale by a datum mark, against which the pointer registers when the power supply is disconnected.

A bridge circuit has the disadvantage that the out-of-balance current also depends on the voltage of the power supply. Hence, errors in indicated readings can occur if the voltage differs from that for which the system was initially calibrated.

Ratiometer system A ratiometer system also consists of a resistive sensor and a moving-coil indicator which, unlike the conventional type, has two coils moving together in a permanent magnetic field of non-uniform strength. The coil arrangements and methods of obtaining the non-uniform field depend on the manufacturer's design, but three typical methods are shown in Fig. 13.9.

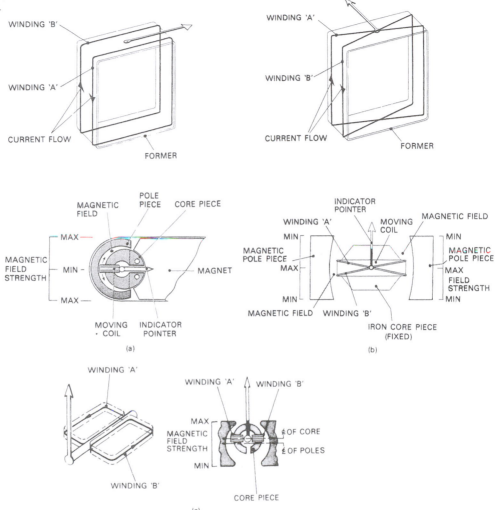

Figure 13.9 Ratiometer coil and field arrangements. (*a*) Crossed-coil; (*b*) parallel coil; (*c*) twin former.

Figure 13.10 shows the circuit in basic form, and from this it will be noted that two parallel resistance arms are formed, one containing a coil and a fixed calibrating resistance R_1, and the other containing a coil in series with a calibrating resistance R_2 and the sensor R_x. Both arms are supplied with either pure dc or rectified ac, and the coils are so wound that current flows through them in opposite directions. As in any moving-coil indicator, rotation of the measuring element, i.e. coil former and pointer, is produced by forces which are proportional to the product of the current and field strength, and the direction of rotation depends on the direction of current relative to the field. In a ratiometer, therefore, it follows that the force produced by one coil will always tend to rotate the measuring element in the opposite direction to the force produced by the second coil, and

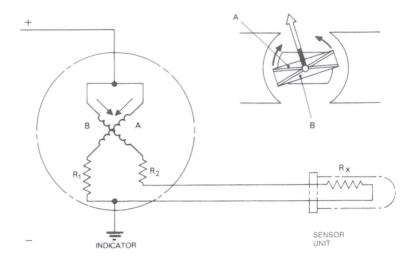

Figure 13.10 Basic ratiometer circuit.

furthermore, as the magnetic field is of non-uniform strength, the coil carrying the greater current will always move towards the area of the weaker field, and vice versa.

For purposes of explanation, let us assume that the basic circuit of Fig. 13.10 employs an indicator which has the crossed-coil method (see (a) of Fig. 13.9), that winding 'B' is in the variable-resistance or sensor arm, and winding 'A' is in the fixed-resistance arm. The resistances of the arms are so chosen that at the zero position of the indicator scale the forces produced by the currents flowing in each winding are in balance. Although the currents are unequal at this point, and indeed at all other points except mid-scale, the balancing of the torques is always produced by the strength of the field in which the windings are positioned.

When the temperature of the sensor R_x increases, then in accordance with the temperature/resistance relationship of the material used for the sensor, its resistance will increase and so cause a decrease in the current flowing in winding 'B' and a corresponding decrease in the force created by it. The current ratio is therefore altered, and the force in winding 'A' will rotate the measuring element so that both windings are carried round the air gap; winding 'B' is advanced further into the stronger part of the field, while winding 'A' is being advanced into a weaker part. When the sensor temperature stabilizes at its new value, the forces produced by both windings will once again balance, at a new current ratio, and the angular deflection of the measuring element will be proportional to the temperature change.

When the measuring element is at the mid-position of its rotation, the currents in both windings are equal since this is the only position where the two windings can be in the same field strength simultaneously.

In a conventional moving-coil indicator, the controlling system is

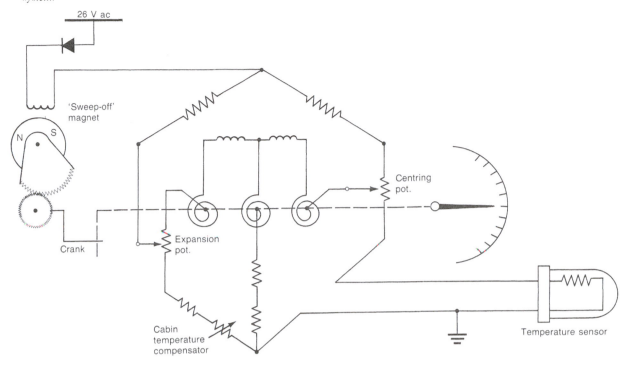

Figure 13.11 Ratiometer type of temperature indicating system.

26 V ac

'Sweep-off' magnet

N S

Crank

Expansion pot.

Cabin temperature compensator

Centring pot.

Temperature sensor

made up of hairsprings which exert a controlling torque proportional to the current flowing through the coil. Therefore, if the current decreases due to a change in the power supply applied to the indicator, the deflecting torque will be less than the controlling torque of the springs and so the coil will move back to a position at which equilibrium between torques is again established. The pointer will thus indicate a lower reading. A ratiometer, on the other hand, does not require hairsprings for exerting a controlling torque, since this is provided solely by the appropriate coil winding and non-uniform field arrangements. Should variations in the power supply occur they will affect both coils equally so that the ratio of currents flowing in the coils remains the same, and tendencies for them to move to positions of differing field strength are counterbalanced.

Having noted this point, however, a spring is, in fact, used in practical applications, but its sole function is to return the measuring element to the 'off-scale' position when the power supply is disconnected. Since it exerts a very much lower torque than a conventional control spring, its effects on the indicating accuracy of a ratiometer in response to power supply changes are very slight.

Figure 13.11 schematically illustrates the circuit arrangement of a type of ratiometer used for the measurement of oil temperature in some series of Boeing 737. It utilizes a twin-coil former system and operates from a 26 V ac supply which is rectified within the indicator. Two potentiometers are included in the circuit, and are pre-adjusted, one to set the range (expansion) and the other (centralizing) to set the mid-scale point at which the currents in the coils should be equal. A thermoresistor, referred to as a cabin temperature compensator, is also included in one arm of the circuit; its purpose is to compensate for the effects that changes of ambient temperature at the indicator's location could otherwise have on the current ratios produced by oil temperature changes.

Another feature of the indicator is the method adopted for returning the coils and pointer to an off-scale position when electrical power is disconnected. This is done by a 'sweep-off' magnet and a crank which mechanically moves the measuring element. Adjacent to the magnet is a coil which is connected in the indicator circuit, so that, when power is switched on, there is an interaction of magnetic fields causing the crank to move clear of the measuring element, thus allowing its normal range of movement.

Thermo-emf systems

These systems play an important part in monitoring the structural integrity of vital components of air-cooled piston engines and turbine engines when operating at high temperatures. In the former class of engine the components concerned are the cylinders, while in turbine engines they are the turbine rotor discs and blading. In basic form, the systems consist of a thermocouple type of sensor which, depending on the application, is secured to an engine cylinder head or exposed to exhaust gases, and an indicator connected to the sensor by special leads.

Thermocouple principle   Thermo-emf systems depend for their operation on electrical energy which is produced by the direct conversion of heat energy at the source of measurement. Thus, unlike variable resistance systems, they are independent of any external electrical supply.

The form of energy conversion known as the *Seebeck effect* is based on the fact that when two wires made of dissimilar metals are joined at their ends, so as to form separate junctions as in Fig. 13.12, a thermo-emf is produced, and if the junctions are maintained at different temperatures, a current is caused to flow around the circuit. This arrangement is called a *thermocouple*, the junction at the higher temperature being conventionally termed the *hot* or measuring junction, and that at the lower temperature the *cold* or reference junction. In practice, the temperature sensor forms the hot junction, and since it is a separate unit, it is generally regarded as the thermocouple proper.

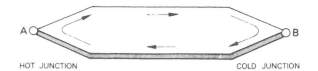

Figure 13.12 Thermocouple principle.

HOT JUNCTION   COLD JUNCTION

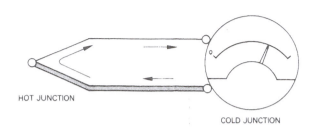

HOT JUNCTION

COLD JUNCTION

Table 13.1   Thermocouple combinations

| Group | Metals and composition | | Maximum temperature °C (continuous) | Application |
	Positive wire	Negative wire		
Base metal	Copper (Cu)	Constantan (NI, 40%; Cu, 60%)	400	Cylinder head temperature measurement
	Iron (Fe)	Constantan (Ni, 40%; Cu, 60%)	850	
	Chomel (Ni, 90%; Cr. 10%)	Alumel (Ni, 90%; Al, 2% + Si + Mn)	1,100	Exhaust gas temperature measurement
Rare metal	Platinum (Pt)	Rhodium-platinum (Rh, 13%; Pt, 87%)	1,400	Not utilized in aircraft temperature-indicating systems

Cr, chromium; Ni, nickel; Al, aluminium; Si, silicon; Mn, manganese

Thermocouple materials and combinations   The materials selected for use as thermocouple sensors fall into two main groups, *base metal* and *rare metal*, and are listed in Table 13.1. The choice of a particular thermocouple combination is dictated by the maximum temperature to be encountered in service.

In order to utilize the thermocouple principle for temperature measurement, it is obviously necessary to measure the emfs generated at the various temperatures. Typical values are listed in the Tables 4, 5 and 6 given at the end of the book. In the basic form of system, measurement is accomplished by connecting a moving-coil millivoltmeter, calibrated in degrees Celsius, in series with the circuits so that it forms the cold junction. The introduction of the indicator into the circuit involves the presence of additional junctions which produce their own emfs and so introduce errors in

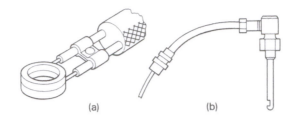

Figure 13.13 Thermocouple
sensors. (*a*) Surface contact;
(*b*) immersion or probe.

(a)

(b)

measurement. However, the effects are taken into consideration when
designing practical thermocouple circuits, and any errors resulting
from 'parasitic emfs', as they are called, are eliminated.

Types of thermocouple sensor   The sensors employed are in general
of two basic types: (i) *surface contact* and (ii) *immersion*. Typical
examples are shown in Fig. 13.13.

The surface contact type is designed to measure the temperature of
a solid component and is used as the sensor in air-cooled engine
cylinder head temperature indicating systems. The material
combinations used are either copper/constantan or iron/constantan,
and their junction may be in the form of a 'shoe' bolted in good
thermal contact with a cylinder head representative of the highest
temperature condition, or in the form of a washer bolted between a
cylinder head and a spark plug.

The immersion, or *probe*, type of thermocouple is designed for the
measurement of gas temperatures, and is therefore adopted as the
sensor in turbine engine exhaust gas temperature (EGT) indicating
systems. Since EGT is a parameter very closely associated with
others required for the monitoring and control of engine power,
further details of probes and EGT indicators will be given in
Chapter 15.

Cold junction temperature compensation   As we have already
learned, the emf produced by a thermocouple depends upon the
difference between the temperature of the hot and cold junctions. It is
thus apparent that, if the ambient temperature of an indicator should
change while the hot junction temperature remains constant, then, by
virtue of the indicator being the cold junction of a circuit, a change
in emf will result causing the indicated readings to be in error. Since
it is essential for the readings to be representative of the hot junction
only, means must be provided for the automatic detection of cold
junction temperature changes, and compensation of resulting errors.
Before going into the details of detection and compensation, it is
useful to consider first how the changes in emf actually arise.

The various combinations of thermocouple materials specified for
use in aircraft conform to standard temperature/emf relationships, and
the associated indicators are calibrated accordingly. The emfs

obtained correspond to a cold junction temperature which is usually maintained at either 0°C or 20°C (see Tables 4, 5 and 6 given at the end of the book).

Let us assume, for example, that the cold junction is maintained at 0°C and that the hot junction temperature has reached 500°C. At this temperature difference a standard value of emf generated by a chromel/alumel combination, say, is 20.64 mV. If now the cold junction temperature increases to 20°C while the hot junction remains at 500°C, the temperature difference decreases to 480°C and the emf equivalent to this difference is now 20.64 mV minus the emf at 20°C; as a standard value this corresponds to 0.79 mV. Thus, the indicator measuring element will respond to an emf of 19.85 mV and move 'down scale' to a reading of 480°C.

A change, therefore, in ambient temperature decreases or increases the emf generated by a thermocouple and makes an indicator read high or low by an amount equal to the change of ambient temperature.

A method commonly adopted for the compensation of these effects in some types of moving-coil indicator is quite simple and is, in fact, an adaptation of the bimetallic strip principle as applied to direct-reading air temperature indicators (see page 61). In this case, however, a strip of dissimilar metals is coiled in the shape of a flat spiral spring. It has one end anchored to a bracket which forms part of the indicator measuring element support, while the other end (free end) is connected by an anchor tag to one of the controlling hairsprings to form a fixture for this spring.

When the indicator is on 'open circuit', i.e. disconnected from the thermocouple system, the spring responds to ambient temperature changes such that the indicator functions as a direct-reading air temperature indicator. With the thermocouple system connected to complete the circuit, then if the two junctions are at the temperatures earlier assumed, namely 0°C and 500°C, the emf will position the measuring element to read 500°C. If the temperature at the indicator increases to 20°C, then, as already illustrated, the emf is reduced but the tendency for the measuring element to move down scale is now directly opposed by the compensating spring as it unwinds in response to the 20°C temperature change. The indicator reading therefore remains at 500°C, the true hot junction temperature.

The compensation of cold junction temperature changes can also be accomplished electrically, but since this relates particularly to the application of servo-operated EGT indicating systems, details of the principle involved will be covered in Chapter 15.

Compensation of moving-coil resistance changes Changes of ambient temperature can also have an effect on the resistance of the moving coil of an indicator, and so this too must be compensated. This can be accomplished by incorporating either a *thermoresistor* or a *thermomagnetic shunt* within the indicator.

A thermoresistor, or thermistor as it is generally known, has a large temperature coefficient which is usually negative. If, therefore, it is connected in a moving-coil circuit, it will, under the same temperature conditions, oppose coil resistance changes and maintain a constant current appropriate only to that produced by the thermocouple hot junction.

A thermomagnetic shunt is a strip of nickel-iron alloy which is sensitive to temperature changes, and is clamped across the poles of the permanent magnet of a moving-coil type of indicator, so that it diverts some of the airgap magnetic flux through itself. For example, if the ambient temperature of an indicator increases, the reluctance of the alloy strip will also increase so that less flux is diverted, or 'shunted', from the airgap. Since the deflecting torque exerted on a moving coil is proportional to the product of current and flux, the increased airgap flux counterbalances the reduction in coil current resulting from the increase in ambient temperature; thus, a constant torque and indicated hot junction temperature reading is maintained.

External circuit and resistance The external circuit of a thermo-emf indicating system consists of the thermocouple and its leads, and the leads from the junction box at an engine bulkhead, or 'firewall', to the indicator terminals. From this point of view, it might therefore be considered as a simple and straightforward electrical instrument system. However, whereas the latter may be connected up by means of the cables normally used in an aircraft, it is not acceptable to do so in a thermo-emf system.

This may be explained by taking the case of a copper/constantan thermocouple which is to be connected to a cylinder head temperature indicator. If a length of normal copper twin-core cable is connected to the thermocouple junction box, then one copper lead will be joined to its thermocouple counterpart, but the other one will be joined to the constantan lead of the thermocouple. It is thus apparent that this introduces another effective hot junction which will respond to temperature changes occurring at the junction box, and so cause an unbalance in the temperature/emf relationship and errors in hot junction temperature readings. Similarly, all terminal connections which may be necessary for routeing the leads through an aircraft, and connections at the indicator itself, will create additional hot junctions and so further aggravate indicator errors.

In order to eliminate these hot junctions, it is the practice to use leads made of the same materials as the thermocouples themselves; such leads are generally known as *extension leads*. It may sometimes be the practice to use leads made of materials having similar thermo-emf characteristics in combination; for example, a chromel/alumel thermocouple may be joined to its indicator by copper/constantan leads, the latter then being referred to as *compensating leads*.

Additional hot junctions at indicators and junction boxes are eliminated by making the terminals of the same materials as those of the appropriate thermocouple combination, i.e. a positive terminal of copper or chromel, and a negative terminal of constantan or alumel.

Another important factor in connection with the external circuit is its resistance, which must be kept not only low but also constant for a particular installation. Indicators are, therefore, normally calibrated for use with external circuits of specific resistance values, e.g. 8 Ω or 25 Ω, and they are identified accordingly. Thermocouples, their leads, and harnesses where appropriate, are made up in fixed low-resistance lengths. Similarly, extension or compensating leads are also made up in lengths and of uniform resistance to suit the varying distances between thermocouple hot junction and indicator locations.

Adjustment of the total external circuit resistance following replacement of thermocouples, leads or indicators is provided by a trimming resistance circuit connected in series with the extension or compensating leads. Although the circuit introduces additional junctions, the temperature coefficients of resistance of the materials used for the resistors (typically Manganin or Eureka) ensure that the circuit has negligible thermo-emf effect.

14 Fuel quantity indicating systems

An indication of the quantity of fuel in the tanks of an aircraft's fuel system is, of course, an essential requirement, and, in conjunction with measurements of the rate at which the fuel flows to the engine or engines, permits an aircraft to be flown at maximum efficiency compatible with its specified operating conditions. Furthermore, a knowledge of both parameters enables not only assessments of remaining flight time to be made, but also comparisons between present engine performance and past or calculated performance.

The method most commonly adopted for quantity indication is one which measures changes in fuel level in terms of the changes in electrical capacitance of special tank units or 'probes', which then transmit corresponding signals to their associated indicators.

Capacitance-type system

In its basic form, a system consists of a variable capacitor located in a fuel tank, an amplifier and an indicator. The complete circuit forms an electrical bridge which is continuously being rebalanced as a result of differences between the capacitances of the tank capacitor and a reference capacitor. The signal produced is amplified to operate a motor, which positions a pointer to indicate the capacitance change of the tank capacitor and, thus, the change in fuel level or quantity.

Before going into the operating details of more practical systems, however, it is first necessary to have some understanding of the fundamental principles of capacitance and its effects in electrical circuits.

Electrical capacitance

Whenever a potential difference is applied across two conducting surfaces separated by a non-conducting medium, called a *dielectric*, they have the property of storing an electric charge; this property is known as *capacitance*. The device comprising the surfaces, or plates, and a dielectric is called a *capacitor*.

The flow of a momentary current into a capacitor establishes a potential difference (pd) across its plates. Since the dielectric contains no free electrons the current cannot flow through it, but the pd sets

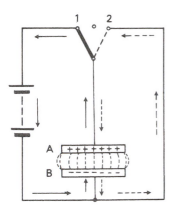
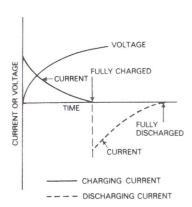

Figure 14.1 Capacitor principle.

up a state of stress in the atoms comprising it. For example, in the circuit shown in Fig. 14.1, when the switch is placed in position 1 a rush of electrons, known as the *charging current*, takes place from plate A through the battery to plate B and ceases when the pd between the plates is equal to that of the battery.

When the switch is opened, the plates remain positively and negatively charged since the atoms of plate A have lost electrons while those at plate B have a surplus. Thus, electrical energy is stored in the electric field between the plates.

Placing the switch in position 2 causes the plates to be short-circuited and the surplus electrons at plate B rush back to plate A until the atoms of both plates are electrically neutral and no potential difference exists between them. This discharging current is in the reverse direction to the charging current, as shown in Fig. 14.1.

Units of capacitance

The capacitance or 'electron-holding ability' of a capacitor is the ratio between the charge and pd between the plates and is expressed in *farads*, one farad representing the ability of a capacitor to hold a charge of one coulomb (6.24×10^{18} electrons) which raises the pd between its plates by one volt.

Since the farad is generally too large for practical work, a sub-multiple of it is normally used called the *microfarad* ($1 \ \mu F = 10^{-6}F$). In the application of the capacitor principle to a fuel quantity indicating system, an even smaller unit, the *picofarad* ($1 \ pF = 10^{-12}F$), is the standard unit of measurement.

Factors on which capacitance depends

Capacitance depends on the area *a* of the plates, the distance *d* between the plates, and the capacitance (or absolute permittivity) ϵ of a unit cube of the dielectric material between the plates.

327

Permittivity ϵ is usually quoted as being relative to that of a vacuum. Relative permittivity, also called *dielectric constant*, is often denoted by K which is the ratio of the capacitance of a capacitor with a given dielectric to its capacitance with air between its plates. The relative permittivities of some pertinent substances are as follows:

Air	1.000
Water	81.07
Water vapour	1.007
Aviation gasolene	1.95
Aviation kerosene	2.10

Capacitors in alternating current circuits

As already mentioned, when direct current is applied to a capacitor there is, apart from the initial charging current, no current flow through the capacitor. In applying the capacitance principle to fuel quantity indicating systems, however, a flow of current is necessary to make the indicator respond to the capacitance changes arising from changes in fuel level. This is accomplished by supplying the tank probes with an alternating voltage, because whenever such voltage across a capacitor changes, electrons flow toward and away from it without crossing the plates and a resultant current flows which, at any instant, depends on the rate of change of voltage.

Basic indicating system

For fuel quantity measurement, the capacitors to be installed in the tanks must, of course, differ in construction from those normally adopted in electronic equipment. The plates therefore take the form of two tubes mounted concentrically with a narrow airspace between them, and extending the full depth of a fuel tank. Constructed in this manner, two of the factors on which capacitance depends are fixed, while the third factor, relative permittivity, is variable since the medium between the tubes is made up of fuel and air. The manner in which changes in capacitance take place is illustrated in Fig. 14.2, and is described in the following paragraphs.

Figure 14.2 Capacitance changes due to fuel and air.

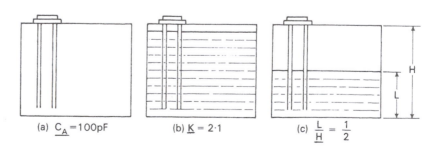

(a) $C_A = 100\text{pF}$ (b) $\underline{K} = 2\cdot1$ (c) $\dfrac{L}{H} = \dfrac{1}{2}$

At (a) a capacitance probe is fitted in an empty fuel tank and its capacitance in air is 100 pF, represented by C_A.

At (b) the tank is filled with a fuel having a K value of 2.1, so that the capacitor is completely immersed and its sensing surfaces are fully 'wetted'. As noted earlier, K is equal to the ratio of capacitance using a given dielectric (in this case C_T) to that using air; therefore

$$K = C_T/C_A \qquad [1]$$

From equation [1] $C_T = C_A K$, and it is thus clear that the capacitance of the tank probe at (b) is equal to 100×2.1, i.e. 210 pF. The increase of 110 pF is the added capacitance due to the fuel and may be represented by C_F. The tank probe may therefore be represented electrically by two capacitors in parallel, and of a total capacitance,

$$C_T = C_A + C_F \qquad [2]$$

In Figure 14.2(c), the tank is only half full and so the total capacitance is $100 + 55$, or 155 pF. The added capacitance due to fuel is determined by transposing equation [2], so that $C_F = C_T - C_A$, and by substituting $C_A K$ for C_T, we thus obtain $C_F = C_A K - C_A$, which may be simplified as:

$$C_F = C_A (K-1) \qquad [3]$$

the factor $(K-1)$ being the increase in the K value over that of air.

The fraction of the total possible fuel quantity in a 'linear tank' at any given level is given by L/H, where L is the height of the fuel level and H the total height of the tank. Thus by adding L/H to equation [3] the complete formula becomes:

$$C_F = \frac{L}{H} (K-1) C_A \qquad [4]$$

The circuit of a basic system is shown in Fig. 14.3. It is divided into two sections or loops by a resistance R, both loops being connected to the secondary winding of a power transformer. Loop A contains the tank probe C_T and may therefore be considered as the sensing loop of the bridge since it detects current changes due to changes in capacitance. Sensing loop voltage V_S remains constant.

Loop B, which may be considered as the balancing loop of the bridge, contains a reference capacitor C_R of fixed value, and is connected to the transformer via the wiper of a balance potentiometer so that voltage V_B is variable.

The balance potentiometer is contained within the indicator together with a two-phase motor which drives the potentiometer wiper and indicator pointer. The reference phase of the motor is continuously energized by the power transformer, and the control phase is only energized when an unbalanced condition exists in the bridge.

Figure 14.3 Basic capacitance system.

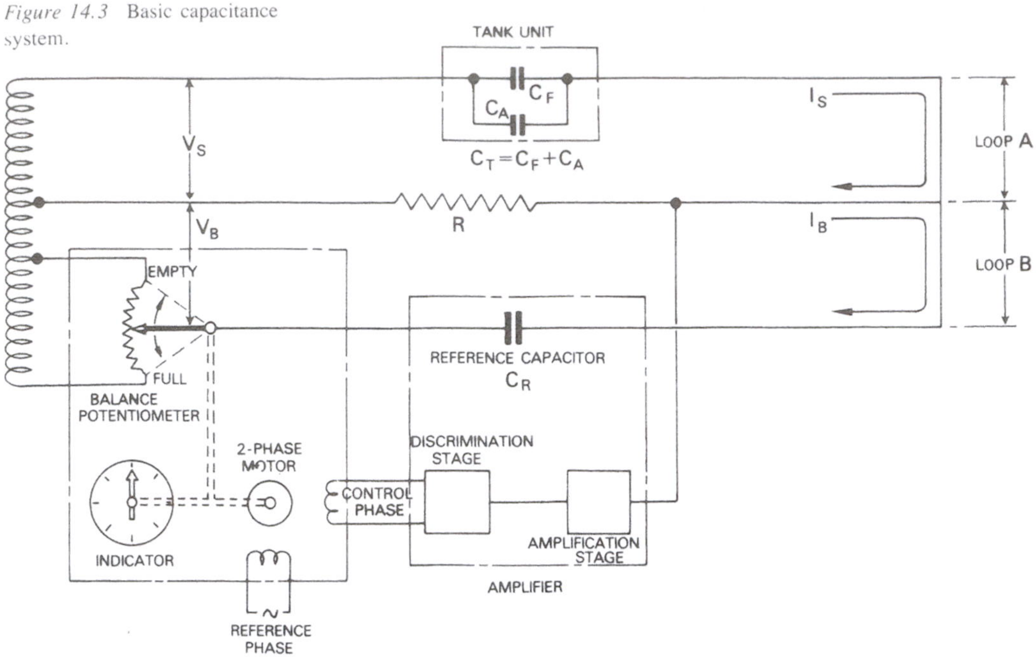

The amplifier has two main stages: one for amplifying the signal produced by bridge unbalance, and the other for discriminating the phase of the signal which is then supplied to the motor.

Let us consider the operation of the complete circuit when fuel is being drawn off from a full tank. Initially, and at the constant full-tank level, the sensing current I_S is equal to the balancing current I_B; the bridge is thus in balance and no signal is produced across R.

As the fuel level drops, the tank probe has less fuel around it; therefore the added capacitance (C_F) has decreased. The capacitance of the probe decreases and so does the sensing current I_S, the latter creating an unbalanced bridge condition with balancing current I_B predominating through R.

A signal voltage proportional to $I_B R$ is developed across R and is amplified and its phase detected before being applied to the control phase of the indicator motor. The discriminator output signal is converted from a half-wave pulse into a full-wave signal, and supplied to the control phase winding of the motor. As in normal two-phase motor operation, the control phase current will either lead or lag that in the reference phase. Thus, in the condition we are considering, the control phase current is lagging since balancing loop current I_B is the predominating one. The motor and balance potentiometer wiper are therefore driven in such a direction as to reduce the current I_B. When this current equals I_S, the bridge is once again in balance, the motor ceases to rotate, and the indicator pointer registers the new lower value of fuel quantity.

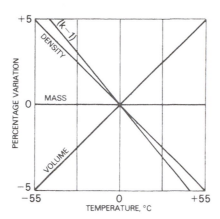

Figure 14.4 Temperature
effects on fuel characteristics.

Effects of fuel temperature changes

With changes in temperature the volume, density and relative
permittivity of fuels are affected to approximately the same degree as
shown in Fig. 14.4, which is a graph of the approximate changes
occurring in a given mass of fuel. From this it should be noted that
$K - 1$ is plotted, since for a system measuring fuel quantity by
volume, the indicator pointer movement is directly dependent on this.
It should also be noted that, although it varies in the same way as
density, the percentage change is greater.

Thus a volumetric system will be subject to a small error due to
variations in fuel temperature. Furthermore, changes in K and density
also occur in different types of fuel having the same temperature. For
example, a system which is calibrated for a K value of 2.1 has a
calibration factor of $2.1 - 1 = 1.1$. If the same system is used for
measuring a quantity of fuel having a K value of 2.3, then the
calibration factor will have increased to 1.3 and the error in
indication will be approximately:

$$\frac{1.3}{1.1} \times 100 - 100\% = 18\%,$$

i.e. the indicator would overread by 18 per cent.

Measurement of fuel quantity by weight

A more useful and accurate method of measuring fuel quantity is to
do so in terms of its mass or weight. This is because the total power
developed by an engine, or the work it performs during flight,
depends not only on the volume of fuel but on the energy it contains,
i.e. the number of molecules that can combine with oxygen in the
engine. Since each fuel molecule has some weight, and also because
one pound of fuel has the same number of molecules regardless of
temperature and therefore volume, the total number of molecules

(total available energy) is best indicated by measuring the total fuel weight.

In order to do this, the volume and density of the fuel must be known, and the product of the two determined. The measuring device must, therefore, be sensitive to changes in both volume and density so as to eliminate undesirable effects due to temperature.

This will be apparent by considering the example of a tank holding 1000 gallons of fuel having a density of 6 lb/gal at normal temperature. Measuring this volumetrically we should of course obtain a reading of 1000 gallons, and from a mass measurement, 6000 lb. If a temperature rise should increase the volume by 10 per cent, then the volumetric measurement would increase to 1100 gallons, but the mass measurement would remain at 6000 lb because the density of the fuel (weight/volume) would have decreased when the temperature increased.

For the calibration of systems in terms of mass of fuel, the assumption is made that the relationship between the relative permittivity (K) and the density (ρ) of a given sample of fuel is constant. This relationship is called the *capacitive index* and is defined by

$$\frac{K-1}{\rho}$$

An indicating system calibrated to this expression is still subject to indication errors, but they are very much reduced. This may be illustrated by a second example. Assuming that the system is measuring the quantity of a fuel of nominal $K = 2.1$ and of nominal density $\rho = 0.779$, then its capacitive index is:

$$\frac{2.1-1}{0.779} = 1.412$$

If now the same system measures the quantity of another fuel for which the nominal K and ρ values are respectively 2.3 and 0.85, then its capacitive index will increase; thus,

$$\frac{2.3-1}{0.85} = 1.529$$

However, the percentage error is now

$$\frac{1.529}{1.412} \times 100 - 100\% = 8\%$$

and this is the amount by which the indicator would overread.

Compensated indicating systems
Although the assumed constant permittivity and density relationship results in a reduction of the indicator error, tests on the properties of

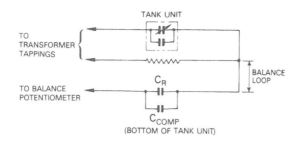

Figure 14.5 Compensator capacitor.

TANK UNIT

TO TRANSFORMER TAPPINGS

TO BALANCE POTENTIOMETER

BALANCE LOOP

C_R

C_{COMP}
(BOTTOM OF TANK UNIT)

fuels showed that, while the capacitive index could vary from one fuel to another, the variation tended to follow the permittivity. Thus, if an indicating system can also detect changes in the permittivity of a fuel as it departs from its nominal value, then the density may be inferred to a greater accuracy, resulting in an even greater reduction of indication errors. Indicating systems currently in use are therefore of the permittivity or inferred-density compensated type, the compensation being effected by a reference capacitor added to the balance loop of the measuring circuit and in parallel with the capacitor C_R, as shown in Fig. 14.5.

A compensator is similar in construction to a standard tank probe. In some systems it is fitted to the bottom end of a standard probe so that it is always immersed in fuel, while in others it forms a separate probe. Located in this manner, its capacitance is determined solely by the permittivity, or K value, of the fuel, and not by its quantity as in the case of the standard tank probe. In addition to variable voltage, the balancing loop will also be subjected to variable capacitance, which means that balancing current I_B will be affected by variations in K as well as sensing-loop current I_S.

Let us assume that the bridge circuit (see Fig. 14.3) is in balance and that a change in temperature of the fuel causes its K value to increase. The tank probe capacitance will increase and so current I_S will predominate to unbalance the bridge and to send a signal to the amplifier and control phase of the motor. This signal will be of such a value and phase that an increase in balancing-loop current is required to balance the bridge, and so the motor must drive the wiper of the balance potentiometer to decrease the resistance. Since this is in the direction of the 'tank full' condition, the indicator will obviously register an increase in fuel quantity. The increase in K, however, also increases the compensator capacitance so that loop current I_B is increased simultaneously with, but in opposition to, the increasing sensing-loop current I_S. A balanced bridge condition is therefore obtained which is independent of the balance potentiometer.

In practice, there is still an indication error due to the fact that the density also varies with temperature, and this is not directly measured. But the percentage increase of density is not as great as that of $K - 1$, and so by careful selection of the compensator

333

capacitance values in conjunction with the reference capacitor, the greatest reduction in overall indication error is produced.

Densitometers Although the application of compensator probes provides for greater accuracy of measurement in terms of mass measurement, the fact that this is based on an assumed relationship between the K and ρ values of fuel results in indications that still fall short of those desired, i.e. true weight of fuel. Such indications can, however, be achieved by measuring variations in ρ separately, and then applying the corresponding values as continuous correction signals to the measuring and indicating circuits. This is a method which is applied to fuel quantity indicating systems installed in several types of current-generation aircraft, and from the point of view of techniques involved, the one adopted in such aircraft as the Boeing 757, 767 and 747–400 serves as an interesting example.

In addition to the normal probes and the compensator probes, a device known as a *densitometer* is installed in each fuel tank. It is comprised of two principal units: (i) an emitter which contains a collimator, a radioactive capsule, and a microswitch, and (ii) an electronics unit containing two detector tubes filled with xenon gas, pre-amplifier, signal processor and power supply circuit card modules. The complete unit is installed on the rear spar of an integral tank such that the emitter is on the inside of the tank. The principle of its operation is shown in Fig. 14.6.

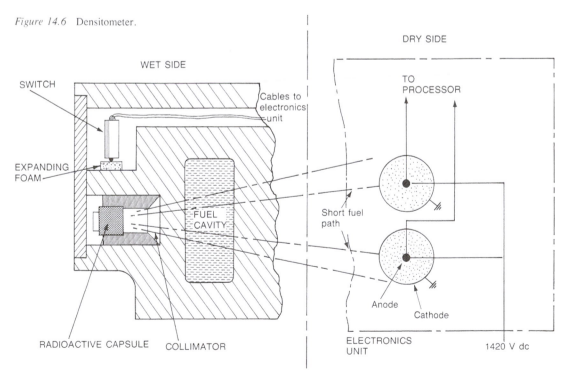

Figure 14.6 Densitometer.

The gamma radiation from the emitter passes through the collimator and the fuel, the density of which will vary the level of radiation and the paths taken from the collimator to the detector tubes. The collimator directs the radiation to the tubes in a conical pattern. One of the tubes detects radiation from what is termed a short fuel path, while the other detects long fuel path radiation so that fuel density is determined as a ratio of radiation.

Each tube is made up of a cathode and an anode which passes through the centre, and is supplied with 1420 V dc generated by the power supply module of the densitometer electronics unit. As noted earlier, the tubes are filled with xenon gas, which on being exposed to gamma radiation causes electrons to be released. This, in turn, generates a pulsed signal which, after amplification and processing, is transmitted from the electronics unit to the processor unit that controls the whole operation of the fuel quantity indicating system adopted for the types of aircraft already mentioned. The number of pulses per unit of time are counted, and are a function of fuel density. In order to obtain the true fuel weight for display on the quantity indicators, the system's processor unit then multiplies the fuel volume values measured by the tank probes in the normal manner, by the fuel density values received from the densitometers.

As will be noted from Fig. 14.6, the collimator, radioactive capsule and microswitch are in a 'dry' and sealed section of a densitometer. In the event that fuel should leak into this section, it causes a piece of foam material, located under the switch, to 'swell' and so exert a pressure on the switch plunger. The resulting actuation of the switch contacts causes the densitometer signals to be transmitted to, and stored in, a fault memory circuit within a built-in test section of the quantity indicating system's processor unit.

Construction of probes

In the majority of applications, probes utilize tubes made of aluminium suitably protected against the effects of corrosion, short-circuiting and grounding. In some cases, the outer tube may be of aluminium, while the inner tube is a non-conducting plastic material that is coated with a metallized film on its outer surface to serve as a capacitor plate.

The tubes are held apart concentrically by insulated cross-pins, and for the purpose of initial calibration and characterization (see page 339) to suit individual tank shapes and sizes, the concentricity is varied throughout the length of a probe. This is accomplished by having an inner tube whose diameter varies along its length, or by off-centering it by means of spacers which differ in length. The spacers are made of teflon, and are colour-coded corresponding to length. In some types of probe, characterization is effected by

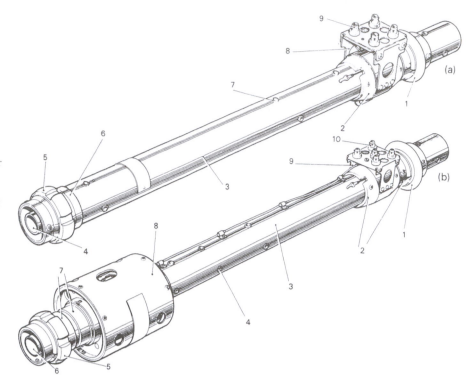

varying the area of an inner tube's conducting surface at various points over its length.

Mounting arrangements vary depending on whether the probes are to be secured to the top and bottom sections of a tank, as in the case of the probes shown as examples in Fig. 14.7, or are to be secured to tank baffles, ribs or spars; in the latter case, mounting brackets are provided. Probes are connected to the indicating system wiring harnesses via terminal blocks which, depending on design, provide for either co-axial or screw-type terminal connections.

The probe illustrated in Fig. 14.8 is of the independent compensator type containing three tubes. The outer and inner tubes are of low impedance, and the middle one is of high impedance. A 'tuning' plate is provided for calibration of the probe, and also for controlling what is termed capacitance fringing. Since the probe is always immersed in fuel, there is a possibility of contaminant build-up which would cause the probe to sense a stagnant sample of fuel. To prevent this, therefore, a tube is provided so that during refuelling operations, a stream of fuel is directed up through the bottom of the probe, thereby washing it out.

Location and connection of tank probes

In practical indicating systems, a number of tank probes are positioned within the fuel tanks as illustrated in Fig. 14.9, and are

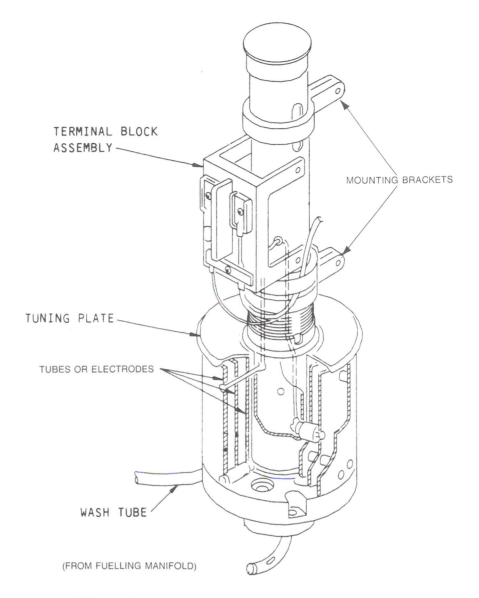

Figure 14.8 Compensator probe.

TERMINAL BLOCK ASSEMBLY

MOUNTING BRACKETS

TUNING PLATE

TUBES OR ELECTRODES

WASH TUBE

(FROM FUELLING MANIFOLD)

connected in parallel to their respective indicators. The reasons for this are to ensure that indications remain the same regardless of the attitude of an aircraft and its tanks, to take into account the effects of wing flexing and, of course, to give a high total capacitance value. This may be understood by considering a two-unit system as shown in Fig. 14.10. If the tank is half-full and in a level attitude, each probe will have a capacitance of half its maximum value; since they are in parallel the total capacitance measured will produce a 'half-full' indication. When the tank is tilted, and because the fuel level remains the same, probe A is immersed deeper in the fuel by the distance d and gains some capacitance, tending to make the indicator overread. Probe B, however, has moved out of the fuel by the same distance d and loses an equal amount of capacitance. The total

337

Figure 14.9 Location of tank probes.

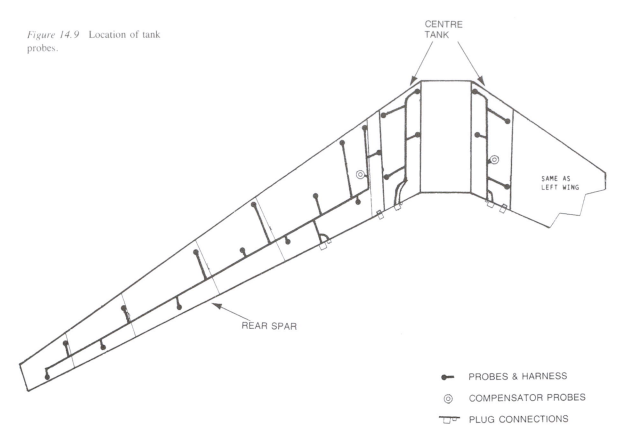

CENTRE TANK

SAME AS LEFT WING

REAR SPAR

●— PROBES & HARNESS

◎ COMPENSATOR PROBES

⊔ᵚ PLUG CONNECTIONS

Figure 14.10 Attitude compensation.

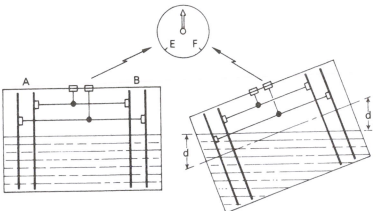

capacitance therefore remains the same as for the level-tank attitude and the indication is unchanged.

The number of probes required is governed by the fuel capacity requirements and tank configurations of any one type of aircraft. In the one on which Fig. 14.9 is based, there are left and right wing main tanks each containing 14 probes, and left and right centre auxiliary tanks each containing four probes. In addition, one compensator probe is installed in each of the tanks.

In the majority of systems, the probe terminals and the wiring harnesses interconnecting them are located within the tanks. The harnesses terminate at single connector boxes, or 'bussing' plugs, which are externally mounted and provide for the connection of signal-processing and indicator units.

Characterization of probes

The fuel tanks of an aircraft may be separate units designed for installation in wings and centre sections, or they may form an integral sealed section of these parts of the structure. This means, therefore, that tanks must vary in contour to suit their chosen locations, with the result that the fuel level is established from varying datum points.

Figure 14.11 represents the contour of a tank located in an aircraft's wing and, as will be noted, the levels of fuel from points A, B and C are not the same. When probes are positioned at these points the total capacitance will be the sum of three different values due to the fuel (C_F), and as the probes produce the same change of capacitance for each inch of 'wetted' length, the indicator scale will be non-linear corresponding to the non-linear characteristic of the tank contour.

The non-linear variations in fuel level are unavoidable, but the effects on the graduation spacing of the indicator scale can be overcome by designing probes so that they can be calibrated, or *characterized*, to measure capacitance changes proportional to tank contour. Some examples of characterization methods have already been described (see page 335).

In addition to tank contouring, account must also be taken of the effects of wing flexing that occur in flight. These effects are normally simulated in full-scale static fuel gauging test rigs, designed for the initial calibration of indicating systems appropriate to the aircraft type.

In indicating systems utilizing digital signal-processing techniques, the need for varying the concentricity of probe inner tubes for the purpose of characterization can be eliminated by incorporating a software program in the processor circuit. The program takes into account all known variables and provides more accurate calculation of

Figure 14.11 Characterization of probes.

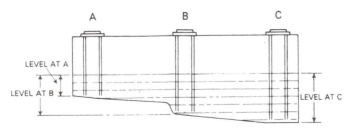

tank volumetric changes resulting from wing bending both on the ground and in the air. The program may be selected to provide four different ground characterizations, and three different airborne characterizations.

'Empty' and 'Full' position adjustments

In systems based on the circuit arrangement shown in Fig. 14.3, it is necessary during calibration to balance the current and voltage of the sensing and balancing loops at the datum corresponding to empty- and full-tank conditions. This is achieved by connecting two potentiometers into the circuit as shown in Fig. 14.12; they are adjusted from the rear of the indicators.

The 'empty' potentiometer is connected at each end to the supply transformer and its wiper is connected to the tank units via a balance capacitor. When a tank is empty, due to the empty capacitance of the probes current will still flow through them. The balance potentiometer wiper will also be at its 'empty' position, but since it is grounded at this point, no current will flow through the reference capacitor. However, current does flow through the balance capacitor and it is the function of the 'empty' potentiometer to balance this out. The balancing signal from the potentiometer is fed to the amplifier, the output signal of which drives the indicator servomotor and pointer to the empty position of the scale.

The 'full' adjustment may be regarded as a means of changing the position of the point on the balance potentiometer at which the balance voltage for any given amount of fuel is found, and also of determining the voltage drop across the potentiometer. Reference to Fig. 14.12 shows that, if the 'full' potentiometer wiper is set at the bottom, the maximum transformer secondary voltage will be applied to the balance potentiometer. Therefore, the distance the potentiometer wiper needs to move to develop a given balance voltage can be varied.

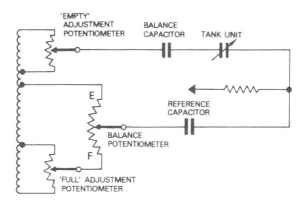

Figure 14.12 'Empty' and 'Full' adjustments.

Electronic displays

These are adopted in systems which process fuel weight values in digital signal format, and two examples based on the Boeing 767 system are illustrated in Fig. 14.13. The primary indication of weight is given by the LCD indicators (diagram (a)), while data related to system operation, e.g. fault occurrences, low fuel state, etc., are displayed as advisory or cautionary messages on a CRT display unit as in diagram (b). The display unit in this case is associated with an engine indicating and control system, details of which will be covered in Chapter 15.

Fail-safe and test circuits

These circuits are incorporated in all indicating systems, and their arrangements can vary dependent on the type of system. In those which operate on the basic principle of bridge balancing, a fail-safe circuit consists of a capacitor and a resistor, connected in the indicator motor circuit so that the reference winding supply is paralleled to the control winding. A small leading current always flows through the parallel circuit, but under normal operating conditions of the system, it is suppressed by tank probe signals flowing from the amplifier through the motor control winding. If a failure of signal flow occurs, the current in the parallel circuit predominates and flows through the control winding to drive the indicator pointer slowly downwards to the 'empty' position.

A test circuit for this type of system incorporates a switch mounted adjacent to its appropriate indicator. When the switch is held in the 'test' position, a signal simulating an emptying tank condition is introduced into the indicator motor control winding, causing it to drive the pointer towards zero if the circuit is functioning correctly. When the switch is released, the pointer should return to its original indication. In multi-indicator installations, a single switch serves to test all indicators simultaneously.

In computer-controlled systems, circuits are provided within the signal processor unit which automatically carry out self-test routines at the time when power is initially applied to a system, and then on a continuous basis. Any faults that occur are stored in memory circuits, and since these can only be recalled when carrying out built-in-test (BIT) checks, an appropriate maintenance message such as 'FUEL QTY BITE' is displayed on the control and monitoring system display unit. Faults are assigned specific status code numbers which are displayed by a two-digit LED display on the processor unit test panel when the 'BIT' switch is operated to recall faults in the order they were stored in memory. When faults have been rectified by the appropriate maintenance actions, the memory is cleared of stored data by the operation of a 'reset' switch.

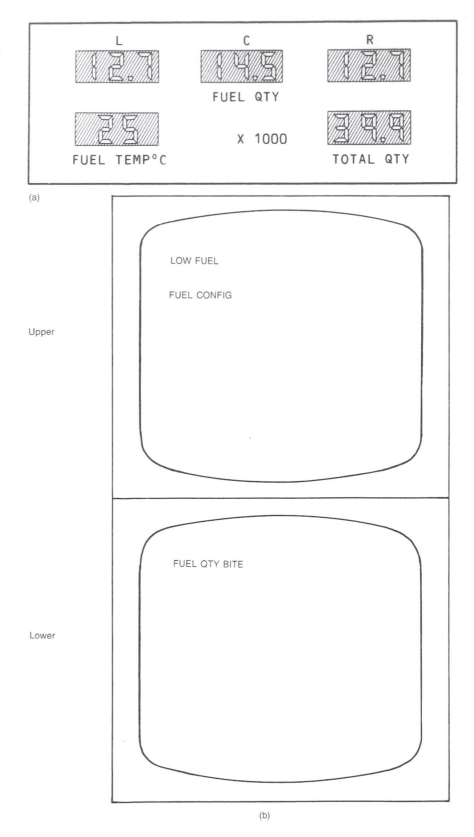

Figure 14.13 Electronic displays. (*a*) LCD indicators; (*b*) EICAS display units.

(a)

(b)

The circuits and software programming are such that adequate redundancy is provided to ensure the display of valid quantity indications in the event of failures of channel power supplies, tank probes, or densitometers.

Since the quantity indicators are of the LCD type, functional testing is simply a matter of checking that each of the seven segments that make up each digit is activated by operating a test switch on a flight deck panel. Thus, all the digits will be displayed as '8's. A test switch is also provided on the processor unit test panel for carrying out similar checks on the fault status code LED display.

Totalizer indicators

These indicators are provided in some types of aircraft in addition to the normal fuel quantity indicating system. The dial presentations of two typical indicators are shown in Fig. 14.14, and from this it will be noted that in addition to the total quantity of fuel remaining, an aircraft's gross weight is also displayed. Both indicators operate on the same principle, the major difference between them being in the method of display. The indicator at (a) utilizes mechanical-type digital counters, while that at (b) utilizes a segmented electronic display.

Each indicator comprises a resistance network with the same number of channels as there are primary fuel quantity indicators appropriate to the particular aircraft system. Each channel receives a signal voltage from a primary indicator, and the resistance network apportions current according to the amount of fuel that each voltage represents, and then sums the currents to represent total fuel weight remaining. Thus, any change in the signal voltage from one or more of the primary indicators causes a proportional change in 'total fuel quantity current' and an unbalancing of circuit conditions which drives the counter to the corresponding value.

The gross weight counter, which is pre-set to indicate an aircraft's

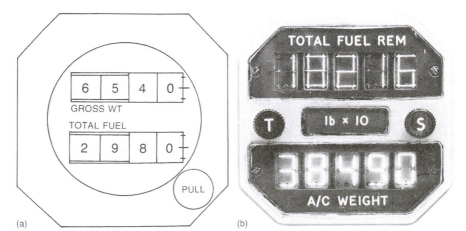

Figure 14.14 Totalizer indicators.

weight prior to flight, is also connected to the primary indicator system, but in such a manner that as fuel is consumed the counter continuously indicates a decreasing gross weight.

Refuelling and load control

In the larger types of aircraft, the location of the primary fuel quantity indicators is, of course, far removed from the tank refuelling points, and so in order that ground handling crews can monitor more precisely the necessary 'fuel-uplift' operations, refuelling and load control panels are located adjacent to the refuelling points.

Figure 14.15 illustrates a panel which is used in conjunction with a volumetric top-off (VTO) system, i.e. one which automatically terminates refuelling at a pre-set fuel level. The indicators are connected into the same sensing circuits as the primary indicators, and so they provide a duplicate indication of the fuel quantity during refuelling. The switches above the indicators control the positions of their respective tank refuelling valves. The indicator lights, which are normally blue, illuminate when the valves are in the open position.

If only a partial load of fuel is required, the indicators are monitored and when the desired quantity has been uplifted, the refuelling valves are closed by selecting this position of the appropriate switch. When, however, a full load of fuel is required, the valves are automatically closed by the VTO system.

A separate system is installed for each fuel tank, comprising a VTO unit and a compensator type of tank probe. Since the system is shared with the primary quantity indicating system, then the latter records the increasing fuel quantity in the normal units of mass, and transmits the corresponding signals to a bridge circuit in the VTO unit. The signals from the VTO compensator probes relate to tank

Figure 14.15 Refuelling panel.

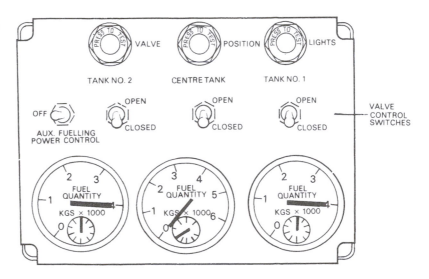

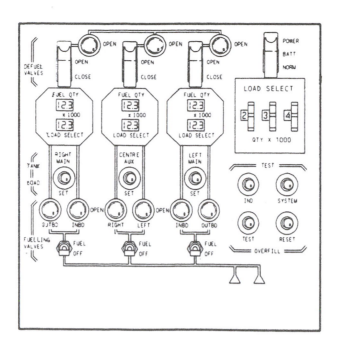

Figure 14.16 Refuelling and load select panel.

unit capacity in volumetric units, and these are also supplied to the bridge circuit which then compares both signals. The resultant signal is amplified and fed through a solid-state switching circuit to energize the fuelling shut-off valve when the pre-set nominal full level, or volume, of the tank is reached. Calibration of the whole VTO system to the desired nominal 'tank full' volume is accomplished by adjusting the current through the VTO compensator probes so that it corresponds with the total current from the tank probes at the particular shut-off level.

On completion of a refuelling operation, the control panel is enclosed by a door which also actuates a switch that isolates all electrical power to the fuelling system.

Figure 14.16 illustrates a refuelling control panel which is used in conjunction with a computer-controlled quantity indicating system. It operates in a similar manner to the one already described but, as will be noted, it also provides for the pre-selection of a desired fuel weight in any tank. The quantity indicators are also of the LCD type.

The desired weight to be loaded is set by means of three thumbwheel rotary switches, each having ten positions and numbered 0 to 9; thus, the ranges are covered in tenths, units and tens. At each numbered position a resistor of differing value is selected so that with power on, and signal currents corresponding to the selected position, multiples are produced and supplied to a load select multiplexer in the processor unit of the main quantity indicating system, A 'SET' switch is provided on the refuelling control panel below each of its quantity indicators, and when pushed in it allows the signals from the processor unit to pass to the lower LCD indicators, which then display the load selected. When this load has been reached, the fuelling valves are automatically closed.

The segments of the six LCD indicators can be checked simultaneously by a push-button 'TEST' switch on the control panel.

15 Engine power and control instruments

The power of an engine refers to the amount of thrust available for propulsion, and depending upon the type of engine, i.e. piston or turbine, it is expressed as power ratings in units of brake or shaft horsepower, or of thrust developed at an exhaust unit. Power ratings of each of the various types of engine are determined, together with operational limitations, during the test-bed calibration runs conducted by the manufacturers for various operating conditions such as take-off, climb, normal cruise, emergency and contingency.

The parameters associated with engine power ratings, and which must be monitored by appropriate instrumentation, are listed in Table 15.1.

RPM measurement

The measurement of engine speed in terms of revolutions per minute is relevant to the three main types of engine and, with the exception of a simple type of unsupercharged piston engine, it is related to the other parameters involved in power control. The power of an unsupercharged engine is directly related to its speed, and so with the throttle at a corresponding operational setting, an rpm indicator system can also serve as a power indicator.

Indicating systems, also generally referred to as tachometers, are of the electrical type and fall into two main categories: (i) generator and indicator, and (ii) tacho probe and indicator.

Table 15.1 Power rating parameters

Parameter	Type of Engine			
	Piston		Turbine	Turboprop
	Unsupercharged	Supercharged		
Rpm	X	X	X	X
Manifold pressure		X		
Torque		X		X
Exhaust gas temperature			X	X
Pressure ratio			X	
Fuel flow		X	X	

Generator and indicator system

A generator is of the ac type, consisting of a permanent magnet rotor rotating within a slotted stator which carries a star-connected three-phase winding. The rotor may be of either two-pole or four-pole construction, and is driven by a splined shaft coupling; the generator is bolted directly to a mounting pad at the appropriate accessories drive gear outlet of an engine. In order to limit the mechanical loads on generators, the operating speed of rotors is reduced by means of either 4:1 or 2:1 ratio gears in the engine drive system.

Two- or four-pole-type generators are utilized in conjunction with a three-phase synchronous motor within the indicator. For the operation of servo-operated indicators, the display of data by electronic CRT indicators, and for supplying signals to automatic power control systems, the appropriate data signals are supplied from generators having 12-pole rotors; these produce a single-phase output at a much higher frequency and sensitivity.

A typical indicator consists of two interconnected elements, a driving element and a speed-indicating element. The *driving element* is a synchronous motor having a star-connected three-phase stator winding, and a rotor which is so constructed that the motor has the self-starting and high torque characteristics of a squirrel-cage motor, combined with self-synchronous properties of a permanent magnet type of motor.

The *speed-indicating element* consists of a permanent magnet device which operates on the eddy-current drag principle, and as indicated in Fig. 15.1 it may utilize either a drag cup or a drag disc. In the former version, the magnet is inserted into a drum so that a small airgap is left between the periphery of the magnet and drum. The drag cup is mounted on a shaft and is supported in such a way that it fits over the magnet rotor to reduce the airgap to a minimum. A calibrated hairspring is attached to one end of the drag cup shaft, and at the other end to the mechanism frame. At the front end of the shaft, a gear train is coupled to two concentrically-mounted pointers; a large one indicating hundreds and a small one indicating thousands of rpm.

System operation

As the generator rotor is driven round inside its stator, the poles sweep past each stator winding in succession so that three waves or phases of alternating emf are generated, the waves being 120° apart. The magnitude of the emf induced depends on the strength of the magnet and the number of turns comprising the phase coils. Furthermore, as each coil is passed by a pair of rotor poles, the induced emf completes one cycle at a frequency determined by the rotational speed of the rotor. Therefore, rotor speed and frequency

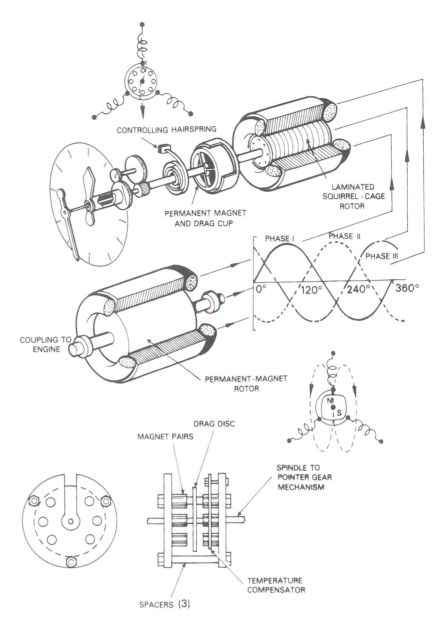

Figure 15.1 Principle of a generator and indicator system.

CONTROLLING HAIRSPRING

PERMANENT MAGNET AND DRAG CUP

LAMINATED SQUIRREL - CAGE ROTOR

PHASE I   PHASE II   PHASE III

0°   120°   240°   360°

COUPLING TO ENGINE

PERMANENT-MAGNET ROTOR

DRAG DISC

MAGNET PAIRS

SPINDLE TO POINTER GEAR MECHANISM

TEMPERATURE COMPENSATOR

SPACERS (3)

are directly proportional, and since the rotor is driven by the engine at some fixed ratio, then the frequency is a measure of the engine speed.

The generator emfs are supplied to the corresponding phase coils of the indicator stator to produce currents of a magnitude and direction dependent on the emfs. The distribution of stator currents produces a resultant magnetic field which rotates at a speed dependent on the generator frequency. As the field rotates it cuts through the bars of the squirrel-cage rotor, inducing a current in them which, in turn, sets up a magnetic field around each bar. The reaction of these fields with the main rotating field produces a torque on the rotor causing it to rotate in the same direction as the main field and at the same speed.

As the rotor rotates it drives the permanent magnet of the speed-indicating unit, and because of relative motion between the magnet and the drag cup, eddy currents are induced in the latter. These currents create a magnetic field which reacts with that of the permanent magnet, and since there is always a tendency to oppose the creation of induced currents (Lenz's law), the torque reaction of the fields causes the drag cup to be continuously rotated in the same direction as the magnet. However, this rotation is restricted by the calibrated hairspring in such a manner that the cup will move to a position at which the eddy-current drag torque is balanced by the tension of the spring. The resulting movement of the drag cup shaft and gear train thus positions the pointers over the dial to indicate the engine speed prevailing at that instant.

Indicators are compensated for the effects of temperature on the permanent magnet of the speed-indicating element by a thermo-magnetic shunt device fitted adjacent to the magnet.

The drag disc version of the speed-indicating element consists of six pairs of small permanent magnets mounted on plates bolted together in such a way that the magnets are directly opposite each other with a small airgap between pole faces to accommodate the disc. Rotation of the disc as a result of eddy-current drag is transmitted to the pointers in a similar manner to that already described.

Percentage rpm indicators
The measurement of engine speed in terms of a percentage is adopted for turbine engine operation, and was introduced so that various types of engine could be operated on the same basis of comparison. The dial presentations of two representative indicators are shown in Fig. 15.2. The main scales are calibrated from 0 to 100 per cent in 10 per cent increments, with 100 per cent corresponding to the optimum turbine speed. In order to achieve this the engine manufacturer chooses a ratio between the actual turbine speed and the generator drive so that optimum speed produces a specific value at the generator drive. A second pointer or digital counter displays speed in one per cent increments.

Figure 15.2 Percentage rpm indicators.

Servo-operated indicators

A schematic diagram of the internal circuit arrangement of a typical indicator is shown in Fig. 15.3, and its modular construction is illustrated in Fig. 15.4; it is used in conjunction with an ac generator.

The generator signals are first converted to a square waveform by a squaring amplifier within the signal-processing module, and in order to obtain suitable positive and negative triggering pulses for each half-cycle of the waveform, it is differentiated by a signal-shaping circuit. The pulses pass through a monostable which then produces a train of pulses of constant amplitude and width, and at twice the frequency of the generator signal. In order to derive the voltage signal to run the dc motor to what is termed the demand speed condition, the monostable output is supplied to an integrator via a buffer amplifier.

The demand signal from the integrator is then applied to a sensing network in a servo amplifier and monitor module, where it is compared with a dc output from the wiper of a position feedback potentiometer. Since the wiper is geared to the main pointer of the indicator, its output therefore represents indicated speed. Any difference between the demand speed and indicated speed results in an error signal which is supplied to the input and output stages of the servo amplifier, and then to the armature winding of the motor; the indicator pointer and digital counter are then driven to the demanded speed position. At the same time, the feedback potentiometer wiper is also repositioned to provide a feedback voltage to back-off the

Figure 15.3 Servo-operated indicator.

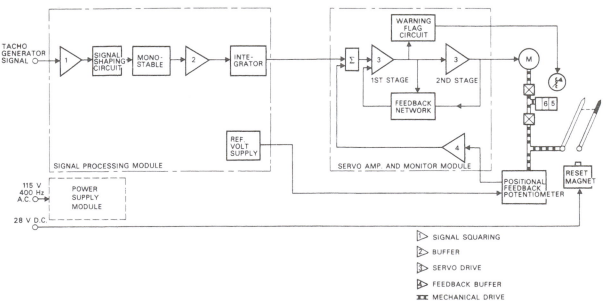

351

Figure 15.4 Construction of a servo-operated indicator.

demanded speed signal until the error signal is zero; at this point, the indicator will then display the demanded speed.

The output voltage from the servo amplifier input stage is also fed to a servo loop monitor, the purpose of which is to detect any failure of the servo circuit to back-off the error signal voltage. In the event of such failure, the monitor functions as an 'on−off' switch, and in the 'off' state it de-energizes a solenoid-controlled warning flag which appears across the digital counter display.

An overspeed pointer is also fitted concentrically with the main pointer, and is initially positioned at the appropriate scale graduation. If the main pointer exceeds this position, the limit pointer is carried with it. When the speed has been reduced the main pointer will move correspondingly, but the limit pointer will remain at the maximum speed reached since it is under the control of a ratchet mechanism. It can be returned to its initial position by applying a separately switched 28 V dc supply to a reset solenoid within the indicator.

Tacho probe and indicator system

This system has the advantage of providing a number of separate electrical outputs additional to those required for speed indication, e.g. automatic power control and flight data acquisition systems.

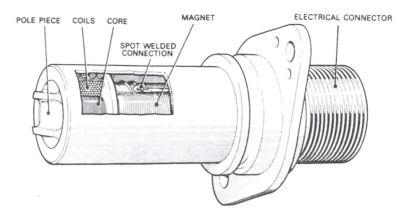

Figure 15.5 Tacho probe.

POLE PIECE COILS CORE

SPOT WELDED
CONNECTION

MAGNET

ELECTRICAL CONNECTOR

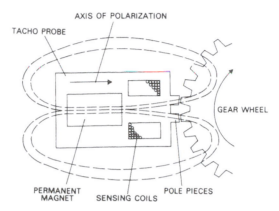

TACHO PROBE

AXIS OF POLARIZATION

GEAR WHEEL

PERMANENT
MAGNET

SENSING COILS

POLE PIECES

Furthermore, and as will be noted from Fig. 15.5, it has the advantage that there are no moving parts for subjection to high rotational loads.

The stainless steel, hermetically-sealed probe comprises a permanent magnet, a pole piece, and a number of cupro-nickel or nickel-chromium coils around a ferromagnetic core. Separate windings (from five to seven depending on the type and application of a probe) provide outputs to the indicator and other processing units requiring engine speed data. The probe is flange-mounted on an engine at a station in the high-pressure compressor section so that it extends into this section. In some turbofan engines, a probe may also be mounted at the fan section for measuring fan speed. When in position, the pole pieces are in close proximity to the teeth of a gear wheel (sometimes referred to as a 'phonic wheel') which is driven at the same speed as the compressor shaft or fan shaft as appropriate. To ensure correct orientation of the probe, a locating plug is provided in the mounting flange.

The permanent magnet produces a magnetic field around the sensing coils, and as the gear wheel teeth pass the pole pieces, the intensity of flux through each pole varies inversely with the width of the air gap between poles and gear wheel teeth. As the flux density

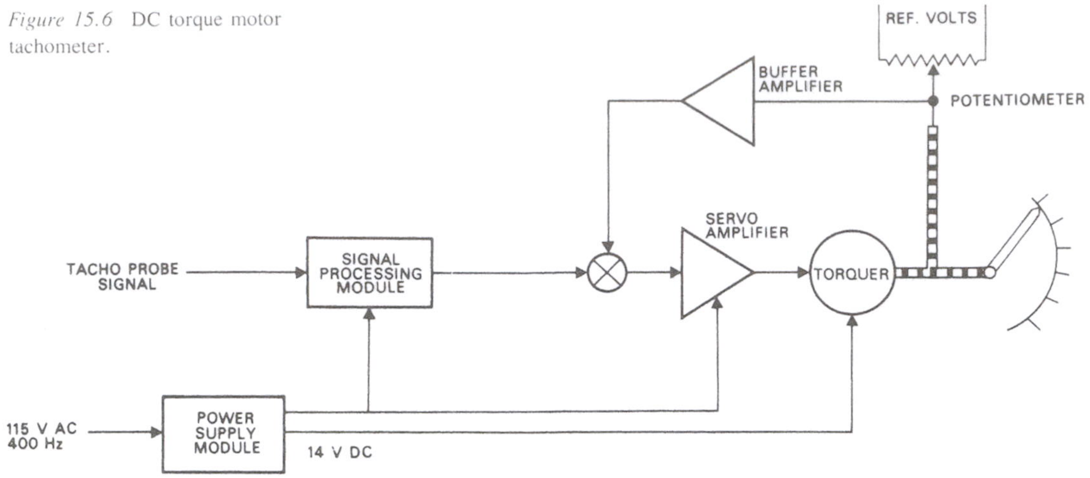

Figure 15.6 DC torque motor tachometer.

changes, an emf is induced in the sensing coils, the amplitude of the emf varying with the rate of flux density change. Thus, in taking the position shown in Fig. 15.5 as the starting position, maximum intensity would occur, but the rate of density change would be zero, and so the induced emf would be at zero amplitude.

When the gear teeth move from this position, the flux density first begins to decrease, reaching a maximum rate of change and thereby inducing an emf of maximum amplitude. At the position in which the pole pieces align with the 'valleys' between gear teeth, the flux density will be at a maximum, and because the rate of change is zero the emf is of zero amplitude. The flux density will again increase as the next gear teeth align with the pole pieces, the amplitude of the induced emf reaching a maximum coincident with the greatest rate of flux density change. The probe and gear teeth may therefore be considered as a magnetic flux switch that induces emfs directly proportional to the gear wheel and compressor or fan shaft speed.

Figure 15.6 illustrates an example of indicator circuit used in a tacho probe system. The probe output signals pass through a signal-processing module and are summed with an output from a servo potentiometer and a buffer amplifier. After summation the signal passes through a servo amplifier to the torquer motor which then rotates the pointer to indicate the change in probe signals in terms of speed. The servo potentiometer is supplied with a reference voltage, and since its wiper is also positioned by the motor, the potentiometer will control the summation of signals to the servo amplifier to ensure signal balancing at the various constant speed conditions. In the event of a power s upply or signal failure, the pointer of the indicator is returned to an 'off-scale' position under the action of a pre-loaded helical spring.

In some applications, the torquer motor is of the ac type, which drives a digital counter in addition to the pointer. Indication of power failure also differs in that a flag is energized to obscure the counter display.

Manifold pressure indicators

These indicators, colloquially termed 'boost gauges', are of the direct-reading type and are calibrated to measure absolute pressure in inches of mercury, such pressure being representative of that produced at the induction manifold of a supercharged piston engine. Before considering a typical example, it is useful to have a brief understanding of the principle of supercharging.

The power output of an internal combustion engine depends on the density of the combustible mixture of fuel and air introduced into its cylinders at that part of the operating cycle known as the induction stroke. On this stroke, the piston moves down the cylinder, an inlet valve opens, and the fuel/air mixture, or charge prepared by the carburettor, enters the cylinder as a result of a pressure difference acting across it during the stroke. If, for example, an engine is running in atmospheric conditions corresponding to the standard sea-level pressure of 14.7 lbf/in^2, and the cylinder pressure is reduced to, say, 2 lbf/in^2, then the pressure difference is 12.7 lbf/in^2, and it is this pressure difference that 'pushes' the charge into the cylinder.

An engine in which the charge is induced in this manner is said to be normally aspirated; its outstanding characteristic is that the power it develops steadily falls off with decrease of atmospheric pressure. This may be understood by considering a second example in which it is assumed that the engine is operating at an altitude of 10 000 ft. At this altitude, the atmospheric pressure is reduced by an amount which is about a third of the sea-level value, and on each induction stroke the cylinder pressures will decrease in roughly the same proportion. We thus have a pressure of about 10 lbf/in^2 surrounding the engine and 1.5 lbf/in^2 in each cylinder, leaving us with a little more than 8.5 lbf/in^2 with which to 'push' in the useful charge. This means then that at 10 000 ft only a third of the required charge gets into the cylinders, and since power is governed by the quantity of charge, we can only expect a third of the power developed at sea-level.

This limitation on the high-altitude performance of a normally-aspirated engine can be overcome by artificially increasing the available pressure so as to maintain as far as possible a sea-level value in the induction system. The process of increasing pressure and charge density is known as *supercharging* or *boosting*, and the device employed is, in effect, an elaborate form of centrifugal air pump fitted betwen the carburettor and cylinders and driven from the engine crankshaft through step-up gearing. It pumps by giving the air a very high velocity, which is gradually reduced as it passes through diffuser

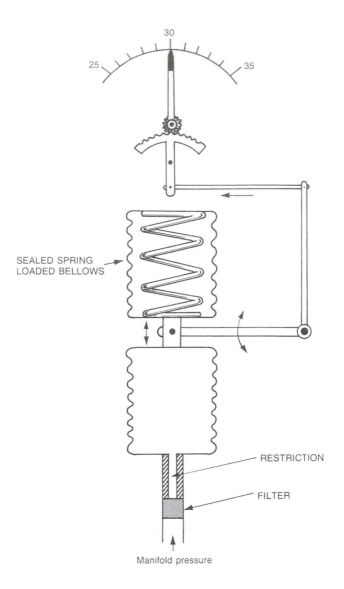

Figure 15.7 Principle of manifold pressure indicator.

SEALED SPRING
LOADED BELLOWS

RESTRICTION

FILTER

Manifold pressure

vanes and a volute, the reduction in speed giving the required increase in pressure.

In order to measure the pressure delivered by the supercharger and so obtain an indication of engine power, it is necessary to have an instrument which indicates absolute pressure. The mechanism of a typical indicator is schematically illustrated in Fig. 15.7. The measuring element is made up of two bellows, one open to the induction manifold and the other evacuated and sealed. A controlling spring is fitted inside the sealed bellows and distension of both bellows is transmitted to the pointer via a lever, quadrant and pinion mechanism. A filter is located at the inlet to open the bellows, where there is also a restriction to smooth out any pressure surges.

When pressure is admitted to the open bellows the latter expands causing the pointer to move over the scale (calibrated in inches of mercury) and so indicate a change in pressure from the standard sea-level value of 29.92 (zero 'boost'). With increasing altitude, there is a tendency for the bellows to expand a little too far because the decrease in atmospheric pressure acting on the outside of the bellows offers less opposition. However, this tendency is counteracted by the sealed bellows, which also senses the change in atmospheric pressure but expands in the opposite direction. Thus a condition is reached at which the forces acting on each bellows are equal, cancelling out the effects of atmospheric pressure so that manifold pressure is measured directly against the spring.

Torque monitoring

The monitoring of torque relates particularly to the power control of certain types of piston engines, to turbopropeller engines, and also to the control of engines in some types of helicopter. In all cases it involves the use of a torquemeter which is essentially an engine component, and is normally built in with the gear transmission assembly between the main drive shaft and the propeller shaft, or the main rotor shaft in respect of a helicopter. The construction of torquemeters depends on the type of engine, but in most cases they are of hydro-mechanical form, operating on the principle whereby any tendency for some part of the gear transmission to rotate is resisted by pistons working in hydraulic cylinders secured to the gear casing. The principle as applied to a piston engine is shown in Fig. 15.8.

Oil, which is supplied from the engine lubricating system to the

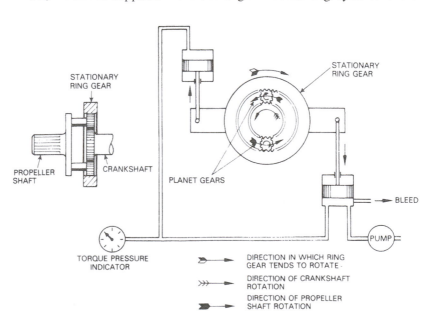

Figure 15.8 Torquemeter principle.

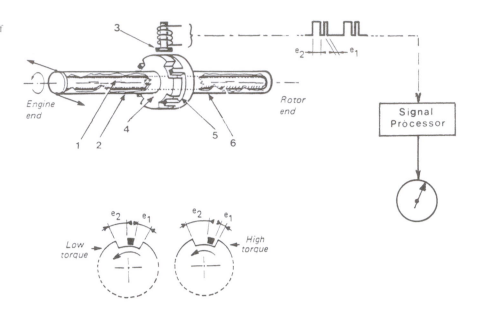

Figure 15.9 Electrical type of
torque-indicating system.
1 Power shaft, 2 sleeve,
3 sensor, 4 notched wheel,
5 toothed wheel, 6 sleeve.

cylinders via a torquemeter pump, absorbs any loads due to
movement of the pistons. The oil is thus subjected to pressures which
are proportional to the applied loads or torques, and are transmitted
to the torque pressure-indicating system which is normally of the
synchronous transmission type.

The power, in this case brake horsepower (bhp), is calculated from
the formula

$$\text{bhp} = pN/K$$

where p is the oil pressure, N the engine speed (rev/min) and K is a
torquemeter constant derived from the reduction ratio between engine
and propeller shaft gearing, length of torque arm, and number and
area of pistons.

Turbopropeller engines are, as far as power is concerned, similar
to large supercharged piston engines: most of the propulsive force is
produced by the propeller, only a very small part being derived from
the exhaust unit thrust. They are, therefore, fitted with a torquemeter
and pressure-indicating system of which the readings are an indication
of the shaft horsepower (shp). The torquemeter pressure indicator is
used in conjunction with the rpm and exhaust gas temperature
indicators.

Figure 15.9 schematically illustrates an electrical torque-indicating
system which is used in one type of helicopter currently in service
for measuring the torsion of the main power shaft in relation to the
effects of engine torque and the drive resistance set up by the main
rotor. The torquemeter consists of two wheels: one wheel is notched
and is attached at the engine end of the power shaft by a sleeve, and

it interfaces with teeth on the second wheel which is also attached by a sleeve at the rotor end of the power shaft. An electromagnetic sensor is mounted in close proximity to the peripheries of both wheels.

Under operating conditions, there is opposition between engine torque and main rotor drive resistance, resulting in torsion of the power shaft. This, in turn, results in relative displacement of the two wheels and variation in the gap widths e_1 and e_2 between teeth and notches. As the gaps pass the sensor unit they cause variations in its magnetic field which induce signal pulses the shape of which represent gap width and, therefore, the torque. The signals are transmitted to the signal processor unit which then produces a dc voltage signal proportional to the torque. After amplification, the signal is supplied to the indicator for the purpose of driving its motor and pointer drive mechanism.

| Exhaust gas temperature | The measurement of exhaust gas temperature (EGT) is based on the thermo-emf principle already described in Chapter 13 (see page 320), and requires the use of chromel/alumel thermocouple probes immersed in the gas stream at the selected points appropriate to the type of engine. |

Types of probe

Probes are generally classified as *stagnation* and *rapid response*, their application depending upon the velocity of gases. In pure jet engines the gas velocities are high, and for this reason stagnation-type probes as indicated in Fig. 15.10(a) are employed. The gas entry and exit holes, usually called sampling holes, are staggered and of unequal size, thus slowing up the gases and causing them to stagnate at the hot junction, thus giving it time to respond to changes in temperature.

Rapid-response thermocouple probes are normally adopted in the EGT systems of turbopropeller engines since their exhaust gas velocities are lower. As can be seen from (b) of Fig. 15.10, the sampling holes are diametrically opposite each other and of equal size; the gases can, therefore, flow directly over the hot junction enabling it to respond more rapidly.

Typical response times for stagnation and rapid-response probes are $1-2$ sec and $0.5-1$ sec respectively.

Probes may also be designed to contain double, triple, and in some cases up to eight hot junctions within a single probe. A triple arrangement is shown in Fig. 15.10(c). The purpose of such multi-arrangements is to provide signals to other systems requiring exhaust gas temperature data. The thermocouple elements are insulated from

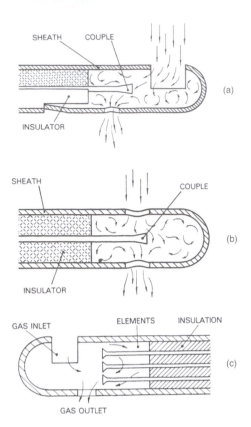

Figure 15.10 Types of
thermocouple probe.
(*a*) Stagnation; (*b*) rapid
response; (*c*) triple-element.

Figure labels: SHEATH, COUPLE, INSULATOR (a); SHEATH, COUPLE, INSULATOR (b); GAS INLET, ELEMENTS, INSULATION, GAS OUTLET (c).

each other and maintained in position by a special compound
material, e.g. compacted magnesium oxide.

When the hot junctions of immersion-type probes are in contact
with the gas stream, then not only will the stream velocity be
reduced, but also the gas will be compressed by the expenditure of
kinetic energy, resulting in an increase of hot-junction temperature. It
is in this connection that the term *recovery factor* is used, defining
the proportion of kinetic energy of the gas recovered when it makes
contact with the hot junction. This factor is, of course, taken into
account in the design of thermocouples so that the 'heat transfer', as
we may call it, makes the final reading as nearly as possible a true
indication to total gas temperature.

Location of probes

The points at which the gas temperature is to be measured are of
great importance, since they will determine the accuracy with which
measured temperature can be related to engine performance. The
ideal location is either at the turbine blades themselves, or at the
turbine entry, but certain practical difficulties are involved which
preclude the application of thermocouple probes at these locations.
Consequently, probes are installed at such locations as exhaust units,

Figure 15.11 Probe locations.

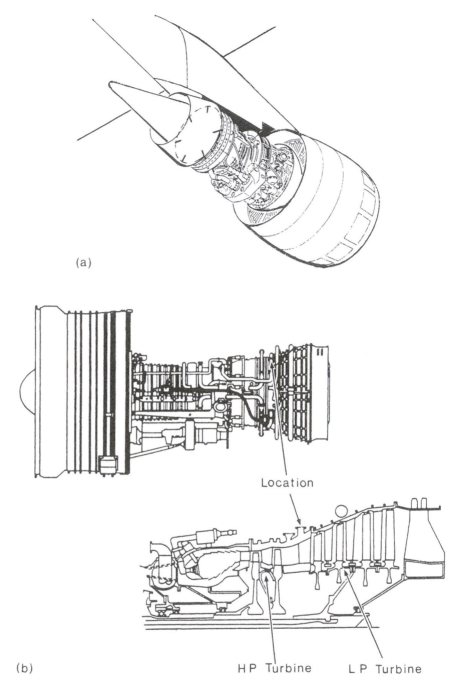

(a)

Location

(b)   HP Turbine   L P Turbine

as at (a) of Fig. 15.11, between high and low pressure turbines as at (b), or in some engines at the leading edge of stator guide vanes between turbine stages.

For accurate measurement it is necessary to sample temperatures from a number of points evenly distributed over a cross-section of the gas flow. This is because temperature differences can exist in various zones or layers of the flow through the turbine section and exhaust

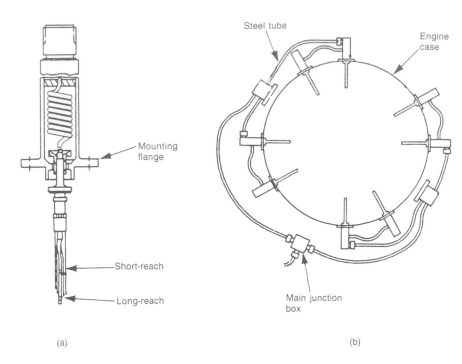

Figure 15.12 Probe grouping.

Steel tube

Engine case

Mounting flange

Short-reach

Long-reach

Main junction box

(a)

(b)

unit, and so measurement at one point only would not be truly representative of the conditions prevailing.

A measuring system, therefore, always consists of a group of thermocouple probes suitably disposed in the gas flow, and connected in parallel so as to measure a good average temperature condition. The probes in a group may contain a single hot junction, or pairs of junctions referred to as *short reach* and *long reach* from the extent to which they reach into the gas stream; an example is shown in Fig. 15.12(a).

Probes and their chromel and alumel cables are made up into a harness assembly of a design appropriate to the type of engine and number of probes required. An eight-probe arrangement comprising 16 hot junctions is shown at (b) of Fig. 15.12. The cables pass through steel tubing and terminate at a main junction box which serves as the 'take-off' point for the connection of indicators and other units requiring EGT data. Terminal studs of junction boxes are also made of chromel and alumel and, in order to ensure correct polarity of cable connections, the diameters of alumel studs are larger than those of the chromel studs.

Indicators

Depending on the instrumentation configurations adopted for a particular type of aircraft, the indication of EGT, as in the case of other power and control parameters, may be provided by servo-operated indicators or by electronic display methods. The modular arrangement of one type of servo-operated indicator is illustrated in Fig. 15.13(a).

The output from the thermocouple probes is supplied first to a cold junction reference bridge circuit, the purpose of which is to compensate for changes in ambient temperature of the indicator. The circuit is shown in more detail in diagram (b). The thermocouple harness and cables are connected to copper leads which are embedded in close proximity to each other within a former which supports a copper coil resistor R_4; thus, together they form the effective cold junction of the system. The bridge circuit is supplied with 7 V dc from a stabilized reference supply module within the indicator, and the bridge output is supplied to a servo amplifier.

As we have already learned, the standard values of emf produced by a thermocouple are related to a selected value of cold junction temperature (see page 322). In this case, the bridge circuit is adjusted by means of a variable resistor RV_1 so that an emf of the correct sense and magnitude is injected in series with that of the thermocouples such that, in combination, the emf is equal to that which would be obtained if the cold junction temperature were 0°C. Since the ambient temperature of the indicator, and hence the cold junction, will in the normal operating environment always be higher than this, then the temperature difference will reduce the thermocouple output. The resistor R_1 will, however, also be subjected to the higher ambient temperature, but because under such conditions the resistance of R_1 decreases, it will modify the bridge circuit conditions so as to restore the combined emf output to the standard value corresponding to a cold junction temperature of 0°C.

The output is termed the *demand* EGT signal and is compared with a dc output from the wiper of a positional feedback potentiometer, and since the wiper is geared to the main pointer and digital counter of the indicator, then the dc output which is fed back to the cold junction reference circuit represents the *indicated* EGT. Any difference between demanded and indicated EGTs results in an error signal being produced by the reference circuit which then supplies the signal to the servo amplifier as shown in (a) of Fig. 15.13. The amplifier output is fed to the armature winding of the dc servomotor which then drives the pointer and digital counter, causing them to display a coarse and fine indication respectively of the EGT. The feedback potentiometer wiper is also repositioned to provide a feedback voltage which backs-off the demanded temperature signal until the error signal is zero; at this point the indicator will then display the demanded temperature.

The output voltage from one stage of the servo amplifier is also fed to a servo loop monitor, the purpose of which is to detect any failure of the loop to back-off the error signal voltage. Should such failure occur, the monitor functions as an 'on–off' switch, and in the 'off'

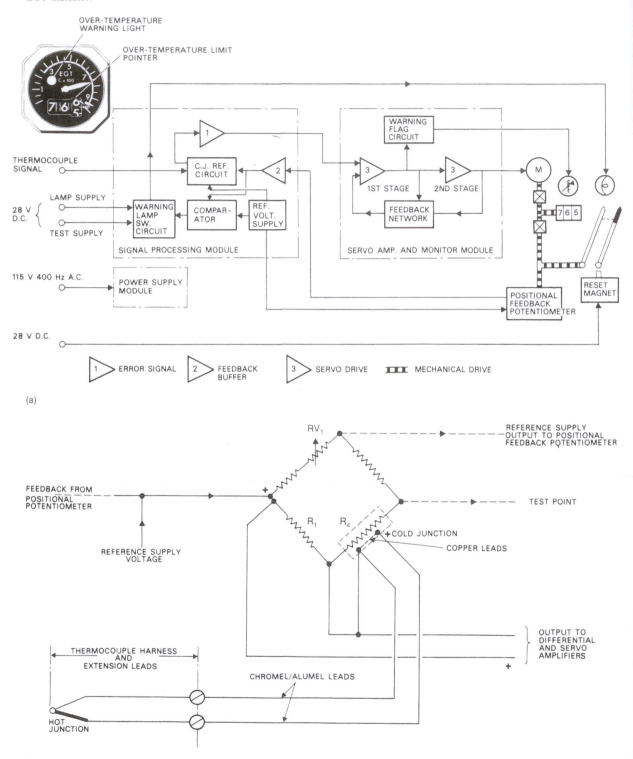

Figure 15.13 Servo-operated EGT indicator.

(a)

1 ERROR SIGNAL	2 FEEDBACK BUFFER	3 SERVO DRIVE

IIIII MECHANICAL DRIVE

(b)

state de-energizes a solenoid-controlled warning flag which appears across the digital counter display. The flag will also appear in the event of the 115 V ac supply to the indicator falling below 100 V.

An over-temperature warning light is incorporated in the indicator, and is controlled by a relay, a comparator, and a solid-state switching circuit. The function of the comparator is to compare the feedback voltage from the positional potentiometer with a pre-set voltage the level of which is equivalent to a predetermiend over-temperature limit for the particular type of engine. In the event of this limit being exceeded, the feedback voltage will exceed the reference voltage level, and the switching circuit will cause the relay to energize, thereby completing the circuit to a warning light. A separate supply voltage may be connected to the light by means of an 'override' facility as a means of testing its filament at any point over the temperature range of the indicator.

An over-temperature pointer is also fitted concentrically with the main pointer, and is initially positioned at the appropriate scale graduation. It operates in a similar manner to the over-speed pointer of a servo-operated tachometer indicator (see page 351).

Examples of EGT indications by means of electronic display systems will be covered in Chapter 16.

Engine pressure ratio (EPR) measurement

EPR is an operating variable which, together with rev/min, EGT and fuel flow, provides an indication of the thrust output of turbine engines, and involves the measurement of the ratio between the pressures at the compressor intake and the turbine outlet or exhaust.

In general, a measuring system consists of an engine inlet pressure probe, a number of pressure-sensing probes projected into the exhaust unit of an engine, a pressure ratio transmitter, and an indicator. The interconnection of these components based on a typical system is schematically shown in Fig. 15.14.

The inlet pressure-sensing probe is similar to a pitot probe, and is mounted so that it faces into the airstream in the engine intake or, as in some power plant installations, on the pylon and in the vicinity of the air intake. The probe is protected against icing by a supply of warm air from the engine anti-ice system.

The exhaust pressure-sensing probes are interconnected by pipelines which terminate at a manifold, thus averaging the pressures. In some engine systems, pressure-sensing is done from chambers contained within the EGT sensing probes. A pipeline from the manifold, and another from the inlet pressure probe, are each connected to the pressure ratio transmitter which comprises a bellows type of pressure-sensing transducer, a linear voltage differential transformer (LVDT), a two-phase servomotor, an amplifier and a potentiometer. The

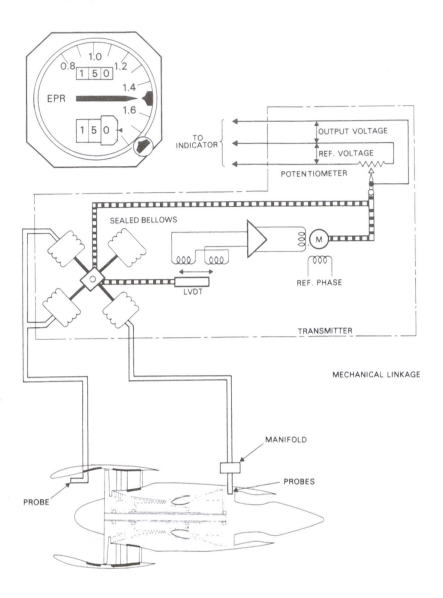

Figure 15.14 EPR system.

transducer bellows are arranged in two pairs at right angles and supported in a frame which, in turn, is supported in a gimbal and yoke assembly. The gimbal is mechanically coupled to the servomotor via a gear train, while the yoke is coupled to the core of the LVDT. The servomotor also drives the wiper of the potentiometer which adjusts the output voltage signals to the indicator in terms of changes in pressure ratio.

The indicator shown in the diagram is of the servo-operated type. In electronic display systems (see Chapter 16) the transmitter output signals are supplied direct to the appropriate system computer.

From Fig. 15.14 it will be noted that the intake pressure is admitted to two of the bellows in the transducer, exhaust gas pressure is admitted to the third bellows, while the fourth is evacuated and sealed. Thus the system, together with its frame, gimbal and yoke assembly, forms a pressure balancing and torsional system.

When a pressure change occurs, it causes an unbalance in the bellows system, and the resultant of the forces acting on the transducer frame acts on the yoke such that it is pivoted about its axis. The deflection displaces the core of the LVDT to induce an ac signal which is amplified and applied to the control winding of the servomotor. The motor, via a gear train, alters the potentiometer output signal to the indicator so that its pointer and digital counter are servo-driven to indicate the new pressure ratio. Simultaneously, the motor drives the transducer gimbal and LVDT core in the same direction as the initial yoke movement, so that the relative movement now produced between the core and coils starts reducing the servo-motor drive signal, until it is finally 'nulled' and the system stabilized at the new ratio.

The lower counter shown in the diagram is for the purpose of indicating a reference EPR value; it is set manually by rotating the setting knob.

If a circuit malfunction occurs, an integrity monitoring circuit within the indicator activates a warning flag circuit, causing the flag to obscure the digital counter display.

In some types of aircraft, a maximum allowable EPR limit indicator is also provided, and is integrated with a TAT indicator (see page 64) and also with an ADC; its purpose being to indicate limits related to air density and altitude values from which thrust settings have been predetermined for specific operating conditions. The conditions are climb, cruise, continuous and go-around, and are selected as appropriate by means of a mode selector switch connected to a computing and switching circuit which generates a datum signal corresponding to each selected condition. The signal is then supplied to a comparator, which also receives temperature signals from the TAT sensor and altitude signals from the ADC. These signals are compared with the datum signal and the lower value of the two is automatically selected as the signal representing the maximum EPR limit for the selected operating condition. The comparator transmits this signal to an amplifier and a servomotor which then drives a digital counter to display the limiting value.

Fuel flow measurement

Fuel flow measuring systems vary in operating principle and construction, but principally they consist of two units: a transmitter or flowmeter, and an indicator. Transmitters are connected in the delivery lines of an engine fuel system, and are essentially electro-mechanical devices producing output signals proportional to flow rate which in a basic system is indicated in either volumetric or mass units. In many of the systems currently in use, an intermediate amplifier/computer is also included to calculate a fuel flow/time ratio and to transmit signals to indicators which can display not only flow rate but also the amount of fuel consumed.

Basic system

Figure 15.15 is a sectioned view of a transmitter that forms the measuring unit of a simple flow rate indicating system. It has a cast body with inlet and outlet connections in communication with a spiral-shaped metering chamber containing the metering assembly. The latter consists of a vane pivoted so that it can be angularly displaced under the influence of fuel passing through the chamber. A small gap is formed between the edge of the vane and the chamber wall which, on account of the volute form of the chamber, increases in area as the vane is displaced from its zero position. The variation in gap area controls the rate of vane displacement which is faster at the lower flow rates (gap narrower) than at the higher ones. The vane is mounted on a shaft carried in two plain bearings, one in each cover plate enclosing the metering chamber.

At one end, the shaft protrudes through its bearing and carries a two-pole ring-type magnet which forms part of a magnetic coupling between the vane and the electrical transmitting unit, which may be a precision potentiometer or an ac torque synchro. The shaft of the transmitting unit carries a two-pole bar-type magnet which is located inside the ring magnet. The interaction of the two fields provides a 'magnetic lock' so that the potentiometer wiper (or synchro rotor) can follow any angular displacement of the metering vane free of friction.

The other end of the metering vane shaft carries the attachment for the inner end of a specially calibrated control spring. The outer end

Figure 15.15 Rotating vane fuel flowmeter.

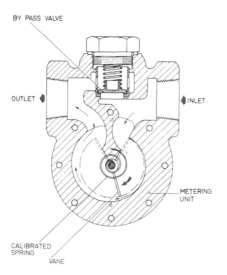

of the spring is anchored to a disc plate which can be rotated by a pinion meshing with teeth cut in the periphery of the plate. This provides for adjustment of the spring torque during flowmeter calibration.

Any tendency for the metering assembly to oscillate under static flow conditions is damped out by a counterweight and vane, attached to the metering vane shaft, and operating in a separate fuel-filled chamber secured to one side of the transmitter body.

When fuel commences to flow it passes through the metering chamber and deflects the metering vane from its zero position and tends to carry it round the chamber. Since the vane is coupled to the calibrated spring, the latter will oppose movement of the vane, permitting it to take up only an angular position at which spring tension is in equilibrium with the rate of fuel flow at any instant. Through the medium of the magnetic-lock coupling the vane will also cause the potentiometer wiper, or synchro rotor, to be displaced. In the former case, and with a steady direct voltage across the potentiometer, the voltage at the wiper is directly proportional to the fuel flow. The voltage is fed to an amplifier, whose output current drives a milliammeter pointer to indicate the current in terms of fuel flow in gal/hr or lb/hr.

In a system employing synchros, the current flow due to differences in angular positions of the rotors will drive the indicator synchro rotor directly to the 'null' position and thereby make the pointer indicate the fuel flow.

In meters of this type it is also necessary to provide a bypass for the fuel in the event of jamming of the vane or some other obstruction causing a build-up of pressure on the inlet side. As may be seen from Fig. 15.15, this is accomplished by a spring-loaded valve incorporated in the metering chamber. The spring tension is adjusted so that the valve lifts from its seating and allows fuel to bypass the metering chamber when the pressure difference across the chamber exceeds 2.5 lbf/in^2.

Integrated flowmeter systems

An integrated flowmeter system may broadly be defined as one in which a fuel consumed measuring element is combined with that of fuel flow, thus permitting the display of both quantities in a single indicator.

In order to accomplish this it is necessary to introduce an integrating system to work out fuel consumed in the ratio of fuel flow rate to time. Such a system may be mechanical, forming an integral part of an indicator mechanism, or as in electronic fuel flow measuring systems, it may be a special dividing stage within the amplifier, or even a completely separate integrator unit.

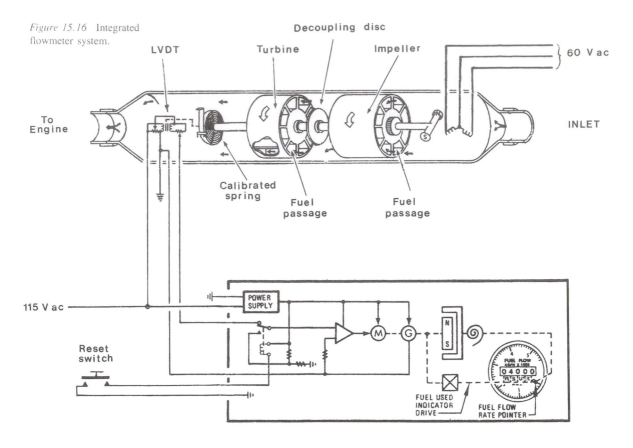

Figure 15.16 Integrated flowmeter system.

The components of a typical system are shown in Fig. 15.16. The transmitter comprises an impeller driven by a two-phase ac motor, a turbine which is interconnected with a calibrated restraining spring, and an LVDT sensor. A decoupling disc is located between the impeller and turbine, its purpose being to prevent an 'hydraulic transmission' effect on both units when operating at low rates of fuel flow. The indicator is servo-operated with the drive to the flow rate pointer being effected by means of an eddy-current drag type of mechanism similar to that adopted in some rpm indicators (see page 350). The fuel consumed indicator is a digital counter which is mechanically integrated with the servomotor via a gear transmission, the ratio of which is preselected to establish the requisite relationship between the motor speed, which is proportional to flow rate, and time.

The system is supplied with 115 V single-phase 400 Hz ac from an aircraft's power system and this is utilized by a power supply unit within the indicator, the primary coil of the LVDT in the transmitter, and by a separate power supply unit (not shown in the diagram). This unit contains a temperature stable oscillator connected to a voltage/frequency converter which converts the main supply into a two-phase 60 V 8 Hz output; this, in turn, is supplied to the transmitter impeller motor. The rotating field set up in the rotor windings interacts with its permanent magnet rotor which rotates in synchronism and drives the impeller at a constant speed.

Fuel flow rate is, in the first instance, always established by an engine's fuel control unit which is calibrated or 'trimmed' to control rates commensurate with the varying operational conditions and the other associated power parameters, i.e. rpm, EGT and EPR. When fuel enters the transmitter it passes through passages in the impeller which, on account of its rotation, causes the fuel to swirl at a velocity governed by the flow rate. The fuel is then diverted around the decoupling disc, and in passing through passages in the turbine, it imparts a rotational force which tends to continuously rotate the turbine in the same direction as the impeller. This tendency is, however, restrained by the calibrated spring such that the rotation is limited and balanced at an angular position proportional to the flow rate of fuel passing through the transmitter.

The movement of the turbine and its shaft alters the position of the LVDT sensor core, so that a signal voltage (up to 5 V at maximum flow rate) is induced in the secondary winding and supplied to the indicator servomotor via the closed contacts of the reset relay and amplifier. The servomotor rotates at a speed proportional to the flow rate, and by means of the eddy-current drag mechanism positions the pointer to indicate this rate.

The reset switch is separately located on a flight deck panel, and when pressed it energizes the relay in the indicator to supply 115 V ac to the servo amplifier and motor, causing it to drive 'downscale' rapidly in order to reset the fuel consumed counter display to zero.

Figure 15.17 illustrates the components of another type of integrated system. It differs from the one just described in that the transmitter utilizes two electromagnetic pick-off elements, and the processing of signals relevant to flow rate and fuel consumed is carried within a separate electronic unit.

The transmitter consists of a light-alloy body containing a flow-metering chamber, a motor-driven impeller assembly, and the externally-mounted coils of the pick-off elements. The impeller assembly consists of an outer drum which is driven through a magnetic coupling and reduction gear by a synchronous motor, and an impeller incorporating vanes and fuel passages to impart swirl and angular velocity to the fuel flowing through the metering chamber. The drum and impeller are coupled to each other by a calibrated spring. The motor is contained within a fixed drum at the inlet end and rotates the impeller at a constant speed (a typical value is 100 rpm). Straightening vanes are provided in the fixed drum around the motor to remove any angular velocity already present in the fuel before it passes through the impeller assembly. A point to note about the use of a magnetic coupling between the motor and impeller

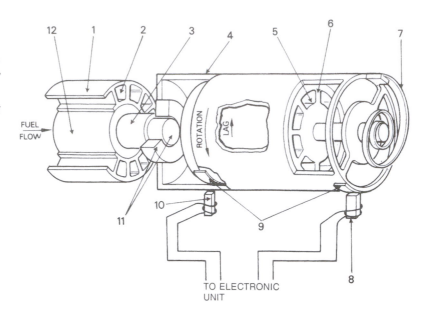

Figure 15.17 Electronic integrated flowmeter system. 1 Fixed drum, 2 fuel passages, 3 motor shaft, 4 rotating drum, 5 fuel passages, 6 impeller, 7 calibrated spring, 8 pick-off (drum), 9 magnets, 10 pick-off (impeller), 11 magnetic coupling, 12 motor.

FUEL FLOW

ROTATION

LAG

TO ELECTRONIC UNIT

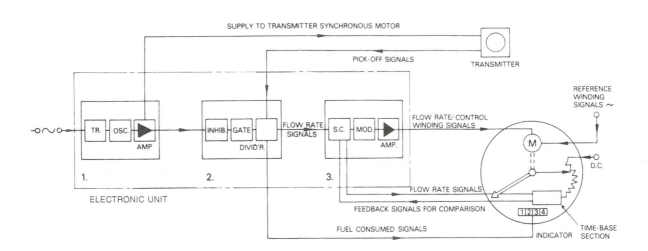

SUPPLY TO TRANSMITTER SYNCHRONOUS MOTOR

PICK-OFF SIGNALS

TRANSMITTER

REFERENCE WINDING SIGNALS ~

TR. | OSC. | AMP

INHIB. | GATE | DIVID'R.

FLOW RATE SIGNALS

S.C. | MOD. | AMP.

FLOW RATE/CONTROL WINDING SIGNALS

M

D.C.

1. 2. 3.

ELECTRONIC UNIT

FLOW RATE SIGNALS

FEEDBACK SIGNALS FOR COMPARISON

FUEL CONSUMED SIGNALS

1 2 3 4

INDICATOR

TIME-BASE SECTION

assembly is that it overcomes the disadvantages which in this application would be associated with rotating seals. The motor and its driving gear are isolated from the fuel by enclosing them in a chamber which is evacuated and filled with an inert gas before sealing.

Each of the two pick-off assemblies consists of a magnet and an iron-cored inductor. One magnet is fitted to the outer drum while the other is fitted to the impeller, thus providing the required angular reference points. The magnets are so positioned that under zero flow conditions they are effectively in alignment with each other. The coils are located in an electrical compartment on the outside of the transmitter body, together with solid-state circuit units which amplify and switch the signals induced.

The electronic unit performs the overall function of providing the power for the various circuits of the system, detecting the number of pulses produced at the transmitter, and computing and integrating the fuel flow rate and amount of fuel consumed. It consists of a number of stages interconnected as shown in Fig. 15.17. The power supply section (1) controls the voltage and frequency of the supply to the transmitter synchronous motor, and consists of a transformer, crystal oscillator, output and power amplifier units.

From the diagram it will be noted that section (2) is made up of three stages: inhibitor, gate and divider. The respective functions of these stages are: to suppress all transmitter signals below a certain flow rate; to control or gate the pulse signals from the power supply oscillator; and to produce output signals proportional to true flow rate, and to provide the time dividing factor and output pulses representing unit mass of fuel consumed. Section (3) is also made up of three stages: signal comparator, modulator and servo amplifier. The respective functions of these stages are: to compare the transmitter output signals with time-base signals fed back from the indicator; to combine the comparator output with 400 Hz ac and produce a new output; and to provide an operating signal to the indicator servomotor control winding.

The indicator employs a flow indicating section consisting of an ac servomotor which drives a pointer, a digital counter display, and a potentiometer wiper through a reduction gear train. The reference winding of the motor is supplied with a constant alternating voltage, while the control winding receives its signals from the servo amplifier in the electronic unit. The potentiometer is supplied with dc and its wiper is electrically connected to a solid-state time-base circuit, also within the indicator. Transmitter output signals are fed into the time-base circuit via a pre-set potentiometer which forms part of the electronic unit's comparator stage. The difference between the time-base and the indicated fuel flow signal voltages is fed to the servomotor, which operates to reduce the error voltage to zero and so to correct the indicated fuel flow.

The fuel consumed section of the indicator consists of a solenoid-actuated five-drum digital counter and a pulse amplifier. The amplifier receives a pulse from the divider stage of the electronic unit for each unit mass of fuel consumed and feeds its output to the solenoid, which advances the counter drums appropriately. A separately located reset switch is also provided for returning the counter to zero; it operates in a similar manner to that described on page 371.

Figure 15.18 Operation of pick-offs.

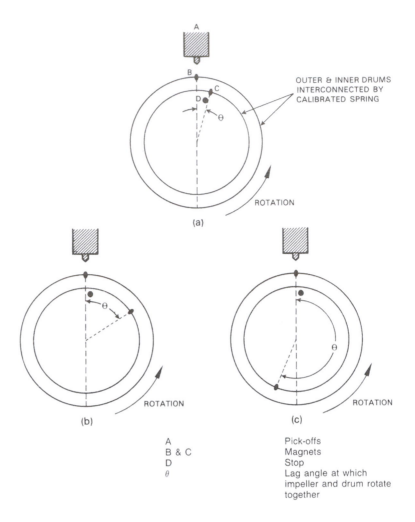

A	Pick-offs
B & C	Magnets
D	Stop
θ	Lag angle at which impeller and drum rotate together

When electrical power is switched on to the system, the transmitter impeller motor is, as already mentioned, rotated at constant speed. Under zero fuel flow conditions, the magnets of the pick-offs are effectively in line with each other, although in practice a small angular difference (typically $3°-5°$) is established to maintain a deflection representing a specific minimum flow rate. This is indicated in Fig. 15.18(a).

As the fuel flows through the metering chamber, a constant angular velocity is imparted to the fuel by the rotating impeller and drum assembly, and since the two are interconnected by the calibrated spring, a reaction torque is created which alters the angular displacement between the impeller and drum and their corresponding magnets. Thus, angular displacement produces a time difference between signal pulses in the pick-offs, both being proportional to flow rate. Diagrams (b) and (c) illustrate the displacement for typical cruising and maximum fuel flow rates respectively.

The position of each magnet is sensed by its own pick-off coil, and the primary pulses induced as each magnet moves past its coil are fed to the dividing stage in the electronic unit (see Fig. 15.17). The output from this stage is fed to the control winding of the indicator servomotor via section (3) of the electronic unit, and the indicator pointer is driven to indicate the fuel flow. At the same time, the motor drives the potentiometer wiper, producing a signal which is fed back to the signal comparator stage for comparison with the output produced by the transmitter. Any resultant difference signal is amplified, modulated and power amplified to drive the indicator motor, pointer and digital counter to a position indicating the actual fuel flow rate.

The divider stage of the electronic unit also uses the transmitter signals to produce pulse 'time' signals for the operation of the fuel consumed counter of the indicator. During each successive revolution of the transmitter impeller assembly the pulses are added and divided by a selected ratio, and then supplied to the counter as an impulse corresponding to each pound of fuel consumed.

| **Engine vibration monitoring** | Engine vibration is a feature of engine operation which cannot be eliminated entirely even with turbine engines, which, unlike piston engines, have no reciprocating parts. Thus, by accurate balancing of such components as crankshafts, compressor and turbine rotor discs, vibration must be kept down to the lowest levels acceptable under all operating conditions. In respect of turbine engine operation, however, there is always the possibility of these levels being exceeded as a result of certain mechanical failures occurring. For example, a turbine blade may crack or 'creep', or an uneven temperature distribution around turbine blades and rotor discs may be set up; either of these will give rise to unbalanced conditions of the main rotating assemblies and possible disintegration. In order, therefore, to indicate when the maximum amplitude of vibration of an engine exceeds the pre-set level, monitoring systems, which come within the control group of instrumentation, are provided. |

A block diagram of a typical system is shown in Fig. 15.19. It consists of a vibration pick-off, or sensor, mounted on an engine at right angles to its axis, an amplifier monitoring unit, and a moving-coil milliammeter calibrated to show vibration amplitude in thousandths of an inch (mils).

The sensor is a linear-velocity detector that converts the mechanical energy of vibration into an electrical signal of proportional magnitude. It does this by means of a spring-supported permanent magnet suspended in a coil attached to the interior of the case.

As the engine vibrates, the sensor unit and core move with it; the magnet, however, tends to remain fixed in space because of inertia.

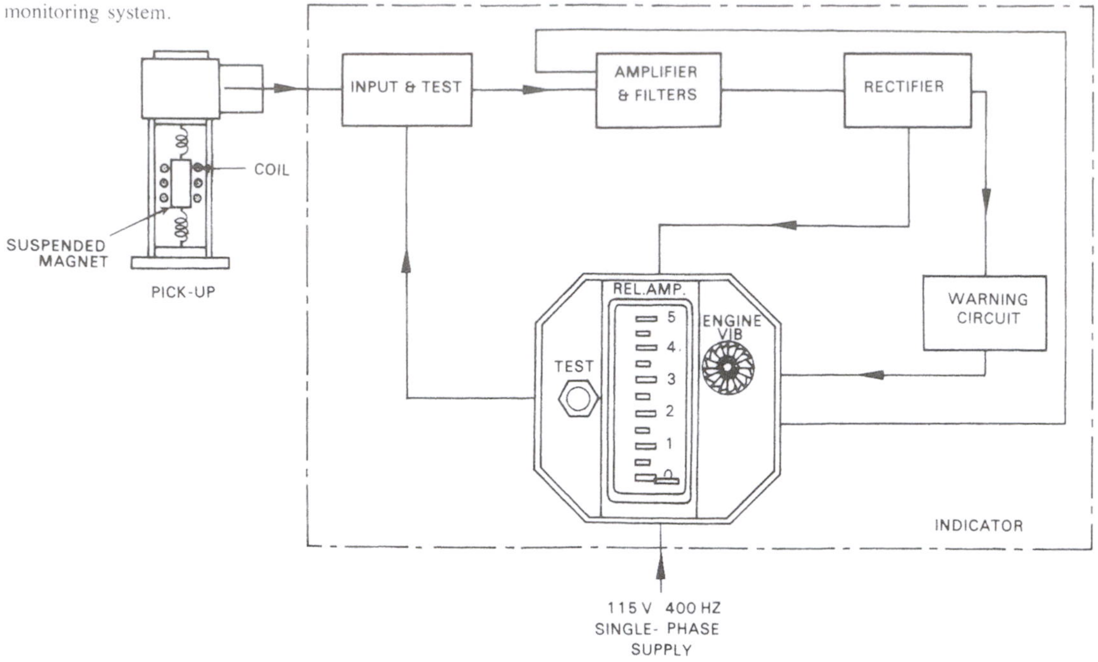

Figure 15.19 Vibration
monitoring system.

In other words, its function is similar to that of an accelerometer. The motion of the coil causes the turns to cut the field of the magnet, thus inducing a voltage in the coil and providing a signal to the amplifier unit. The signal, after amplification and integration by an electrical filter network, is fed to the indicator via a rectifying section.

An amber indicator light also forms part of the system, together with a test switch. The light is supplied with dc from the amplifier rectifying section and it comes on when the maximum amplitude of vibration exceeds the pre-set value. The test switch permits functional checking of the system's electrical circuit.

In some engine installations, two sensors may be fitted to an engine: for example, in a typical turbofan engine, one monitors vibration levels around the fan section, and the other around the engine core section.

In systems developed for use in conjunction with LCD and CRT display indicators, the vibration sensors are of the type whereby vibration causes signals to be induced in a piezoelectric stack (see also page 165). A CRT display of vibration is shown in Fig. 16.2.

16 Electronic instruments for engine and airframe systems control

The display of the parameters associated with engine performance and airframe systems control by means of CRT-type display units has, like those of flight instrument systems, become a standard feature of many types of aircraft. The display units form part of two principal systems designated as *engine indicating and crew alerting system* (EICAS) and *electronic centralized aircraft monitoring* (ECAM) system, which were first introduced in Boeing 757 and 767 aircraft and the Airbus A310 respectively. At the time of their introduction, there were differing views on the approach to such operating factors as flight deck layouts and crews' controlling functions, the extent to which normal, alerting and warning information should be displayed, and in particular, whether engine operating data required to be displayed for the whole of a flight, or only at various phases.

In respect of EICAS, engine operating data is displayed on its CRT units, thereby eliminating the need for traditional instruments. The data, as well as those relevant to other systems, are not necessarily always on display but in the event of malfunctions occurring at any time, the flight crew's attention is drawn to them by an automatic display of messages in the appropriate colours. The ECAM system, on the other hand, displays systems' operation in checklist and schematic form, and as this was a concept based on the view that engine data need to be displayed during the whole of a flight, traditional instruments were retained in the Airbus A310. It is of interest to note, however, that in subsequent types produced by this manufacturer, e.g. A320, the ECAM system is developed to include the display of engine data in one of its display units.

EICAS

The basic system comprises two display units, a control panel, and two computers supplied with analog and digital signals from engine and system sensors as shown in the schematic functional diagram of Fig. 16.1. The computers are designated 'Left' and 'Right', and only one is in control at a time; the other is on 'standby', and in the event of failure it may be switched in either manually or automatically.

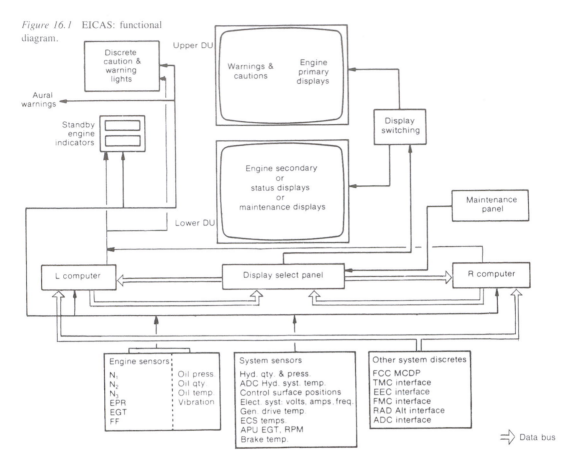

Figure 16.1 EICAS: functional diagram.

Upper DU

Discrete
caution &
warning
lights

Warnings &
cautions

Engine
primary
displays

Aural
warnings

Display
switching

Standby
engine
indicators

Engine secondary
or
status displays
or
maintenance displays

Maintenance
panel

Lower DU

L computer

Display select panel

R computer

Engine sensors		System sensors	Other system discretes
N₁	Oil press.	Hyd. qty. & press.	FCC MCDP
N₂	Oil qty.	ADC Hyd. syst. temp.	TMC interface
N₃	Oil temp.	Control surface positions	EEC interface
EPR	Vibration	Elect. syst: volts, amps, freq.	FMC interface
EGT		Gen. drive temp.	RAD Alt interface
FF		ECS temps.	ADC interface
		APU EGT, RPM	
		Brake temp.	

⟹ Data bus

Operating in conjunction with the system are discrete caution and warning lights, standby engine indicators and a remotely-located panel for selecting maintenance data displays. The system provides the flight crew with information on primary engine parameters (full-time), with secondary engine parameters and advisory/caution/warning alert messages displayed as required.

Display units

These units provide a wide variety of information relevant to engine operation, and operation of other automated systems, and they utilize colour shadow mask CRTs and associated card modules whose functions are identical to those of the EFIS units described in Chapter 12. The units are mounted one above the other as shown in Fig. 16.2.

The upper unit displays the primary engine parameters N_1 speed, EGT, and warning and caution messages. In some cases this unit can also display EPR depending on the type of engines installed and on the methods of processing data by the thrust management control system. The lower unit displays secondary engine parameters, i.e. N_2

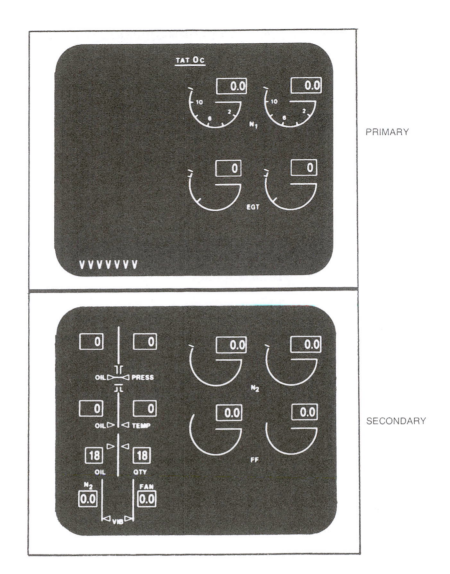

Figure 16.2 EICAS: engine data displays.

PRIMARY

SECONDARY

speed, fuel flow, oil quantity, pressure and temperature, and engine vibration. In addition, the status of non-engine systems, e.g. flight control surface positions, hydraulic system, APU, etc., can also be displayed together with aircraft configuration and maintenance data. The rows of 'V's shown on the upper display unit only appear when secondary information is being displayed on the lower unit.

Seven colours are produced by the CRTs and they are used as follows:

White   All scales, normal operating range of pointers, digital readouts.

Red   Warning messages, maximum operating limit marks on scales, and digital readouts.

379

Green	Thrust mode readout and selected EPR/N_1 speed marks or target cursors.
Blue	Testing of system only.
Yellow	Caution and advisory messages, caution limit marks on scales, digital readouts.
Magenta	During in-flight engine starting, and for cross-bleed messages.
Cyan	Names of all parameters being measured (e.g. N_1 oil pressure, TAT, etc.) and status marks or cues.

The displays are selected according to an appropriate display selection mode.

Display modes

EICAS is designed to categorize displays and alerts according to function and usage, and for this purpose there are three modes of displaying information: (i) *operational*, (ii) *status*, and (iii) *maintenance*. Modes (i) and (ii) are selected by the flight crew on the display select panel, while mode (iii) is selected on the maintenance panel which is for the use of engineers only.

Operational mode
This mode displays the engine operating information and any alerts required to be actioned by the crew in flight. Normally only the upper display unit presents information; the lower one remains blank and can be selected to display secondary information as and when required.

Status mode
When selected this mode displays data to determine the dispatch readiness of an aircraft, and is closely associated with details contained in an aircraft's Minimum Equipment List. The display shows positions of the flight control surfaces in the form of pointers registered against vertical scales, selected sub-system parameters, and equipment status messages on the lower display unit. Selection is normally done on the ground either as part of pre-flight checks of dispatch items, or prior to shut-down of electrical power to aid the flight crew in making entries in the aircraft's Technical Log.

Maintenance mode
This mode provides maintenance engineers with information in five different display formats to aid them in trouble-shooting and verification testing of the major sub-systems. The displays, which are presented on the lower display unit, are not available in flight.

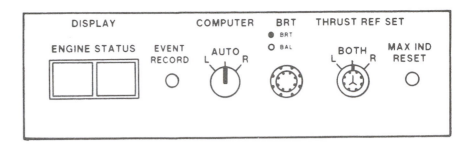

Figure 16.3 EICAS: display select panel.

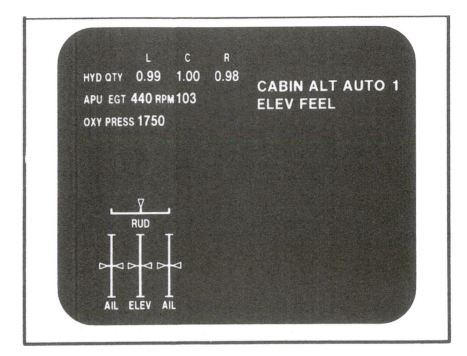

Figure 16.4 Status mode display.

Display select panel

This panel, as indicated in Fig. 16.3, permits control of EICAS functions and displays and can be used both in flight and on the ground. It is normally located on the centre pedestal of an aircraft's flight deck, and its controls are as follows:

1. *Engine display switch* This is of the momentary-push type for removing or presenting the display of secondary information on the lower display unit.

2. *Status display switch* Also of the momentary-push type, this is used to display the status mode information referred to earlier, on the lower display unit. The display is known as a 'status page', an example of which is shown in Fig. 16.4.

3. *Event record switch* This is of the momentary-push type and is used in the air or on the ground, to activate the recording of fault data relevant to the environmental control system, electrical power, hydraulic system, performance and APU. Normally, if any

malfunction occurs in a system, it is recorded automatically (called an 'auto event') and stored in a non-volatile memory of the EICAS computer. The push switch also enables the flight crew to record a suspect malfunction for storage, and this is called a 'manual event'. The relevant data can only be retrieved from memory and displayed when the aircraft is on the ground and by operating switches on the maintenance control panel.

4. *Computer select switch* In the 'AUTO' position it selects the left, or primary, computer and automatically switches to the other computer in the event of failure. The other positions are for the manual selection of left or right computers.

5. *Display brightness control* The inner knob controls the intensity of the displays, and the outer knob controls brightness balance between displays.

6. *Thrust reference set switch* Pulling and rotating the inner knob positions the reference cursor on the thrust indicator display (either EPR or N_1) for the engine(s) selected by the outer knob.

7. *Maximum indicator reset switch* If any one of the measured parameters, e.g. oil pressure, EGT, should exceed normal operating limits, this will be automatically alerted on the display units. The purpose of the reset switch is to clear the alerts from the display when the excess limits no longer exist.

Alert messages

The system continuously monitors a large number of inputs (typically over 400) from engine and airframe systems' sensors and will detect any malfunctioning of systems. If this should occur, then appropriate messages are generated and displayed on the *upper* display unit in a sequence corresponding to the level of urgency of action to be taken. Up to 11 messages can be displayed, and at the following levels:

Level A — Warning requiring immediate corrective action. They are displayed in red. Master warning lights are also illunianted, and aural warnings (e.g. fire bell) from a central warning system are given.
Level B — Cautions requiring immediate crew awareness and possible action. They are displayed in amber, and also by message caution lights. An aural tone is also repeated twice.
Level C — Advisories requiring crew awareness. Also displayed in amber. No caution lights or aural tones are associated with this level.

The messages appear on the top line at the left of the display screen as shown in Fig. 16.5. In order to differentiate between a caution and an advisory, the latter is always indented one space to the right.

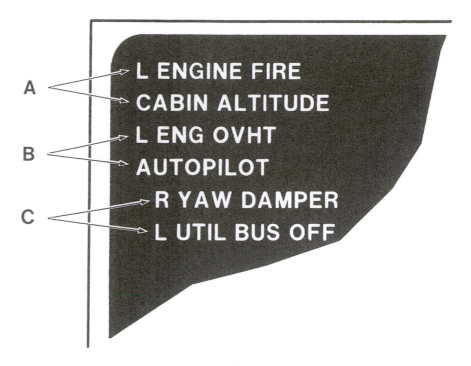

Figure 16.5 Alert message levels.

A L ENGINE FIRE
CABIN ALTITUDE

B L ENG OVHT
AUTOPILOT

C R YAW DAMPER
L UTIL BUS OFF

The master warning and caution lights are located adjacent to the display units together with a 'Cancel' switch and a 'Recall' switch. Pushing the 'Cancel' switch removes only the caution and advisory messages from the display; the warning messages cannot be cancelled. The 'Recall' switch is used to bring back the caution and advisory messages into the display. At the same time, the word 'RECALL' appears at the bottom of the display.

A message is automatically removed from the display when the associated condition no longer exists. In this case, messages which appear below the deleted one each move up a line.

When a new fault occurs, its associated message is inserted on the appropriate line of the display. This may cause older messages to move down one line. For example, a new caution message would cause all existing caution and advisory messages to move down one line.

If there are more messages than can be displayed at one time, the whole list forms what is termed a 'page', and the lowest message is removed and a page number appears in white on the lower right side of the list. If there is an additional page of messages it can be displayed by pushing the 'Cancel' switch. Warning messages are carried over from the previous page.

Display unit failure

If the lower display unit should fail when secondary information is being displayed on it, an amber alert message appears at the top left

Figure 16.6 Compact format.

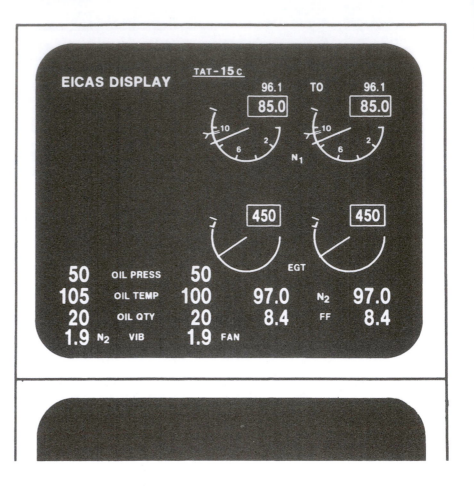

of the upper display unit, and the information is transferred to it as shown in Fig. 16.6. The format of this display is referred to as 'compact', and it may be removed by pressing the 'ENGINE' switch on the display select panel. Failure of a display unit causes the function of the panel 'STATUS' switch to be inhibited so that the status page format cannot be displayed.

Display select panel failure

If this panel fails the advisory message 'EICAS CONTROL PANEL' appears at the top left of the upper display unit together with the primary information, and the secondary information automatically appears on the lower display unit. The 'cancel/recall' switches do not operate in this failure condition.

Standby engine indicator

This indicator provides primary engine information in the event that a total loss of EICAS displays occurs. As shown in Fig. 16.7, the

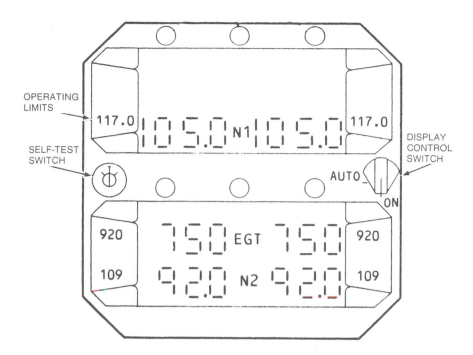

Figure 16.7 Standby engine indicator.

information relates to N_1 and N_2 speeds and EGT; the displays are of the LCD type. Operating limit values are also displayed.

The display control switch has two positions: 'ON' and 'AUTO'. In the 'ON' position, the displays are permanently on. In the 'AUTO' position the internal circuits are functional, but the displays will be automatically presented when the EICAS displays are lost due to failure of both display units or both computers.

The test switch has three positions, and is spring-loaded to a centre off position. It is screwdriver-operated and when turned to the left or right, it changes over power supply units within the indicator to ensure that they each provide power for the displays. The test can be performed with the display control switch in any position.

Maintenance control panel

This panel is for use by maintenance engineers for the purpose of displaying maintenance data stored in system computer memories during flight or ground operations. The layout of the panel and the principal functions of each of the controls are shown in Fig. 16.8.

The five display select switches are of the momentary-push type, and as each one is activated, a corresponding maintenance display page appears on the lower display unit screen. The pages are listed together with two example displays in Fig. 16.9. The upper display unit displays data in the 'compact' format (see Fig. 16.6) with the message 'PARKING BRAKE' in the top left of the screen.

System failures which have occurred in flight and have been

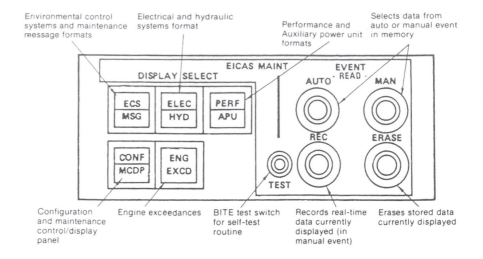

Figure 16.8 Maintenance control panel.

Environmental control systems and maintenance message formats

Electrical and hydraulic systems format

Performance and Auxiliary power unit formats

Selects data from auto or manual event in memory

Configuration and maintenance control/display panel

Engine exceedances

BITE test switch for self-test routine

Records real-time data currently displayed (in manual event)

Erases stored data currently displayed

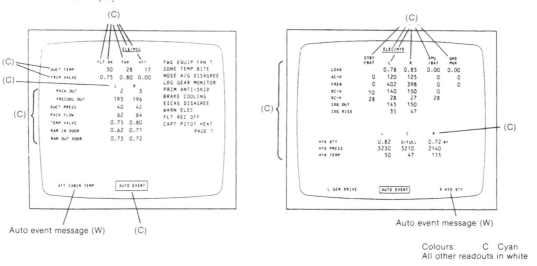

Figure 16.9 Examples of maintenance mode displays.

Auto event message (W)

Colours: C Cyan
All other readouts in white

automatically recorded ('auto event') in computer memory, as also data entered as 'manual event', can be retrieved for display by means of the 'event record' switch on the panel.

A self-test of the whole system, which can only be activated when an aircraft is on the ground and its parking brake set, is performed by means of the 'TEST' switch on the maintenance control panel. When the switch is momentarily pressed, a complete test routine of the system, including interface and all signal-processing circuits, and power supplies, is automatically performed. For this purpose an initial test pattern is displayed on both display units with a message in white to indicate the system being tested, i.e. 'L or R EICAS' depending on the setting of the selector switch on the display select panel. During the test, the master caution and warning lights and aural devices are activated, and the standby engine indicator is turned on if its display control switch is at 'AUTO'.

The message 'TEST IN PROGRESS' appears at the top left of display unit screens and remains in view while testing is in progress. On satisfactory completion of the test, the message 'TEST OK' will appear. If a computer or display unit failure has occurred, the message 'TEST FAIL' will appear followed by messages indicating which of the units has failed.

A test may be terminated by pressing the 'TEST' switch a second time or, if it is safe to do so, by releasing an aircraft's parking brake. The display units revert to their normal primary and secondary information displays.

ECAM

The units comprising this system, and as originally developed for the Airbus A310, are shown in the functional diagram of Fig. 16.10. As far as the processing and display of information are concerned, it differs significantly from EICAS in that data relate essentially to the primary systems of the aircraft, and are displayed in check-list and pictorial or synoptic format. Engine operating data are displayed by

Figure 16.10 ECAM system functional diagram.

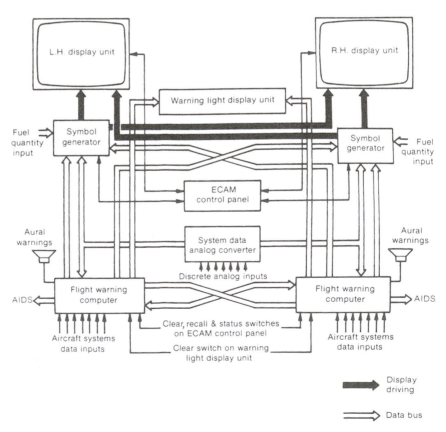

conventional types of instruments as noted in the introduction to this chapter. Other differences relate to display locations and selection of system operating modes.

Display units

These units are mounted side-by-side; the left-hand unit is dedicated to information on the status of systems, warnings and corrective action in a sequenced check-list format, while the right-hand unit is dedicated to associated information in pictorial or synoptic format.

Display modes

There are four display modes, three of which are automatically selected and referred to as flight phase-related, advisory (mode and status), and failure-related modes. The fourth mode is manual and permits the selection of diagrams related to any one of 12 of the aircraft's systems for routine checking, and also the selection of status messages provided no warnings have been 'triggered' for display. The selections are made by means of illuminated push-button switches on the system control panel.

In normal operation the automatic flight phase-related mode is used, and in this case the displays are appropriate to the current phase of aircraft operation, i.e. pre-flight, take-off, climb, cruise, descent, approach, and after landing. An example of a pre-flight phase is shown in Fig. 16.11; the left-hand display unit displays an advisory memo mode, and the right-hand unit displays a diagram of the aircraft's fuselage, doors, and arming of the escape slides deployment system.

The failure-related mode takes precedence over the other two

Figure 16.11 Pre-flight phase-related mode display.

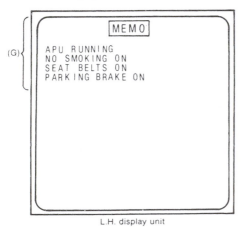
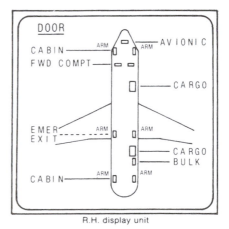

Examples: Doors locked: Door symbols green and name of door white
Doors unlocked: Door symbols and name of door amber

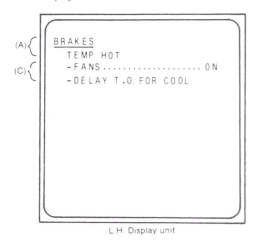

Figure 16.12 Failure-related mode display.

(A) {
BRAKES
 TEMP HOT

(C) {
 -FANS....................ON
 -DELAY T.O. FOR COOL

L.H. Display unit

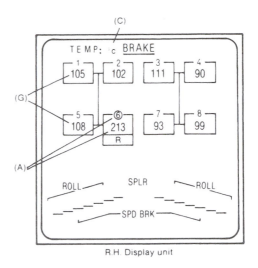

TEMP: °c BRAKE (C)

| 1 | 2 | 3 | 4 |
| 105 | 102 | 111 | 90 |

(G)

5	⑥	7	8
108	213	93	99
	R		

(A)

ROLL   SPLR   ROLL

SPD BRK

R.H. Display unit

Colours: A Amber
          C Cyan
          G Green
          Remainder of display white

modes and the manual mode. An example of a display associated with this mode is shown in Fig. 16.12. In this case, while taxiing out for take-off, the temperature of the brake unit on the rear right wheel of the left main landing gear bogie has become excessive. A diagram of the wheel brake system is immediately displayed on the right-hand display unit, and simultaneously the left-hand unit displays corrective action to be taken by the flight crew. In addition, an aural warning is sounded, and a light (placarded 'L/G WHEEL') on a central warning light display panel is illuminated. As the corrective action is carried out, the instructions on the left-hand display are replaced by a message in white confirming the result of the action. The diagram on the right-hand display unit is appropriately 'redrawn'.

In the example considered, the warning relates to a single system, and by convention such warnings are signified by underlining the system title displayed. In cases where a failure can affect other sub-systems, the title of the sub-system is shown 'boxed', as for instance in the display shown in Fig. 16.13. Warnings and the associated lights are cleared by means of 'CLEAR' push-button switches on either the ECAM control panel or a warning light display panel.

Status messages, which are also displayed on the left-hand display unit, provide the flight crew with an operational summary of the aircraft's condition, possible downgrading of autoland capability, and as far as possible, indications of the aircraft status following all failures except those that do not affect the flight. The contents of an example display are shown in Fig. 16.14.

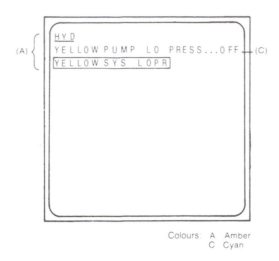

Figure 16.13 Display of failure affecting a sub-system.

```
HYD
YELLOW PUMP  LO  PRESS...OFF
YELLOW SYS  LO PR
```

(A)   (C)

Colours: A Amber
         C Cyan

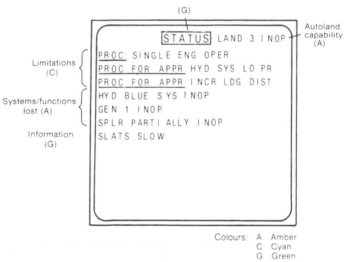

Figure 16.14 Example of status display.

```
                    STATUS  LAND 3 INOP
PROC: SINGLE ENG OPER
PROC FOR APPR: HYD SYS LO PR
PROC FOR APPR: INCR LDG DIST
HYD BLUE SYS INOP
GEN 1 INOP
SPLR PARTIALLY INOP
SLATS SLOW
```

(G)

Autoland capability (A)

Limitations (C)

Systems/functions lost (A)

Information (G)

Colours: A Amber
         C Cyan
         G Green

Control panel

The layout of this panel is shown in Fig. 16.15; all switches, with the exception of those for display control, are of the push-button, illuminated caption type.

1. *SGU selector switches* Control the respective symbol generator units, and the lights are off in normal operation of the system. The 'FAULT' caption is illuminated amber if a failure is detected by an SGU's internal self-test circuit. Releasing a switch isolates the corresponding SGU, and causes the 'FAULT' caption to extinguish, and the 'OFF' caption to illuminate white.

2. *Synoptic display switches* Permit individual selection of synoptic diagrams corresponding to each of 12 systems, and illuminate white when pressed. A display is automatically cancelled whenever a warning or advisory occurs.

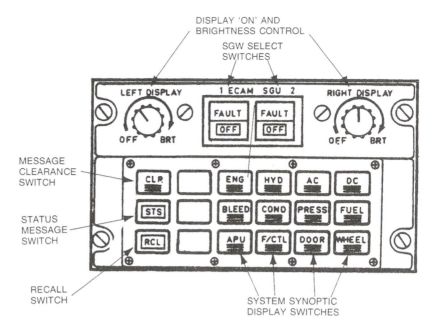

Figure 16.15 Control panel.

3. *CLR switch* Light illuminated white whenever a warning or status message is displayed on the left-hand display unit. Pressed to clear messages.
4. *STS switch* Permits manual selection of an aircraft status message if no warning is displayed; illuminated white. Pressing the switch also causes the CLR switch to illuminate. A status message is suppressed if a warning occurs or if the CLR switch is pressed.
5. *RCL switch* Enables previously cleared warning messages to be recalled provided the failure conditions which initiated them still exist. Pressing the switch also causes the CLR switch light to illuminate. If a failure no longer exists the message 'NO WARNING PRESENT' is displayed on the left-hand display unit.

System testing

Each flight warning computer of the system is equipped with a monitoring module which automatically checks data acquisition and processing modules, memories, and the internal power supplies as soon as the aircraft's main power supply is applied to the system. A power-on test routine is also carried out for correct operation of the symbol generator units. During this test the display units remain blank.

In the event of failure of the data acquisition and processing modules, or of the warning light display panel, a 'failure warning system' light on the panel is illuminated. Failure of a computer causes a corresponding annunciator light on the maintenance panel,

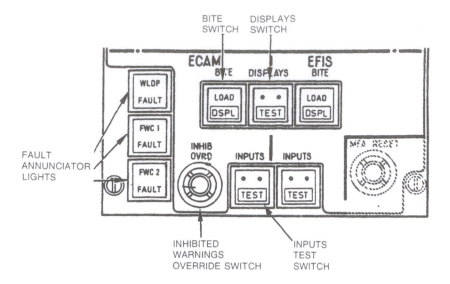

Figure 16.16 Maintenance panel.

captioned 'FWC FAULT', to illuminate. A symbol generator unit failure causes a 'FAULT' caption on the appropriate push-button switch on the system control panel to illuminate.

Manual self-test checks for inputs and displays are carried out from a maintenance panel shown in Fig. 16.16. When the 'INPUTS' switch is pressed, a 'TEST' caption is illuminated white, and most of the inputs to each computer are checked for continuity. Any incorrect inputs appear in coded form on the left-hand display unit. The right-hand display unit presents a list of defective parameters at the system's data analog converter. The diagrams of systems appear on the right-hand display unit with the caption 'TEST' beside the system title, as each corresponding push-button switch is pressed. Calibrated outputs from the data analog converter are also displayed. Any defective parameters are identified by a flag display.

A 'DISPLAYS' push-button switch is provided on the maintenance panel and when pressed it initiates a check for correct operation of the symbol generator units, and the optical qualities of the display units by means of a test pattern display. The 'LOAD' caption is illuminated each time a failure is memorized in the relevant test circuits of the SGUs.

The annunciator lights on the maintenance panel illuminate white simultaneously with a failure warning system light on the central warning light display panel when a corresponding computer fails.

The 'INHIB OVRD' switch enables inhibited warnings to be displayed.

17 **Flight management systems**

Computerized systems designed in various forms to carry out performance advisory or comprehensive flight management functions are an essential feature of a number of types of public transport category aircraft, their development having stemmed from the need to ensure the most efficient use of fuel, to reduce flight crew workload, and to reduce operating costs overall.

In performing an advisory function a system merely advises the flight crew of the optimum settings of various control parameters, such as engine pressure ratio (EPR) and climb rate under varying flight conditions, in order to achieve the most economical use of the available fuel. Such systems require adjustments of controls on the part of the flight crew if they are to be utilized to maximum advantage.

A system performing a combined function is one in which the computer and display units are interfaced with a thrust or power management control system and an automatic flight control system. Thus, in isolating the flight crew from the control loop, an integrated automatic flight management system (FMS) is formed to provide greater precision of engine power and flight path control.

Early forms of such systems, whether purely advisory or of combined function, were limited to supervising control parameters affecting the vertical flight path. In order to ensure maximum fuel economy it is, however, also necessary to integrate this optimized flight path management with the lateral flight path; in other words, a system must also be provided with a navigation capability. This requires interfacing the computer with such navigation systems as Doppler, INS/IRS, DME and VOR. The inputs from these systems permit continuous monitoring of an aircraft's track in relation to a flight plan which may be pre-stored in the computer memory and an immediate identification of deviations. Furthermore, it allows flight plan variants to be constructed and evaluated. It is thus apparent that by combining these inputs with those controlling the vertical flight path parameters mentioned earlier, an FMS can integrate the functions of navigation, performance management, flight planning and three-dimensional guidance and control along a pre-planned flight path.

In addition to changing data inputs from such systems as those mentioned above, an FMS also requires data bases for storing bulk navigation data, and the characteristics of an aircraft and its engines, in order that the system will operate in a full three-dimensional capacity. The navigation data base is capable of storing the necessary flight environmental data associated with a typical airline's entire route structure, including pertinent navigation aids and waypoints, airports and runways, published terminal area procedures, etc. The memory bank also contains flight profile data for a variety of situation modes such as take-off, climb, cruise, descent, holding, go-around and 'engine-out'. The cruise mode is also sub-divided into sub-mode variants such as economy, long-range, manual and thrust-limited. The integration of all the foregoing data, plus other variable inputs such as wind speeds and air traffic control clearances, permit the automatic generation or modification of flight plans to meet the needs of any specific flight operation.

Typical systems

Performance data computer system

This system provides advisory data in alphanumeric format on a CRT display, in addition to the positioning of target command 'bugs' on a Mach/airspeed indicator and EPR indicators, such indicators operating on electrical servomechanism principles. Provision is also made for interfacing the system with autothrottle and automatic flight control systems. A schematic diagram of the system, which consists primarily of a control and display unit, computer and mode annunciator, is shown in Fig. 17.1.

Abbreviations are extensively used for the display of data by the control and display units of this and other flight management computer systems, and these abbreviations/acronyms and their definitions are given in Appendix 3.

Control and display unit (CDU)
The CDU provides the major input link to the system and allows the flight crew to make inputs to obtain EPR and airspeed displays and can also be used for obtaining decision-making data in relation to an aircraft's flight profile. The CRT has a 2 in × 3 in screen and enables data to be displayed over a 13 (column) × 6 (row) matrix. The selection of EPR and airspeed data for various phases of flight is accomplished by a flight mode select switch, the modes and associated displays being as follows:

TO   Take-off EPR limits for the outside air temperature entered by the flight crew

CLB   EPR and speeds for the desired climb profile: best economy, maximum climb rate, or crew-selected speeds

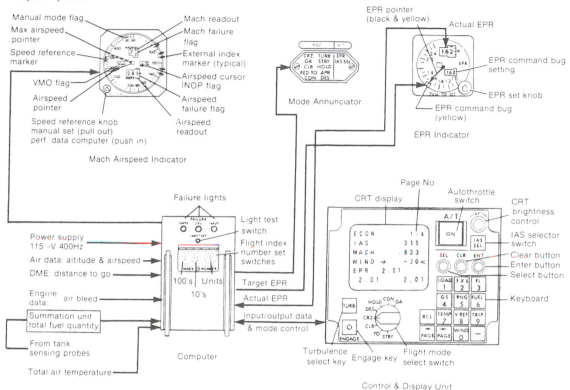

Figure 17.1 Performance data computer system.

CRZ   EPR and speeds for the desired cruise schedule: best economy, long-range cruise, or crew-selected speeds

DES   Descent speed, time and distance for best economy

HOLD   EPR, speed and endurance for holding

CON   Maximum continuous EPR limits for existing altitude, temperature and speed

GA   Go-around EPR limit for existing altitude, temperature and speed

The standby (STBY) position of the select switch is used for data entry and for an automatic check-out of the system.

The function of the 'ENGAGE' key is to couple the target command 'bugs' of the Mach/airspeed indicator and EPR indicators to computer command signals which drive the bugs to indicate the speed and EPR values corresponding to those displayed on the CRT screen. If the data is verified by the computer to be valid, engageable and different from the data presently engaged, the engage key illuminates and is extinguished after engagement takes place; at the same time the appropriate light of the mode annunicator is illuminated.

The key marked 'TURB' is for use only in cruise and when

turbulent flight conditions are to be encountered. When pressed it causes the CRT to display the appropriate turbulence penetration data, i.e. airspeed in knots (also Mach number at high altitudes), pitch attitude and the N_1 percentage rpm. In the turbulence mode, the target command speed and EPR 'bugs' engage automatically. This mode is disengaged by pressing the key a second time or else engaging another flight mode.

In order that the flight crew may load keyboard-selected data into the system, three push-button switches are provided above the keyboard for SELecting, CLeaRing and ENTERing data. In connection with the selection and entering of data, question marks and two symbols are displayed at the right-hand end of a data line: a caret ($<$) and an asterisk (*). The caret signifies that the computer is ready to accept data, while the asterisk signifies that the data next to it may be entered or changed if necessary.

The keyboard primarily serves a dual function in that it (1) permits the flight crew to enter pure numeric data into the computer and (2) permits desired performance function data to be called up from the computer for display. The data appropriate to the keys is given in Table 17.1 and is displayed in the form of pages, each page being numbered in the top right-hand corner. For example, the page shown on the CDU in Fig. 17.1 is page 1 of a set of four relating to 'economy fuel' in the cruise mode. In order to call up each of the remaining pages the $\overrightarrow{\text{PAGE}}$ key is successively pressed. Similarly, the $\overleftarrow{\text{PAGE}}$ key permits cycling of the pages in reverse order. When a flight mode or performance function is first selected, the first page of a set is always automatically displayed.

The RCL key is used whenever a performance function is being displayed and if it is required to recall a display corresponding to a selected flight mode.

The two switches in the upper right-hand corner of the CDU are associated with autothrottle system operation. When the A/T annunciator switch is pressed, an internal light is illuminated to indicate connection of the auto-throttle system, and at the same time an 'EPR' light in the mode annunciator is illuminated. The PDCS then adjusts the throttles to track the EPR target values displayed on the CDU and by the command bugs of the associated EPR indicators. In order for the autothrottle system to adjust engine power in relation to indicated airspeed, the second switch 'IAS SEL Annunicator' is operated; the system then drives the throttles so as to track the speed target values displayed on the CDU and by the command bug of the Mach/airspeed indicator.

Computer

The computer is of the hybrid type, and the inputs, outputs and unit interfaces are as shown in Fig. 17.1. Program storage is by means of

Table 17.1 Performance function data

Key	Data pages displayed
LOAD 1	Outside air temperature, destination airport elevation, reserve and alternate fuel and zero fuel weight for the intended flight plan
↑ × ↓ 2	Flight level intercept data for use in solving time and distance problems in climb and descent
FL 3	Economy cruise and long-range cruise speeds at appropriate flight level
GS 4	Present ground speed computed from a known true airspeed (TAS) and wind component
RNG 5	Total flight endurance as well as distance/time solutions to fuel reserve at any flight level
FUEL 6	Fuel for engaged cruise speed (economy, long-range and manually entered speeds). Used only in the CRZ and CON modes
TEMP 7	Ambient temps (TAT & SAT) TAS, and temperature deviation from ISA
V REF 8	Reference landing speeds for various flap settings, based on aircraft's correct gross weight
TRIP 9	Optimum initial cruise flight level for inserted trip distances
WIND 0	Data in respect of automatically computed and manually inserted wind components
—	Negative value data and also 'test pages' while in 'STBY' mode

a PROM and an additional non-volatile memory for retaining all entered data during any interruption of the power supply. Built-in test equipment circuits and software operate continuously to check all critical circuits of the system. The fuel summation unit, which is a component of the PDCS, develops an ac voltage signal that is proportional to the total fuel on board the aircraft, the signal being a combination of those produced by the fuel quantity indicating system sensing probes which are located in the fuel tanks.

Failure lights on the front of the computer indicate whether a fault is in the computer, CDU or input signals. The INDEX NUMBER switches, which are of the rotary type, are used for programming a flight index number from 0 to 200 into the computer so that maximum economy flight modes are modified according to time-related costs compared to fuel costs. The switches are guarded to eliminate the possibility of inadvertent changing of the index number.

Mode annunciator

This unit indicates the flight mode driving the command bugs of the EPR and Mach/airspeed indicators. The legend appropriate to a flight mode is illuminated when the mode is selected, and the 'ENGAGE' or 'TURB' button switches are depressed.

Typical display

The number of data pages associated with all the performance functions and flight modes it is possible to select on the CDU is quite considerable, and limitations on space do not permit a full description of each to be given here. However, the fundamentals of presentation and data entry methods may be understood from the example given in Fig. 17.2.

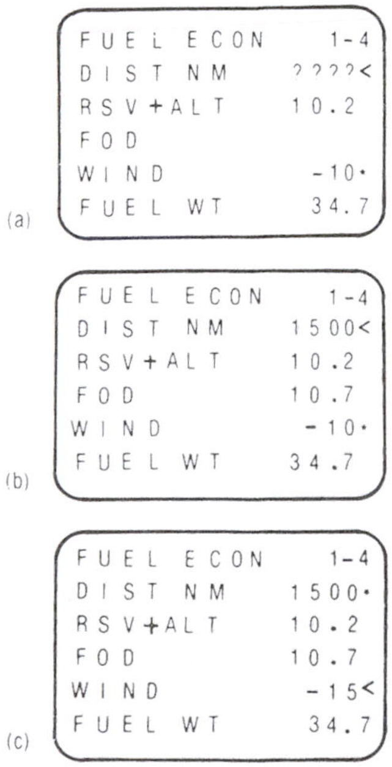

Figure 17.2 Examples of PDCS displays.

```
      FUEL ECON      1-4
      DIST NM      ?????<
      RSV+ALT      10.2
      FOD
      WIND          -10·
      FUEL WT       34.7
(a)
```

```
      FUEL ECON      1-4
      DIST NM      1500<
      RSV+ALT      10.2
      FOD          10.7
      WIND          -10·
      FUEL WT       34.7
(b)
```

```
      FUEL ECON      1-4
      DIST NM      1500·
      RSV+ALT      10.2
      FOD          10.7
      WIND          -15<
      FUEL WT       34.7
(c)
```

In this case, the first page of the fuel performance function for the maximum economy speed schedule is displayed, and the question marks on the second line (Fig. 17.2(a)) indicate that the computer is asking for the distance to go from the present position of the aircraft. The caret indicates the computer's readiness to accept the information. The CLR button switch is then pressed and this causes erasure of the question marks and the caret to flash on and off to indicate that the numbers appropriate to the distance should be 'keyed in' to the display from the keyboard — 1500 nautical miles in this case (Fig. 17.2(b)). When this has been done the ENTER button switch is then pressed, following which the computer goes through an input validity check routine. If the input is valid, the caret will stop flashing, thereby advising that the data has been accepted. An INVALID message is displayed if the input exceeds any limitation. The computer also computes the fuel over destination (FOD) value and causes it to be displayed. If the computer requires more data on another line the caret drops to that line automatically. The asterisk against the wind component of -10 knots signifies that any change to its value may be effected. If, for example, the wind speed has increased to 15 knots (the minus sign indicates a headwind) then it may be entered by first pressing the SEL key which causes the caret and the asterisk to exchange positions (Fig. 17.2(c)). The CLR key is then pressed, the new wind speed value is keyed in and is finally displayed by pressing the ENTER button switch.

Flight management system

A flight management system (FMS) is currently the most advanced of systems, providing as it does full integration of all the functions referred to earlier (see page 393) and which are necessary to fly optimized flight profiles either in manual or fully automatic control modes. The system is a union of autonomous and generally asynchronous units interconnected by a network of ARINC data busses to satisfy specific functional needs. In many cases redundant units are present to meet requirements for functional availability, flight safety or aircraft coverage.

Figure 17.3 illustrates schematically the computing units which are typically a formal part of an FMS and also how by means of the data busses they communicate with the principal elements of the system, namely the flight management computer (FMC) and its associated control and display unit (CDU). By virtue of their communication link these two units are together designated as a flight management computer system and this provides the primary interface between the flight crew and the aircraft. Inputs from other interfacing systems and sensors are also transmitted to the data busses, but for reasons of clarity have been omitted from Fig. 17.3.

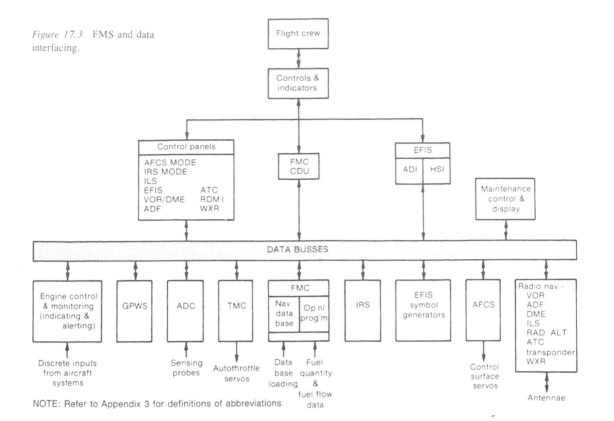

Figure 17.3 FMS and data interfacing.

NOTE: Refer to Appendix 3 for definitions of abbreviations

Flight management computer

Typically, a computer incorporates three different types of memory: a bubble memory for holding the bulk navigation and aircraft performance characteristics data bank; a C-MOS RAM for holding specific navigation and performance data, and the active and secondary flight plan, all 'down-loaded' from the bubble memory; and a UV-PROM for the operation program, which may be reprogrammed at card level.

The data base which is used for all computations contains numberous types of records in memory and these are given in Table 17.2. All the data are unique to each aircraft operator, depending on the routes flown, and are initially programmed on magnetic tape. The tape cartridge is inserted into a portable data base loader unit which, after connection to the computer, is operated so as to transfer the data to the bubble memory. Any subsequent changes in navigation aids and procedures, and route structure changes, are also incorporated in the data base by means of the data loader, in accordance with a specified time schedule, e.g. a 28-day cycle.

Control and display unit

The CDU of one example of an FMCS which is currently in use is shown in Fig. 17.4. It is basically similar to that adopted for the

Table 17.2 Records in data base memory

Record	How identified and defined
Radio-nav aids: VOR, DME, VORTAC, ILS, TACAN	Identifier ICAO region, latitude and longitude, frequency, magnetic declination, class (VOR, DME, etc.), company defined figure of merit,* elevation for DME, ILS category, localizer bearing
Waypoints	Each waypoint defined by its ICAO region, identifier, type (en-route, terminal), latitude and longitude
En-route airways	Identified by route identifier, sequence number, outbound magnetic course
Airports	Each identified by ICAO four-letter code, latitude and longitude, elevation, alternate airports
Runways	Each identified by ICAO identifier, number, length, heading, threshold latitude and longitude, final approach fix identifier, threshold displacement
Airport procedures	Each identified by its ICAO code, type (SID, STAR, profile descent, ILS, RNAV), runway number/transition, path and termination code
Company routes	Origin airport, destination airport, route number, via code (SID, airway, direct, STAR, profile descent, approach), via identifier (SID, name, airway identifier, etc.), cruise altitude, cost index

* The figure of merit is a number assigned to each navigational aid to indicate the maximum distance at which it can be tuned.

Figure 17.4 FMS control and display unit.

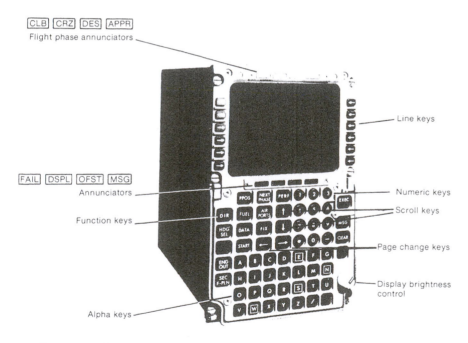

performance data computing system described earlier, but in keeping with the role to be played by an FMCS the unit is much more sophisticated in respect of data selection and corresponding displays.

The operation of the keyboard function keys is summarized in Table 17.3. As in the case of a PDCS display unit, annunciators are also provided but form an integral part of the unit.

Table 17.3 Summary of operation of keyboard function keys

Key	Selection and data displayed
P POS	Returns display to show the active navigation leg page, i.e. the aircraft's present position
NEXT PHASE	Changes a navigation leg display to the beginning of the next phase of the flight plan
PERF	Selection of performance pages
DIR	Permits direct entry of revisions to flight plan from present position to any waypoint
FUEL	Selection of fuel pages
AIR-PORTS	Displays the navigation legs page which includes the next airport along the current flight plan
HDG SEL	Selection of headings to be flown automatically via the FMS
DATA	Displays data index pages relevant to: lateral, vertical, performance, key waypoints, sensors, maintenance, navigation, aircraft configuration, history
FIX	To check or up-date aircraft's position
START	Selects 'START DATA REQUIRED' pages for flight crew to initiate and construct flight plans
ENG OUT	Presents performance data pages relating to engine out operation
SEC.F-PLAN	Selects secondary flight plan facility for re-clearance or return-leg planning
EXEC	To promote a temporary plan to active status. The bar illuminates when the FMCS has enough data to create an active plan, and remains illuminated until the temporary plan has been executed, or cancelled by pressing PPOS
MSG	Informs computer that any message displayed on CDU has been acknowledged and the message will either be stored or erased
CLEAR	To delete incorrect scratch pad entries

While on the ground, the flight crew can construct a detailed flight plan by inserting data in selected data pages and, in conjunction with the comprehensive data base stored in the computer, the plan is raised to active control status. In flight, the system receives data from the relevant aircraft sensors and radio-navigational aids, and then as the flight proceeds it presents the flight plan in a progressive 'scroll' form. Pitch, roll and thrust demands are also computed, and in communicating these to an AFCS and a TMS accurate control of the flight profile and maximum fuel economy can be provided. Various pages of data can be selected for review by the flight crew at any stage of a flight, and predictions concerning its future phases can be assessed by inserting detailed revisions, the future implications of which are computed and displayed. In addition to its own display unit, the FMCS also has the unique capability of presenting navigation data in 'changing map' form on the display unit of an electronic flight instrument system (see also Chapter 12).

The flat-faced CRT of the display unit gives a dual character size presentation with 24 stroke-written characters per 14 lines. Small-size characters signify data with default values, or computer-predicted values which can be changed by the flight crew when the data is being supplied by the computer; large-size characters signify data entered by the flight crew.

Data pages and flight plan construction

Many pages of data can be accessed and displayed, and space does not permit them all to be shown here. However, some indication of character presentation and method of entering data in general may be understood by considering the example of the display shown in Fig. 17.5, which is used to initiate the construction of a flight plan when the 'START' key is pressed.

The data lines adjacent to the line keys constitute the 'operational area' of the display and can be accessed by line key selection for entry or revision of flight plan information or for selection of displayed options. The first 12 character spaces on the bottom line (line 14) are used as a scratch pad for information entered by the flight crew via the alphanumeric keys. The next 10 spaces are reserved for FMS messages to the crew and the last two spaces are reserved for scroll cues signified by upward- and downward-pointing arrows. When the appropriate scroll key on the keyboard is pressed the display moves up or down for the purpose of reviewing the display.

Arrows which appear against characters of a display indicate flight crew options, which may be the choice of display data or may result in some functional activity of the system such as aligning the inertial reference system (IRS) as indicated in Fig. 17.5. The choice of option is signified by pressing the line key adjacent to an arrow; this

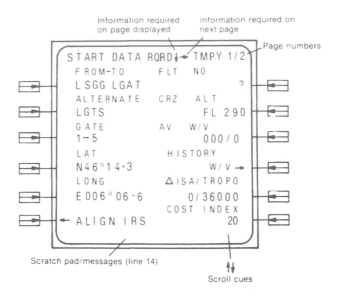

Figure 17.5 Example of a page display.

Information required on page displayed

Information required on next page

Page numbers

```
START DATA RQRD ↓→ TMPY 1/2
FROM-TO        FLT  NO
LSGG LGAT                    ?
ALTERNATE      CRZ  ALT
LGTS                  FL 290
GATE           AV   W/V
1-5                   000/0
LAT            HISTORY
N46°14·3              W/V →
LONG           Δ ISA/TROPO
E006°06·6          0/36000
               COST  INDEX
← ALIGN IRS               20
```

Scratch pad/messages (line 14)

↑↓
Scroll cues

results in a change to the content of an existing page and erasure of the arrow.

Requests for data entry are indicated by question marks. For example, in Fig. 17.5 the request is for the flight number, and in response this is first entered into the scratch pad by using the alphanumeric keys and then pressing the line key adjacent to the question mark. If the format of the entry is correct, the flight number appears in large-size characters on the appropriate line and the scratch pad clears. If the format is incorrect, the word ERROR appears in the message space, the MSG annunciator illuminates and the incorrect number remains in the scratch pad. After an incorrect entry has been attempted the ERROR legend and MSG annunciation are acknowleged by pressing the MSG key. The CLEAR key is then pressed to delete the entire scratch pad entry.

It is possible to change any data by over-writing with new values from a scratch pad entry. If, for example, it is required to change the displayed data base cruise altitude, the new value is 'keyed' into the scratch pad and then transferred to the altitude line by pressing the adjacent line key. Similarly, if the airport terminal gate from which an aircraft is to depart is changed, the new gate number may be over-written in the display. The FMC will automatically enter the revised latitude and longitude values for the new gate to which the IRS must be realigned.

The TMPY legend on the title line of start data pages signifies that the data being entered by the flight crew relates to the construction of a temporary flight plan.

Flight plan construction and associated changes of data pages are essentially in two sections. First, the navigation section involves insertion and acceptance of route number, airport codes, cruise altitude, latitude and longitude data and IRS alignment. When this has been accepted by the computer the bar in the EXEC key is illuminated and this section of the plan is executed by pressing the key. A page is then displayed requesting data needed to construct the second section of the plan which relates to fuel on board, reserve fuel, trip fuel and time, weights and centre of gravity position. When this has been computed, a 'START COMPLETE' page is displayed to show the relevant fuel, weight data and time components, and the EXEC key is again pressed so that the whole of the flight plan is raised to 'active' status.

Performance and in-flight displays

The next step is to initiate the display of data pages relevant to the performance section of the flight plan, and this is done by pressing the PERF key. These pages are concerned with management of engine thrust, pitch attitude and other alternative modes of performance control limited to the various phases of flight. The normal mode is 'economy' by which the most economical climb, cruise and descent speeds are computed. Each flight plan has its own performance scroll of pages related to those of the navigation legs scroll. The first page is the take-off performance page in which the flight crew must enter the values of the take-off speeds V_1 and V_R.

After take-off and during climb at the appropriate climb speed, the CLB annunciator is illuminated and the take-off performance page is automatically replaced by a climb performance page. At the top of the climb, the CRZ annunciator is illuminated and a cruise performance page is then presented to display continuously updated information related to optimum performance and destination arrival fuel and time. During cruise descent a descent performance page is displayed, but the point at which it is presented depends on the point at which the descent is commenced. For example, if the descent is commenced prior to the planned point in the vertical profile (by authorizing it via the AFCS control panel) the cruise performance page is first replaced by a cruise descent page and the aircraft descends at the selected vertical speed. When the aircraft captures the planned descent profile the display then changes to the descent performance page and the DES annunciator is illuminated. If the descent is commenced at the planned profile point, the cruise performance page is replaced directly by the descent performance page and the CR DES annunciator is illuminated. During approach, and when the leading edge slats are extended, an approach performance page replaces the descent performance page and the APPR annunciator is then illuminated.

If any revisions are made while in flight, the TMPY legend will also appear on the appropriate data page being displayed to indicate that a complete temporary flight plan is generated. The aircraft continues under the control of the active flight plan until the temporary plan is raised to active status by pressing the EXEC key. If it is to be aborted, the P POS key is pressed.

Data index

An index of data is contained on a page which can be called up for display by pressing the data key. When the line keys adjacent to the titles are pressed, further pages are presented to display the information noted in Fig. 17.6.

System configuration

Two FMC systems are installed in an aircraft, each having its own CDU situated on the centre consol and each controlling its associated automatic flight control system, autothrottle system and radio-navigational aids. The basic configuration is shown in Fig. 17.7. In normal operating conditions, both computers operate together and share and compare each other's information, i.e. they 'cross talk' by means of an interconnecting data bus. The pilots can operate their displays independently for review or revision purposes without disturbance to the active flight plan and without affecting the other CDU commands. Typically, a working arrangement would be for the performance pages to be displayed on one pilot's CDU, while the other pilot's CDU would display navigation legs.

When a temporary flight plan is created in one system the other system (referred to as the 'offside' system) has no access to this plan, but it can review the active plan in the normal way. In addition, the offside system is inhibited from creating a temporary flight plan until the previous temporary state is cancelled or raised to active.

Each computer has its own VOR/DME receiver and determines the frequencies it requires for its own purposes. No interconnection between systems is possible except when a lateral revision is effected at the present position of the aircraft.

In the event of failure of one computer, each pilot has the means whereby he can select his own CDU into the other system.

Figure 17.6 Data index page.

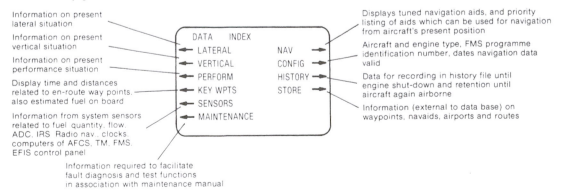

Information on present lateral situation

Information on present vertical situation

Information on present performance situation

Display time and distances related to en-route way points. also estimated fuel on board

Information from system sensors related to fuel quantity. flow. ADC. IRS. Radio nav., clocks. computers of AFCS. TM. FMS. EFIS control panel

Information required to facilitate fault diagnosis and test functions in association with maintenance manual

```
    DATA    INDEX

←  LATERAL        NAV   →

←  VERTICAL       CONFIG →

←  PERFORM        HISTORY →

←  KEY WPTS       STORE  →

←  SENSORS

←  MAINTENANCE
```

Displays tuned navigation aids, and priority listing of aids which can be used for navigation from aircraft's present position

Aircraft and engine type, FMS programme identification number, dates navigation data valid

Data for recording in history file until engine shut-down and retention until aircraft again airborne

Information (external to data base) on waypoints, navaids, airports and routes

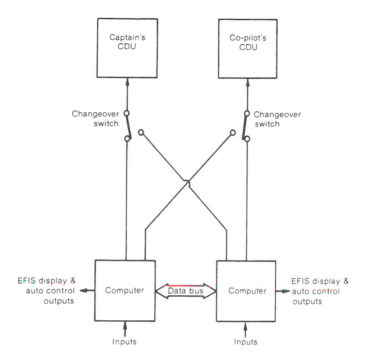

Figure 17.7 FMS configuration.

Tables

Table 1 Standard atmosphere

Altitude	Pressure			Temperature	Velocity of sound
ft	millibars	in hg	lbf/in^2	°C	ft/sec
−1.000	1050·41	31·019	15·234	+16·981	1119.9
0	1013.25	29.921	14.696	15.000	1116.1
1,000	977.17	28.856	14.172	13.019	1112.3
2,000	942.13	27.821	13.664	11.038	1108.4
3,000	908.12	26.817	13.170	9.056	1104.5
4,000	875.10	25.842	12.691	7.075	1100.7
5,000	843.07	24.896	12.226	5.094	1096.7
6,000	811.99	23.978	11.775	3.113	1092.8
7,000	781.85	23.088	11.338	1.132	1088.9
8,000	752.62	22.225	10.913	−0.850	1085.0
9,000	724.28	21.388	10.502	−2.831	1081.0
10,000	696.81	20.577	10.104	−4.812	1077.0
11,000	670.20	19.791	9.718	−6.793	1073.1
12,000	644.41	10.029	0.344	−8.774	1069.1
13,000	691.43	18.292	8.891	−10.756	1065.0
14,000	595.24	17.577	8.630	−12.737	1061.0
15,000	571.82	16.886	8.291	−14.718	1057.0
16,000	549.15	16.216	7.962	−16.699	1052.9
17,000	527.22	15.569	7.643	−18.680	1048.8
18,000	506.00	14.942	7.335	−20.662	1044.7
19,000	485.47	14.336	7.038	−22.643	1040.6
20,000	465.63	13.750	6.750	−24.624	1036.5
21,000	446.45	13.184	6.472	−26.605	1032.4
22,000	427.91	12.636	6.203	−28.686	1028.2
23,000	410.00	12.107	5.943	−30.568	1024.0
24,000	392.71	11.597	5.692	−32.549	1019.8
25,000	376.01	11.104	5.450	−34.530	1015.6
26,000	359.89	10.628	5.216	−36.511	1011.4
27,000	344.33	10.168	4.991	−38.492	1007.1
28,000	329.32	9.725	4.773	−40.474	1002.9
29,000	314.85	9.298	4.563	−42.455	998.6
30,000	300.89	8.885	4.361	−44.436	994.3
31,000	287.45	8.488	4.166	−46.417	990.0
32,000	274.49	8.106	3.978	−48.398	985.6
33,000	262.01	7.737	3.797	−50.380	981.3
34,000	249.99	7.382	3.622	−52.361	976.9
35,000	238.42	7.041	3.455	−54.342	972.5

Table 1 cont'd

Altitude	Pressure			Temperature	Velocity of sound
ft	millibars	in hg	lbf/in²	°C	ft/sec
36,000	227.29	6.712	3.293	−56.323	968.1
37,000	216.63	6.397	3.139		
38,000	206.46	6.097	2.991		
39,000	196.77	5.811	2.851		
40,000	187.54	5.538	2.717		
41,000	178.74	5.278	2.589		
42,000	170.35	5.030	2.468		
43,000	162.36	4.794	2.352		
44,000	154.74	4.569	2.241	−56.500	967.7
45,000	147.48	4.355	2.136		
46,000	140.56	4.150	2.036		
47,000	133.96	3.956	1.940		
48,000	127.67	3.770	1.849		
49,000	121.68	3.593	1.762		
50,000	115.97	3.425	1.679		

Table 2 Mach no./airspeed relationship

Height (ft)	Abs temp °C	Speed of sound ft/sec mph	Mach number						
			0.3	0.4	0.5	0.6	0.7	0.8	0.9
			Airspeed in mph						
0	288.0	761.6	228.5	304.6	380.8	457.0	533.1	609.3	685.4
5000	278.1	748.4	224.5	299.4	374.2	449.0	523.9	598.7	673.6
10000	268.2	734.9	220.5	294.0	367.5	440.9	514.4	587.9	661.4
15000	258.3	721.2	216.4	288.5	360.6	432.7	504.8	577.0	649.1
20000	248.4	703.3	212.2	282.9	353.7	424.4	495.1	565.8	636.6
25000	238.5	691.1	207.9	277.2	346.5	415.9	485.2	554.5	623.8
30000	228.6	678.5	203.5	271.4	339.2	407.1	474.9	542.8	610.6
35000	218.7	663.7	199.1	265.5	331.8	398.2	464.6	531.0	597.3
36000	216.5	660.3	198.1	264.1	330.2	396.2	462.2	528.2	594.3

Table 3 Temperature/resistance equivalents

Nickel sensing elements

°C	0	10	20	30	40	50	60	70	80	90
Minus	—	95.4	90.8	86.4	82.1	77.8	73.6	69.5	65.4	61.4
Plus	100.0	104.7	109.5	114.4	119.4	124.6	129.9	135.3	140.8	146.4
100	152.2	158.2	164.4	170.6	176.9	183.4	190.1	197.0	204.1	211.4
200	218.8	226.4	234.3	242.3	250.6	259.0	267.7	—	—	—

Table 3 cont'd

Platinum sensing elements

°C	0	10	20	30	40	50	60	70	80	90
Minus	—	105.9	101.8	97.7	93.5	89.4	85.2	81.1	76.9	72.7
Plus	110.0	114.1	118.1	122.2	126.3	130.3	134.3	138.3	142.4	146.3
100	150.3	154.2	158.2	162.1	166.1	170.0	173.9	177.8	181.7	185.6
200	189.4	193.2	197.0	200.9	204.7	208.5	212.3	216.0	219.8	223.5
300	227.3	231.0	234.7	238.4	242.1	245.8	249.5	253.1	256.8	260.4
400	264.1	267.6	271.2	274.8	278.4	281.9	285.5	289.0	292.5	296.0
500	299.5	303.0	306.5	309.9	313.4	—	—	—	—	—

Table 4　Temperature/millivolt equivalent of typical iron v constantan thermocouples

Cold junction at 0°C

°C	0	10	20	30	40	50	60	70	80	90	100
0	0	0.48	0.96	1.46	1.96	2.46	2.97	3.48	3.99	4.50	5.02
100	5.02	5.54	6.06	6.59	7.12	7.65	8.18	8.71	9.24	9.76	10.28
200	10.28	10.80	11.33	11.86	12.33	12.93	13.47	14.01	14.55	15.08	15.62
300	15.62	16.16	16.70	17.24	17.79	18.33	18.87	19.40	19.93	20.45	20.98
400	20.93	21.51	22.04	22.57	23.10	23.62	24.15	24.68	25.21	25.74	26.28
500	26.28	26.32	27.36	27.90	28.44	28.98	29.52	30.07	23.62	31.18	31.75
600	31.75	32.32	32.90	33.48	34.06	34.66	35.28	35.90	36.53	37.15	37.77
700	37.77	38.39	39.01	39.63	40.25	40.87	41.49	42.11	42.73	43.35	43.97
800	44.97	—	—	—	—	—	—	—	—	—	—

Table 5　Temperature/millivolt equivalents of typical copper v constantan thermocouples

Cold junction at 0°C

°C	0	10	20	30	40	50	60	70	80	90	100
0	0	0.37	0.75	1.16	1.58	2.01	2.44	2.89	3.34	3.80	4.27
100	4.27	4.75	5.23	5.71	6.19	6.69	7.20	7.71	8.22	8.73	9.25
200	9.25	9.77	10.29	10.83	11.37	11.92	12.47	13.03	13.60	14.17	14.75
300	14.75	15.33	15.91	16.49	17.08	17.68	18.28	18.88	19.48	20.08	20.68
400	20.68	21.30	21.92	22.54	23.16	23.78	24.40	25.02	25.64	26.26	26.88
500	26.88	—	—	—	—	—	—	—	—	—	—

Table 6　Temperature/millivolt equivalents of typical chromel v alumel thermocouples

Cold junction at 0°C

°C	0	10	20	30	40	50	60	70	80	90	100
0	0	0.40	0.80	1.20	1.61	2.02	2.43	2.85	3.26	3.68	4.10
100	4.10	4.51	4.92	5.33	5.73	6.13	6.53	6.93	7.33	7.73	8.13
200	8.13	8.53	8.93	9.34	9.74	10.15	10.56	10.97	11.38	11.80	12.21
300	12.21	12.62	13.04	13.45	13.87	14.29	14.71	15.13	15.55	15.97	16.39
400	16.39	16.82	17.24	17.66	18.08	18.50	18.93	19.36	19.78	20.21	20.64

Table 6 cont'd

Cold junction at 0°C

°C	0	10	20	30	40	50	60	70	80	90	100
500	20.64	21.07	21.49	21.92	22.34	22.77	23.20	23.62	24.05	24.48	24.90
600	24.90	25.33	25.75	26.18	26.60	27.03	27.45	27.87	28.29	28.72	29.14
700	29.14	29.56	29.98	30.40	30.82	31.23	31.65	32.07	32.48	32.90	33.31
800	33.31	33.71	34.12	34.53	34.94	35.35	35.75	36.16	36.56	36.96	37.36
900	37.36	37.76	38.16	38.56	38.96	39.35	39.75	40.14	40.53	40.92	41.31
1000	41.31	41.70	42.08	42.47	42.86	43.24	43.62	44.00	44.38	44.76	54.14
1100	45.14	45.52	45.89	46.27	46.64	47.01	47.38	47.75	48.12	48.48	48.85
1200	48.85	49.21	49.57	49.94	50.29	50.65	51.00	51.36	51.71	52.06	52.41
1300	52.41	52.75	53.10	53.45	53.79	54.13	54.47	54.81	55.15	55.48	55.81
1400	55.81	—	—	—	—	—	—	—	—	—	—

Table 7   Nominal dielectric constants and densities of fuels

Fuel Type	Dielectric constant K			Density D (lb/gal)		
	Temperature °C					
	+55	0	−55	+55	0	−55
91/98	1.914	1.990	2.066	5.636	6.025	6.414
100/130	1.912	1.991	2.070	5.597	5.988	6.379
115/145	1.895	1.971	2.047	5.517	5.913	6.308
JP−1	2.071	2.145	2.219	6.493	6.835	7.177
JP−3	2.017	2.098	2.179	6.100	6.464	6.827
JP−4	2.007	2.083	2.159	6.160	6.520	6.880

Principal symbols and abbreviations

Symbols for quantities are in *italic* type, and abbreviations for the names of units (unit symbols) are in ordinary type.

A	ampere	N	newton (force)
a	speed of sound	*N*	number of turns of a coil
B	magnetic flux density	Nm	Newton Metre (torque)
bhp	brake horsepower	Pa	pascal (N/m^2)
C	capacitance	pF	picofarad
	radiation constant	*R*	resistance
°C	degree Celsius		gas constant
c/s	cycle per second	rad	radian
F	farad	rev/min	revolution per minute
ft/min	foot per minute	*T*	period, periodic time
ft/h	foot per hour	V	volt
ft/s^2	foot per second per	*V*	velocity
	second	*W*	weight
g	acceleration due to	Wb	weber (magnetic flux)
	gravity	*X*	reactance
G	conductance	*Z*	impedance
H	magnetic field strength	α	temperature coefficient of
Hz	hertz (frequency)		resistance
I	electric current		angle
	moment of inertia	μ	permeability
in Hg	inch of mercury	μF	microfarad
K	Kelvin	ρ	density
k	coefficient of heat		resistivity
	transmission	Φ	magnetic flux
L	inductance		heat current or flow
lb/gal	pound per gallon	Ω	ohm
lbf/in^2	pound force per square	ω	angular velocity
	inch		(radians per second)
M	Mach number; torque	λ	thermal conductivity
m	magnetic moment	ϵ	permittivity
	mass		dielectric constant
mA	milliampere	γ	conductivity
mb	millibar		cubic coefficient of expansion
mV	millivolt		ratio of specific heat
mph	mile per hour		
mm H$_2$O	millimetre of water		

Appendix 1
Conversion factors

1. Pressure

To convert	into	multiply by
Atmospheres	inches Hg (0°C)	29.921
	inches H_2O	406.9
	kilogrammes per sq cm	1.0333
	millibars	1013.25
	millimetres Hg (0°C)	760.00
	piézes	101.331
	pounds per sq in	14.696
	*pascals	101325.000
Inches Hg	atmospheres	0.03342
	inches H_2O	13.60
	kilogrammes per sq cm	0.03453
	millibars	33.8639
	millimetres Hg	25.40
	pounds per sq in	0.4912
Inches H_2O	atmospheres	2.458×10^{-3}
	inches Hg	0.07355
	kilogrammes per sq cm	2.540×10^{-3}
	millibars	2.490
	pounds per sq in	0.03613
	pascals	249.089
Kilogrammes per sq cm	atmospheres	0.000987
	inches Hg	28.96
	millimetres Hg	735.54
	piézes	1.0197
	pounds per sq in	14.223

*The pascal (Pa) is an SI unit and is the pressure produced by a force of 1 newton applied, uniformly distributed, over an area of 1 square metre.
Note: It is common practice to refer to a pressure as so many 'pounds per square inch'. Since, however, pressure is more exactly pounds-weight or force, acting per square inch, the symbol 'lbf/sq in' is now adopted in lieu of 'lb/sq in'.

Millibars	atmospheres	0.01450
	inches Hg	0.02953
	millimetres Hg	0.7450
	pounds per sq in	0.0145
	pascals	100.00
Millimetres Hg	inches Hg	0.03937
	kilogrammes per sq cm	0.0013596
	pascals	133.322
	pounds per sq in	0.019337
Pounds per sq in	atmospheres	0.06804
	inches Hg	2.03596
	inches H_2O	27.68
	kilogrammes per sq cm	0.0703
	millibars	68.9476
	millimetres Hg	51.713
	pascals	6896.55
Pascals	pounds per sq in	0.0001450

2. Velocity

To convert	into	multiply by
Feet per minute	feet per second	0.01667
	kilometres per hour	0.01829
	knots	35.524
	metres per minute	0.3048
	miles per hour	0.01136
Feet per second	kilometres per hour	1.0973
	knots	0.5921
	metres per minute	18.29
	miles per hour	0.6818
Kilometres per hour	feet per minute	54.68
	feet per second	0.9113
	knots	0.5396
	metres per minute	16.67
	miles per hour	0.6214
Knots	feet per minute	101.34
	feet per second	1.689
	kilometres per hour	1.8532
	miles per hour	1.1516

Metres per minute	feet per minute	3.281
	feet per second	0.05468
	kilometres per hour	0.06
	knots	0.03238
	miles per hour	0.03728
Miles per hour	feet per minute	88.00
	feet per second	1.4666
	kilometres per hour	1.60934
	knots	0.8684

3. Volumetric

To convert	into	multiply by
Cubic cemtimetres	cubic feet	3.531×10^{-5}
	cubic inches	0.06102
	imperial gallons	2.1997×10^{-4}
	litres	0.001
	pints	1.7598×10^{-3}
	quarts	8.7988×10^{-4}
Cubic feet	cubic centimetres	28316.85
	cubic inches	1728.00
	imperial gallons	6.2288
	litres	28.32
	pints	59.84
	quarts	29.92
	US gallons	7.481
Cubic inches	cubic centimetres	16.39
	cubic feet	5.787×10^{-4}
	imperial gallons	3.6047×10^{-3}
	litres	0.01639
	pints	0.5688
	quarts	0.2844
Imperial gallons	cubic centimetres	4546.087
	cubic feet	0.160544
	cubic inches	277.42
	litres	4.54596
	US gallons	1.201

Litres	cubic centimetres	1000.00
	cubic feet	0.03532
	cubic inches	61.025
	imperial gallons	0.21998
	pints	1.7598
	quarts	0.8799
	US gallons	0.2642
Pints	cubic centimetres	568.26
	cubic feet	0.02007
	cubic inches	34.68
	litres	0.5682
	quarts	0.5
Quarts	cubic centimetres	1136.522
	cubic feet	0.04014
	cubic inches	69.3548
	litres	1.13649
US gallons	cubic centimetres	3786.44
	cubic feet	0.1337
	cubic inches	231.00
	imperial gallons	0.8327
	litres	3.785

4. Angular Measure

To convert	into	multiply by
Degrees	minutes	60.00
	quadrants	0.0111
	radians	0.0175
Minutes	degrees	0.0166
	quadrants	1.852×10^{-4}
	radians	2.909×10^{-4}
Quadrants	degrees	90.00
	minutes	5400.00
	radians	1.571
Radians	degrees	57.2957
	minutes	3437.75
	quadrants	0.6366

5. Angular Velocity

To convert	*into*	*multiply by*
Degrees per second	radians per second	0.01745
	revolutions per second	2.788×10^{-3}
	revolutions per minute	0.1667
Radians per second	degrees per second	57.30
	revolutions per second	0.1592
	revolutions per minute	9.5493
Revolutions per second	degrees per second	360.00
	radians per second	6.283
	revolutions per minute	60.00
Revolutions per minute	degrees per second	6.00
	radians per second	0.10472
	revolutions per second	0.01667

6. Temperature

$$°F = (°C \times 1.8) + 32 \quad or \quad = (°C \times 9/5) + 32$$
$$°C = (°F - 32)/1.8 \quad or \quad = (°F - 32) \times 5/9$$

$$\text{Absolute zero (K)} = -273.15°C \quad 0°C = 273.15K$$
$$= -459.67°F \quad 0°F = 459.67K$$

Appendix 2
Logic gates & truth tables

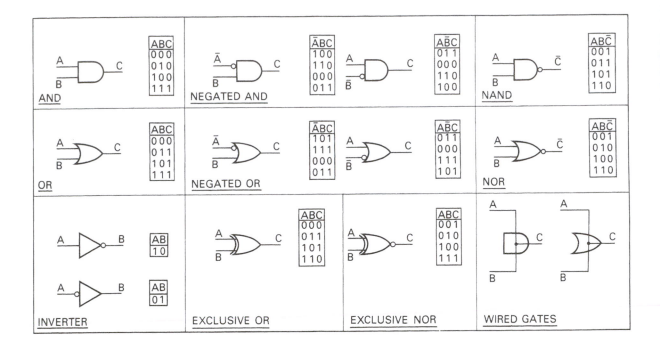

Appendix 3
Acronyms and abbreviations

This Appendix, which is by no means exhaustive, is intended as a guide to the meanings of acronyms and abbreviations found in the documentation dealing with the description, operation, logic signal functions and maintenance of instruments and integrated systems.

ACARS	ARINC Communications Addressing and Reporting System
ACAS	Airborne Collision and Avoidance System
ACCEL	Accelerometer
ACQ	Acquire (prefixed by a condition, e.g. ALT ACQ)
A/D	Analog to Digital
ADC	Air Data Computer
ADEU	Automatic Data Entry Unit
ADF	Automatic Direction Finder
ADI	Attitude Director Indicator
AFCS	Automatic Flight Control System
AFS	Automatic Flight System
AGC	Automatic Gain Control
AGS	Automatic Gain Stabilization
AHRS	Attitude and Heading Reference System
AIDS	Airborne Integrated Data System
ALPHA	Angle of Attack
ALU	Arithmetic Logic Unit
ANN	Annunciator
AP, A/P	Autopilot (suffixed by condition, e.g. ENG, DISC)
APFDS	Autopilot and Flight Director System
APMS	Automatic Performance and Management System
APPR OC	Approach On Course
APS	Altitude Preselect
ARINC	Aeronautical Radio InCorporated
ARM	Armed (prefixed by condition, e.g. LOC ARM, VOR ARM)
AS, A/S	Airspeed
ASA	Autoland Status Annunciator

AT	Autothrottle
ATE	Automatic Test Equipment
ATLAS	Abbreviated Test Language for Avionic Systems
ATOL	Automatic Test-Oriented Language
ATR	Austin Turnbull Radio (formerly Air Transport Radio)
ATS	AutoThrottle System
AT/SC	AutoThrottle/Speed Control
ATT	Attitude (may be followed by condition, e.g. ATT HOLD)
ATT ERR	Attitude Error
AUTO APPR	Automatic Approach
B/A	Bank Angle
BARO	Barometric
BB	Bar Bias
B/B	Back Beam
B/C, BC, B/CRS	Back Course
B/D	Bottom of Descent
BITE	Built-In Test Equipment
BRG	Bearing
CADC	Central Air Data Computer
CAP	Capture (prefixed by a condition, e.g. LOC CAP, NAV CAP)
CAPS	Collins Adaptive Processor System
CAWP	Caution And Warning Panel
CBB	Collective Bar Bias
CDU	Control and Display Unit
CE	Course Error
CG	Character Generator
CLK	Clock
CMD	Command (prefixed by another abbreviation, e.g. FD CMD)
COMP	Compensation, Compass, Comparator
CONT	Controller
CP	Control Panel
CPL	Coupled (prefixed by condition, e.g. ROLL, PITCH, APPR)
CPU	Central Processor Unit
CRS	Course
CRT	Cathode Ray Tube
CSEU	Control Systems Electronic Unit
CT	Control Transformer
CW	Caution and Warning
CWS	Control Wheel Steering

D/A	Digital to Analog
DAD	Data Acquisition Display
DADC	Digital Air Data Computer
DAIS	Digital Avionics Information System
DDI	Dual Distance Indicator
DDS	Digital Display System
DES	Desired (suffixed by condition, e.g. DES TRK, DES CRS)
DEVN	Deviation
DFDAU	Digital Flight Data Acquisition Unit
DFDR	Digital Flight Data Recorder
DG	Directional Gyroscope
DH	Decision Height
DI	Digital Interface
DIFCS	Digital Integrated Flight Control System
DISC	Disconnect
DISPL	Displacement
DME	Distance Measuring Equipment
DMLS	Doppler Microwave Landing System
DMM	Digital Multi-Meter
DMUX	Demultiplexer
DRC	Dual Remote Compensator
DSR TK	Desired Track
DTG	Distance-To-Go
DU	Display Unit
DVM	Digital Volt-Meter
EADI	Electronic Attitude Direction Indicator
ECAM	Electronic Centralized Aircraft Monitor
ECS	Environmental Control System
EDPS	Electronic Data Processing System
EEC	Electronic Engine Control
EFCU	Electronic Flight Control Unit
EFIS	Electronic Flight Instrument System
EGT	Exhaust Gas Temperature
EHSI	Electronic Horizontal Situation Indicator
EHSV	Electro-Hydraulic Servo Valve
EICAS	Engine Indicating and Crew Alerting System
ENG	Engage
EO	Easy-On
EPR	Engine Pressure Ratio
EX LOC	Expanded Localizer
EXT	Extend
FAC	Flight Augmentation Computer
FADEC	Full Authority Digital Engine Control

FAWP	Final Approach Waypoint
FCC	Flight Control Computer
FCES	Flight Control Electronic System
FCEU	Flight Control Electronic Unit
FCU	Flight Control Unit
FD, F/D	Flight Director
FDEP	Flight Data Entry Panel
FGS	Flight Guidance System
FIM	Fault Isolation Monitoring
FIS	Flight Instrument System
FL CH	Flight Level CHange
FMA	Flight Mode Annunciator
FMC	Flight Management Computer
FMCS	Flight Management Computer System
FMCU	Flight Management Computer Unit
FMS	Flight Management System
FODTS	Fiber-Optic Data Transmission System
FPC	Fuel Performance Computer
FPM	Flap Position Module
FS	Fast Slew
FSEU	Flap/Slot Electronic Unit
FTR	Force Trim Release
FVC	Frequency-to-Voltage Converter
FWC	Flight Warning Computer
GA, G/A	Go-Around
GPWS	Ground Proximity Warning System
GS, G/S	Glide Slope
HARS	Heading and Attitude Reference System
HDG	Heading (can be suffixed by condition, e.g. HDG HOLD, HDG SELect)
HLD	Hold
HSI	Horizontal Situation Indicator
HUD	Head-Up Display
HVPS	High Voltage Power Supply
IAS	Indicated Airspeed
IAWP	Initial Approach Waypoint
ICU	Instrument Comparator Unit
ILS OC	Instrument Landing System On Course
IMU	Inertial Measuring Unit
INC−DEC	Increase−Decrease
INS	Inertial Navigation System
INTGL	Integral
INTLK	Interlock

INWP	Intermediate Waypoint
IRMP	Inertial Reference Mode Panel
IRS	Inertial Reference System
IRU	Inertial Reference Unit
ISA	International Standard Atmosphere
ISS	Inertial Sensing System
IVS	Instantaneous Vertical Speed
LAU	Linear Accelerometer Unit
LBS	Lateral Beam Sensor
LNAV	Lateral NAVigation
LOC	Localizer
LRRA	Low-Range Radar Altimeter
LRU	Line Replaceable Unit
LSU	Logic Switching Unit
LVDT	Linear Voltage Differential (also Displacement) Transformer
MALU	Mode Annunciation Logic Unit
MAN	Manual
MAP	Mode Annunciator Panel
MAWP	Missed Approach WayPoint
MCDP	Maintenance Control Display Panel
MCP	Mode Control Panel
MCU	Modular Concept Unit
MDA	Minimum Descent Atitude
MIP	Maintenance Information Printer
MM	Middle Marker
M_{MO}	Maximum Operating Mach No.
MPU	Microprocessor Unit
MSU	Mode Selector Unit
MTP	Maintenance Test Panel
MUX	Multiplexer
MWS	Master Warning System
NAV	Navigation
NC	No Connection or Normally Closed
NCD	No Computed Data
NCU	Navigation Computer Unit
ND	Navigation Display
NDB	Non-Directional Beacon
NM	Nautical Mile
NOC	NAV On Course
OAT	Outside Air Temperature
OC, O/C	On Course

OD	Out of Detent (may be prefixed, e.g. CWS OD)
OM	Outer Marker
ONS	Omega Navigation System
OSS	Over Station Sensor
PAFAM	Perforrmation And Failure Assessment Monitor
PAS	Performance Advisory System
P ATT	Pitch Attitude
PBB	Pitch Bar Bias
PCA	Power Control Actuator
PCB	Printed Circuit Board
PCPL	Pitch Coupled
PCWS	Pitch Control Wheel Steering
PDCS	Performance Data Computer System
PDU	Pilot's Display Unit
PECO	Pitch Erection Cut-Off
PFD	Primary Flight Display
P HOLD	Pitch Hold
PIU	Peripheral Interface Unit
PMS	Performance Management System
PNCS	Performance Navigation Computer System
PRAM	Programmable Analog Module
PSAS	Pitch Stability Augmentation System
PSM	Power Supply Module
PSO	Phase Shift Oscillator
P SYNC	Pitch Synchronization
RA, R/A	Radio (Radar) Altimeter
RALU	Register and Arithmetic Logic Unit
RBA	Radio Bearing Annunciator
RBB	Roll Bar Bias
RCPL	Roll Coupled
RCVR	Receiver
RCWS	Roll Control Wheel Steering
RDMI	Radio Distance Magnetic Indicator
REF	Reference
REV/C	Reverse Course (same as Back Course)
RG	Raster Generator
R/HOLD	Roll Hold
RLS	Remote Light Sensor
RMI	Radio Magnetic Indicator
RN, RNAV	Area Navigation
RN/APPR	Area Navigation Approach
RSAS	Roll Stability Augmentation System
RSU	Remote Switching Unit
R/T	Receiver/Transmitter

RTE DATA	Route Data
RVDT	Rotary Voltage Differential Transmitter
R/W	Read/Write
SAI	Stand-by Attitude Indicator
SAM	Stabilizer Aileron Module
SAS	Stability Augmentation System
SAT	Static Air Temperature
SCAT	Speed Command of Altitude and Thrust
SCM	Spoiler/Speedbrake Control Module
SEL	Select
SELCAL	Selective Calling
SFCC	Slat/Flap Control Computer
SID	Standard Instrument Departure
SG	Symbol Generator (Stroke Generator)
SGU	Symbol Generator Unit
SPD	Speed (Airspeed or Mach hold)
SRP	Selected Reference Point
SS	Slow Slew
SSEC	Static Source Error Correction
STAR	Standard Terminal Arrival Route
STBY	Standby
STCM	Stabilizer Trim Control Module
STS	Status (prefixed by a function, e.g. TRACK STS)
TACAN	Tactical Air Navigation
TAS	True Air Speed
TAT	Total Air Temperature
T/C	Top of Climb
TCC	Thrust Control Computer
TCS	Touch Control Steering
T/D	Top of Descent
TET	Turbine Entry Temperature
TGT	Turbine Gas Temperature
TK CH	Track Change
TKE	Track Angle Error
TMC	Thrust Management Computer
TMS	Thrust Management System
TMSP	Thrust Mode Select Panel
TRP	Thrust Rating Panel
TTL	Tuned to Localizer
VAR	Variable
VBS	Vertical Beam Sensor
VDU	Visual Display Unit
VGU	Vertical Gyro Unit

VLD	Valid (usually suffixing a condition, e.g. VG VLD, FLAG VLD)
V_{MO}	Maximum Operating Airspeed
VNAV	Vertical Navigation
VOR	Very-high-frequency Omnidirectional Range
VOR APPR	VOR Approach
VOR OC	VOR On Course
VORTAC	VOR TACtical (Air navigation)
V_{REF}	Reference Speed
VS	Vertical Speed
VSCU	Vertical Signal Conditioner Unit
WO, W/O	Washout
WPT	Waypoint
WXR	Weather Radar transceiver
XTK DEV	Cross Track Deviation
XTR	Transmitter
YD, Y/D	Yaw Damper
YDM	Yaw Damper Module

Exercises

Chapter 1

1. Explain the difference between quantitative and qualitative displays, and quote some examples of instruments to which they are applied.
2. What is the difference between static and dynamic counter displays?
3. To which types of instrument is a director display applied?
4. Describe the operating principle of an LED, and by means of a diagram show how the principle is applied to produce a segmented numeric display.
5. Describe the operating principle of an LCD.
6. What type of display is formed when the elements are arranged in, say, a 4 × 7 configuration, and to which display element does it specifically apply?
7. Name the flight instruments that comprise the basic 'T' layout, and state their respective positions. Does this layout also apply to electronic displays?
8. What do you understand by the term 'head-up' display? With the aid of a diagram describe how the required basic flight data are displayed.
9. What is the significance of the coloured markings and/or 'memory bugs' applied to certain instruments?

Chapter 2

1. Define the following: (i) troposphere, (ii) tropopause, and (iii) stratosphere.
2. What do you understand by the term 'ISA'? State also the assumptions made.
3. The pressure of the atmosphere:
   (a) increases non-linearly with height;
   (b) decreases non-linearly with height;
   (c) decreases linearly with height.
4. 1 inch of mercury is equal to:
   (a) 14.7 millibars;
   (b) 2.49 millibars;
   (c) 33.87 millibars.
5. What are the principal components and instruments which comprise a basic air data system?
6. With the aid of a simple diagram, describe the construction of a combined pitot and static pressure sensing probe.

7. Explain what is meant by the 'PE' of an air data system, and how its effects are minimized.

8. How are alternative sources of pitot and static pressure normally provided for, and connected to the appropriate instruments?

9. Explain the principle of pitot pressure measurement and how the $1/2\rho V^2$ law is derived.

10. Describe the construction and operation of a typical pneumatic type of airspeed indicator.

11. Computed airspeed is:
    (a) calibrated airspeed corrected for PE;
    (b) indicated airspeed compensated for the square-law response of the airspeed sensor;
    (c) indicated airspeed corrected for PE.

12. Define the term Mach number, and describe how it is indicated by measurement in terms of the ratio $(p_t - p_s)/p_s$.

13. Describe how the functions of a Machmeter and an airspeed indicator can be combined to provide indications of V_{mo}.

14. What is the difference between 'pressure altitude' and 'indicated altitude'?

15. When setting the BP counters of an altimeter to the pressure prevailing at a particular airport, the corresponding 'Q' code is known as:
    (a) QFE and the altimeter will read zero;
    (b) QFE and the altimeter will read the airport height above sea-level;
    (c) QNH and the altimeter will read zero.

16. With the aid of a diagram explain the operating principle of a VSI.

17. An IVSI provides more rapid indications of climb and descent because it utilizes a vertical acceleration pump:
    (a) instead of a metering unit;
    (b) which is connected between the pressure sensing capsule and a metering unit;
    (c) which is connected directly to the metering unit.

18. Why is it customary to sense air temperature in terms of a total value? Briefly describe the construction and operation of a typical sensing probe used for this purpose.

19. Describe the operation of a typical Mach warning system.

20. Explain the operating sequence of an altitude alerting unit when an aircraft descends to a preselected altitude.

21. Explain the operation of a stall warning and stick-shaker system.

Chapter 3

1. Define the following: (i) magnetic meridian, (ii) magnetic variation, (iii) isogonal lines, and (iv) agonic lines.

2. The angle the lines of magnetic force make with the earth's surface is called:
   (a) deviation;
   (b) dip;
   (c) variation.
3. Describe the construction of a typical direct-reading compass.
4. Define the two principal errors that can occur in the readings of a compass under aircraft operating conditions.
5. What do you understand by the terms 'hard-iron' and 'soft-iron' magnetism of an aircraft?
6. Name the components of hard-iron magnetism and the aircraft axes about which they are effective.
7. Which of the hard-iron and soft-iron components are associated with the deviation coefficients B and C?
8. Express the formulae used for the calculation of deviation coefficients A, B and C.
9. Briefly describe how a deviation compensating device neutralizes the fields due to aircraft magnetic components.

Chapter 4

1. Define the two fundamental properties of a gyroscope, and state the factors on which they depend.
2. How are the properties defined in Q.1 utilized in flight instruments?
3. What are the input and output axes of a gyroscope?
4. As the speed of a gyroscope rotor increases, the rate of precession for a given torque:
   (a) remains constant;
   (b) decreases;
   (c) increases.
5. With the aid of a diagram, explain how a gyroscope precesses under the influence of an applied force.
6. How are the spin axes of gyroscopes arranged for the detection of pitch and roll attitude changes, and for establishing directional references?
7. What is meant by 'earth rate', and how would the input axis of a gyroscope have to be aligned to exhibit apparent drift equal to this rate?
8. What is meant by 'transport wander', and does it have the same effects on horizontal-axis and vertical-axis gyroscopes?
9. Briefly describe some methods of controlling drift and transport wander.
10. What do you understand by the terms 'gimbal lock' and 'gimbal error'?
11. Describe how pitch and roll attitudes are displayed by a gyro horizon.

12. How is the gyroscope of an electrically-operated gyro horizon erected to, and maintained in, its normal operating position?
13. Describe the operation of an electrolytic type of levelling switch.
14. What effects do acceleration and turning of an aircraft have on the indications of a gyro horizon, and how are they compensated?
15. Assuming that an aircraft accelerates while in straight and level flight, the effect on a gyro horizon utilizing a levelling torque motor system would cause it to indicate a:
   (a) pitch-up attitude;
   (b) pitch-down attitude;
   (c) pitch-up and a left bank attitude.
16. With the aid of a diagram, describe the operation of an erection cut-out system.
17. Describe how the rate gyroscope principle is applied to indicate the rates at which an aircraft turns.
18. With the aid of diagrams, describe how a ball type of bank indicator displays (a) a correctly-banked turn, and (b) a turn to starboard in which the aircraft is overbanked.
19. Describe how a rate gyroscope may be utilized to sense both banking and rate of turn.

Chapter 5

1. Explain the operation of a torque synchro system.
2. In what type of synchro system is a control transformer utilized?
3. Explain how a synchro is applied to systems involving the measurement of the sine and cosine components of angles.
4. What do you understand by the term 'electrical zero' as applied to a synchro system?
5. The letters 'TDX' designate a:
   (a) combined torque and differential synchro transmitter element;
   (b) differential synchro used in a torque synchro system;
   (c) torque synchro transmitter element when used with a differential synchro.
6. Explain the operating principle of a synchrotel.

Chapter 6

1. What are the elements that constitute the hardware and software of a digital computer?
2. Name the principal sections of a CPU and explain their functions.
3. Briefly explain the functions of the busses comprising a computer highway.
4. Which of the highway busses are bidirectional?

5. What is the name given to the digital code through which a computer carries out instructions?
6. What is the difference between a RAM and a ROM?
7. A 16K memory has a bit storage capacity of:
   (a) 16 000;
   (b) 32 000;
   (c) 16 384.
8. How is data in analog form converted to binary-coded format?
9. In the ARINC 429 format of data transfer, how is data identified according to function?
10. How are any errors in the codes used in the transmission of data detected?

Chapter 7

1. Explain some of the reasons why ADC systems are used in aircraft.
2. Describe the operation of a piezoelectric type of pressure transducer.
3. Explain how an output in terms of vertical speed can be obtained from the altitude module of an ADC.
4. What inputs are required to obtain signals whose values correspond to TAS?
5. How are square-law characteristics and PE compensated and corrected by an ADC?
6. What warning and indicating flags are provided in a typical servo-operated Mach/airspeed indicator?
7. What is the function of the capsule-type sensor element in a pneumatic/servo-operated type of altimeter?
8. How are indications of SAT derived from the signals produced by a TAT sensing probe?
9. Explain how BP settings are made in a servo-operated altimeter which is supplied with signals from a digital type of ADC.

Chapter 8

1. Draw a diagram to show the basic coil and core arrangement of a MHRS detector unit, and explain how fluxes and voltages are induced.
2. Explain how the magnitude of the voltages induced in a detector unit is used as a measure of aircraft heading.
3. The sensing element of a detector unit is pendulously mounted within its casing, and has limited freedom in:
   (a) pitch only so as to reduce the effects of acceleration;
   (b) pitch and roll so as to reduce the effect of the earth's magnetic field component Z;
   (c) azimuth only so as to reduce errors due to turning.

4. Explain what is meant by a 'pre-indexed' type of detector unit, and how it is mounted in an aircraft.

5. With the aid of a diagram explain how a detector unit monitors a DGU and an RMI.

6. What is the purpose of the synchronizing and annunciator system?

7. When a monitored gyroscope type of MHRS is operating in the 'slaved' mode, any drift of the gyroscope is controlled by:
   (a) manually resetting the compass card of the RMI;
   (b) setting the known latitude of the aircraft on the scale of a latitude corrector unit;
   (c) the slaving and servo synchro loops between the DGU and RMI.

8. How are slaving and servo signal transmission circuits controlled in an MHRS which is integrated with an INS?

9. Under what conditions is the DG mode of operation selected?

10. The purpose of the 'SET HDG' facility is to:
    (a) mechanically set a heading 'bug' to the required heading, and also to position the rotor of a CX synchro which provides heading signals to other systems;
    (b) mechanically rotate a compass card into alignment with a heading 'bug';
    (c) position the stator of a servo CT and so provide a servo drive to a compass card.

Chapter 9

1. Draw a diagram to illustrate the display of a typical ADI.

2. With the aid of a schematic diagram, explain how primary attitude changes are displayed by an ADI.

3. What is the significance of the dots against which the localizer and glide slope elements of an ADI and HSI are registered?

4. State the purpose of the ADI command bars, and explain how they are positioned.

5. What is the difference between 'course' and 'heading'? Explain how they are selected and displayed on an HSI.

6. How is an aircraft's position with respect to a VOR station displayed?

7. GS and VOR/LOC deviation indicators are operated by:
   (a) synchros;
   (b) servomotors;
   (c) dc meter movements.

8. What is the purpose of 'TO−FROM' indicators, and in which of the FDS indicators are they incorporated?

9. Explain how 'TO−FROM' indicators are activated.

10. Under what conditions is the 'GA' mode of operation selected? Explain how selection is carried out and the effect it has on the ADI display.

11. At which stage of a flight profile are the amber and green lights of an annunciator panel illuminated? Describe their operating sequence.

12. Describe how the convergence of a GS beam is allowed for during the 'AUTO APPR' mode of operation.

13. What is the purpose of an instrument comparator and warning system?

14. What warning and indicating flags are provided in an ADI and an HSI?

15. What heading information is displayed on the HSI when an FDS is interfaced with an INS and operates in the 'INS' mode?

16. In addition to heading, what other navigational data can be displayed on an HSI when the FDS operates in the 'INS' mode?

17. In dual FD systems, how is it ensured that data displays are available in the event of failure of data supply sources?

Chapter 10

1.   State the functions of the principal units that comprise an INS.

2.   From the operational point of view, what are the differences between an INS and an IRS?

3.   Define the terms 'longitude' and 'latitude'.

4.   What is meant by convergency?

5.   What do the abbreviations TK, TKE, and XTK signify?

6.   Draw a simple diagram to illustrate the significance of the abbreviated term 'WPT'.

7.   If the value of an aircraft's TK is geater than that of its heading:
     (a) no drift will occur;
     (b) drift would be to the left;
     (c) drift would be to the right.

8.   What are the fundamental laws of mechanics on which INS operation depends?

9.   What data must first be entered into an INS computer in order for the system to navigate an aircraft?

10. With the aid of a schematic diagram, describe how accelerometer signals are integrated to provide information on an aircraft's present position.

11. Describe how the X and Y accelerometers are aligned, and state also the coordinate system to which the output signals are related.

12. What do you understand by the term 'Schuler tuning'?

13. The gyroscopes utilized in a stabilized platform type of INU are of the:
     (a) displacement type;
     (b) rate-integrating type;
     (c) laser type.

14. How are the input axes of gyroscopes positioned with respect to an aircraft's axes, and what attitude changes do they sense?

15. What is the function of the Z gyroscope of a stabilized platform type of INU?
16. Describe how the gyroscopes are compensated for earth rate and transport rate.
17. What are the functions of the pitch, outer roll, and azimuth synchros connected to a gimballed platform?
18. Describe the constructional arrangement of a laser gyroscope.
19. Describe how a laser gyroscope senses aircraft attitude changes in terms of an angular rate.
20. How are attitude changes sensed and resolved when an inertial reference unit is installed in a 'strapdown' configuration?
21. What are the principal modes in which an IN/IR system operates, and how are they selected?
22. How are waypoints selected and inserted?
23. What is the purpose of an 'ALERT' annunciator?
24. Briefly describe the alignment procedure essential for IN operation.
25. What do you understand by the term 'gyrocompassing'?
26. Under what conditions would a 'slewing' procedure be carried out?
27. What is the purpose of the code numbering system that is programmed into an IN computer? State how the codes are displayed.

Chapter 11

1. What are the principal elements of a CRT?
2. In a CRT, the electrons are produced and emitted by:
   (a) passing signal currents through deflection coils;
   (b) the heating of an element called a cathode;
   (c) the heating of an element called an anode.
3. When an electron beam strikes the inside surface of a CRT screen, how is it made to produce a spot of light?
4. How is the spot of light made to trace out a line or a pattern on the screen of a CRT?
5. The three primary colours of a CRT are produced by:
   (a) electron beams that are coloured red, blue and green;
   (b) passing the electron beams from three electron 'guns' through a shadow mask containing red, blue and green screens;
   (c) directing the beams from three electron 'guns' through a perforated shadow mask so that they strike three different kinds of phosphor coatings.
6. Describe how colours other than the primary ones are produced.
7. Describe the raster scanning and stroke scanning techniques, stating the types of display provided.

Chapter 12

1. What are the main units that comprise an EFIS?
2. What are the functions of an SGU?
3. To which areas of the indicator displays is the raster scanning technique applied?
4. State the colours assigned to the displays and the type of information to which they correspond.
5. What is the function of a remote light sensor?
6. In dual EFIS installations, how is it ensured that the failure of an SGU will not affect the display of data?
7. State the modes that can be selected for display on the EHSI.
8. How are failures of data signal sources displayed?
9. Under what conditions do the GS and LOC deviation pointers of the EADI change from white to amber?
10. Can weather radar 'returns' be displayed in all modes of EHSI operation?

Chapter 13

1. What types of sensing elements are used for pressure measurement?
2. Describe a method of measuring engine oil pressure based on the principle of synchronous transmission.
3. Explain how an inductor type of pressure transmitter produces the varying currents required for the operation of a ratiometer.
4. For what purposes are pressure switches required?
5. With the aid of a diagram, explain how a pressure switch is made to give a warning of a pressure falling below its normal operating value.
6. Describe how temperature changes can cause variations in the properties of substances. Which of these variations are utilized in engine temperature indicating systems?
7. Name the materials most commonly used for variable resistance-type sensing elements, and describe the construction of a typical element.
8. Describe how a Wheatstone bridge circuit may be utilized for the measurement of temperature.
9. Describe the construction and operation of a ratiometer type of temperature indicating system.
10. What would be the effect of an 'open circuit' between the sensing element and indicator of the ratiometer type?
11. Explain the operating principle of a thermo-emf system, and state the engine parameters measured by such a system.
12. The metal combinations used in an EGT sensing probe are:
    (a) copper and constantan.
    (b) chromel and alumel.
    (c) iron and constantan.
13. What effects can changes in cold junction temperature have on the indications of thermo-emf instruments? Describe a method of compensation.

14. What is the difference between extension leads and compensating leads?
15. The 'external circuit' of a thermo-emf system is that part which extends from the thermocouple probes to the:
   (a) ends of the harness only.
   (b) junction at which conductors and extension leads enter an aircraft's fuselage.
   (c) terminals of the indicator.

Chapter 14

1. Explain the operating principle of a capacitor and state the factors on which it depends.
2. Define: (a) the units in which capacitance is expressed, (b) permittivity.
3. Describe how the capacitance principle is applied to the measurement of fuel quantity.
4. Why is it necessary to install a number of sensing probes in a fuel tank system?
5. What effects do temperature changes have on the fuels used, and how are they compensated?
6. Why is it preferable to measure fuel weight rather than fuel volume?
7. Explain why densitometers provide greater accuracy in fuel weight measurement than compensator probes.
8. Describe the construction and operation of a densitometer.
9. What do you understand by the term 'characterization' as applied to tank sensing probes?
10. What adjustments are normally provided in a capacitance type of system?
11. Explain the function of the test switch incorporated in some fuel quantity indicating systems.
12. How are checks carried out in computer-controlled fuel quantity indicating systems?
13. Briefly explain how total fuel remaining and aircraft's gross weight can be displayed.
14. Why are refuelling control panels provided on some types of aircraft?
15. Describe a method of refuelling control.

Chapter 15

1. Describe the construction and operation of a tacho-generator type of rpm-indicating system.
2. Why is the speed of turbine engines measured as a percentage?
3. What is the purpose of a tacho probe? Describe the operation of an indicating system which utilizes such a sensor.

4. Briefly explain the principle of supercharging, and how the increase in induction manifold pressure is measured.

5. To which types of engine does torque monitoring relate? Describe a method of torque measurement.

6. What do you understand by the terms 'stagnation' and 'rapid response' as applied to EGT sensing probes?

7. How is it ensured that an EGT indicating system measures good average temperature conditions of the exhaust gases?

8. Describe the operation of an electrical method of CJ compensation.

9. Describe the operation of a basic type of fuel flow indicating system.

10. What is the purpose of an EPR indicating system? Explain the fundmental principle of measuring the pressure ratio.

11. What is meant by an 'integrated' flowmeter system? Describe a method of achieving integration.

12. Describe the operation of an EVI system.

Chapter 16

1. What are the meanings of the acronyms 'EICAS' and 'ECAM'?

2. What are the principal units that comprise an EICAS installation?

3. Which engine parameters are classified as primary and secondary information for purposes of display?

4. Primary information is displayed on the:
(a) lower display unit;
(b) standby engine indicator;
(c) upper display unit.

5. Name the three modes of EICAS operation, and state how they are selected.

6. What data are displayed in each of the three operating modes?

7. What is meant by 'auto event' and 'manual event', and how is the associated data called up for display?

8. With an aircraft on the ground, and with electrical power 'on', is it necessary to switch on the display units from the display select panel?

9. Name the three levels of alert messages, and state on which display unit they are presented.

10. If several alert messages were displayed at the same time, and it was noted that one of them was indented one space, what level of message would it signify?

11. If the condition that generated an alert message no longer exists, the message is removed from the display:
(a) by pressing the 'cancel' switch on the DSP;
(b) automatically;
(c) by pressing the 'cancel' switch located near the display units.

12. What happens to the display of secondary information in the event of failure of the lower display unit?

13. The standby engine indicator displays are permanently on when its display control switch is in the:
    (a) 'on' position and the test switch is at the left or right position;
    (b) 'auto' position;
    (c) 'on' position only.
14. Under what conditions can an EICAS self-test routine be carried out, and from which panels is it controlled?
15. What are the principal differences between EICAS and the ECAM system?
16. Name the four modes of an ECAM display, and state which one of these is used in normal operation of the system.
17. Which display mode takes precedence over the others?
18. On which display unit are 'STATUS' messages displayed and what information do they provide?
19. In order to carry out an automatic test routine, is it necessary for a test switch to be operated first?
20. During a manual test of ECAM, when do diagrams appear on the right-hand display unit?

Chapter 17

1. What are the modes that can be selected on the CDU of a performance data computer system?
2. What is the significance of the caret ($<$) and asterisk symbols displayed on the CDU?
3. What is the purpose of the mode annunciator of a PDCS?
4. How are the records for an FMS computer data base entered into the memory?
5. What do the two sizes of characters displayed by the CDU signify?
6. What is the purpose of the line keys either side of the CDU screen?
7. For what purpose is the bottom line of the display used?
8. The 'CLEAR' key of the CDU is used to remove:
    (a) any of the data displays;
    (b) all records from the computer data base;
    (c) only incorrect entries from the scratch pad.
9. Under what conditions is data presented in 'scroll' form?
10. What is the purpose of the flight phase annunciators, and what happens to data display pages when the annunciators illuminate?
11. What is the function of the 'EXEC' key, and at what stage does the bar illuminate?
12. If an arrow appears against a data line of the display, what does it signify?
13. Describe a typical dual FMS configuration.

Solutions to exercises

Index

443